The Rise of Intelligent Machines

A fascinating shift marks the journey of technological evolution. Historically, humans were trained to build and operate machines. This education emphasised mechanical skills, logical thinking, and problem-solving abilities, laying the groundwork for the following industrial revolutions. Early machines, from the steam engine to the assembly line, were designed and maintained by skilled human workers, reflecting a one-way relationship where humans were the creators and controllers of machines.

Today, we stand on the brink of a new paradigm. Advanced artificial intelligence (AI) systems and generative AI tools are not just aiding human tasks but are also capable of educating and guiding humans. These intelligent machines can analyse vast amounts of data, provide personalised learning experiences, and generate creative content. The transformation from humans building machines to machines educating humans signifies a profound shift in our technological landscape, impacting industries.

As editors, we are united by a profound conviction that bridging the gap between industry and higher education is imperative in the face of rapid advancements in generative AI and innovations across various industries. This connection is crucial to equipping graduates and young professionals with the skills to innovate and excel in the future workforce. Through this multidisciplinary exploration, *The Rise of Intelligent Machines* aims to provide readers with an understanding of AI's transformative potential and the strategies needed to harness its benefits for industry and education.

Every chapter reflects our shared passion for AI's potential and close industry-higher education collaboration, and we hope that our work inspires you to embrace the opportunities and challenges of the future with curiosity and confidence.

The Rise of Intelligent Machines

A Multi-disciplinary Perspective from Industry and Impact on Higher Education

Edited by
Nadya Shaznay Patel and Nitin Indurkhya

CRC Press
Taylor & Francis Group
Boca Raton London New York

CRC Press is an imprint of the
Taylor & Francis Group, an **informa** business

A CHAPMAN & HALL BOOK

Designed cover image: Ernest Tey

First edition published 2025
by CRC Press
2385 NW Executive Center Drive, Suite 320, Boca Raton FL 33431

and by CRC Press
4 Park Square, Milton Park, Abingdon, Oxon, OX14 4RN

CRC Press is an imprint of Taylor & Francis Group, LLC

ISBN: 9781032745008 (hbk)
ISBN: 9781032744995 (pbk)
ISBN: 9781003469551 (ebk)

DOI: 10.1201/9781003469551

Typeset in Times
by KnowledgeWorks Global Ltd.

Contents

Preface

The rapid advancement of artificial intelligence (AI) and other emerging technologies has fundamentally transformed our world in recent decades. These innovations have reshaped industries, altered societal norms, and redefined human-machine relationships. This book seeks to explore these profound changes, offering insights from professionals across diverse fields and examining the potential of AI.

Historically, humans were trained to build and operate machines. Education emphasised mechanical skills, logical thinking, and problem-solving abilities, laying the groundwork for the following industrial revolutions. Early machines, from the steam engine to the assembly line, were designed and maintained by skilled human workers, reflecting a one-way relationship where humans were the creators and controllers of machines. As technology progressed, the relationship between humans and machines began to evolve. The advent of computers and early AI systems in the mid-20th century marked the beginning of a new era. Machines could now process information, perform calculations, and execute previously unimaginable tasks. This period saw the emergence of a symbiotic relationship where machines assisted humans in complex tasks, enhancing productivity and enabling new possibilities.

Today, we stand on the brink of a new paradigm. Advanced AI systems and generative AI tools are not just aiding human tasks but are also capable of educating and guiding humans. These intelligent machines can analyse vast amounts of data, provide personalised learning experiences, and generate creative content. The transformation from humans building machines to machines educating humans signifies a profound shift in our technological landscape. Scholars advocate for integrating AI into education by leveraging AI's strengths while maintaining human oversight to enhance learning outcomes. This co-intelligence approach ensures that AI aids rather than replaces human educators (Mollick, 2024). Moreover, as AI becomes more integrated into daily life, it is crucial to emphasise "human-in-the-loop" systems where human judgement and oversight are integral to AI processes. Hence, despite the profound shift, this approach helps maintain ethical standards and enhances the reliability of AI systems. Data privacy, algorithmic bias, and the potential for job displacement have sparked widespread debate. As machines become more capable, questions about accountability and control become increasingly pressing, underscoring the profound existential questions AI raises, such as what it means to be human and the implications for future employment, reflecting anxieties about AI's potential to surpass human abilities in various tasks.

While all these changes may seem overwhelming, the pace of technological change has always been rapid, and the demands of the job market have tended to evolve faster than those of education systems, leading to a mismatch between the skills taught and those required by industries. This has worsened dramatically with the emergence of modern AI and underscores the need for a reimagined approach to education that emphasises lifelong learning and adaptability and highlights the necessity of integrating disciplined experimentation and practical skills in education to bridge this gap effectively. The impact of AI extends beyond skills education for the workforce. It influences various aspects of society, including healthcare, legal systems, creative industries, and leadership. The integration of AI in these domains must be guided by principles that prioritise human well-being and ethical considerations, ensuring that technological advancements do not undermine fundamental human values. How can education systems prepare their graduates to address these issues? This question has no single answer, and opinions vary on what can be done.

Unlike other books on this topic, we have sought to present a multidisciplinary perspective, bringing together insights from diverse fields. Another unique aspect of this book is that we focus on how educational institutions can prepare students for a world shaped by AI, in which graduates need to be equipped with critical design futures (CDF) thinking.

STRUCTURE OF THE BOOK

This is an edited book. It includes the views from Southeast Asia generally and, more specifically Singapore, 'the little red dot', as it is sometimes affectionately referred to by locals. Each chapter offers a unique perspective from local industry professionals who reflect on the question: "How could your industry embrace emerging AI technologies more strongly?" They are organised as follows. Chapter 1 examines how AI revolutionises client-agency dynamics in the advertising industry, enhancing strategic communication and trust. Chapter 2 delves into the integration of generative AI in the creative industry, demonstrating its capacity to streamline operations and drive innovation. Chapter 3 explores the role of AI in customer experience, particularly how AI can enhance design efficiency and creativity through strategic change management. Chapter 4 explores innovation in healthcare, focusing on innovative strategies to enhance senior care and address workforce challenges in eldercare centres. Chapter 5 investigates AI's impact on interdisciplinary workplace communication with oral history, while Chapter 6 examines its integration into legal practices. The following chapters, Chapters 7 to 10, cover AI's influence on leadership, lifestyle, student development, and young adult career coaching. Finally, Chapter 11 addresses the evolution of higher education institutions into lifelong learning ecosystems, highlighting the importance of transdisciplinary learning, real-world applications, and embracing AI. It examines the integration of CDF thinking and generative AI to prepare students for the complexities of the future workforce, pursuing innovation with ethical considerations, and advancing human and machine intelligence at unprecedented levels. This approach fosters critical competence, design dexterity, and futures flexibility, ensuring graduates and professionals navigate and lead in a rapidly changing world. Throughout, the book emphasises the necessity of lifelong learning and adaptability. It illustrates how AI can enhance educational outcomes and foster innovation while highlighting the ethical considerations and practical challenges involved.

AUDIENCE

The book is aimed at students and educators, researchers and administrators in higher education, business leaders, executives and professionals across diverse industries, policy-makers, tech enthusiasts, and the general public interested in the impact of AI in varied disciplines. Most chapters can be read and understood by anyone with a basic knowledge of the domains. If you are looking to apply AI in a novel field and gain a comprehensive understanding of AI's transformative potential and the strategies to harness its benefits for industry and education, then the material in this book can provide some direction by showing how it has been done in various industries. The chapters are based on Masters research completed in February 2024 by industry professionals in a programme on digital management by Hyper Island Asia, making the book a useful reference for similar postgraduate courses.

REFERENCE

Mollick, E. (2024). *Co-intelligence: Living and Working with AI*. Portfolio/Penguin.

Acknowledgements

As editors, we are united by a profound conviction that bridging the gap between industry and higher education is imperative in the face of rapid advancements in generative AI and innovations across various industries. This connection is crucial to equipping graduates and young professionals with the skills to innovate and excel in the future workforce. Through this multidisciplinary exploration, the book aims to provide readers with a comprehensive understanding of AI's transformative potential and the strategies needed to harness its benefits for industry and education.

The rapid evolution of artificial intelligence has reshaped industries, altered societal norms, and redefined human-machine relationships. This transformation demands a reimagined approach to education that prepares students with technical skills and human meta-skills thinking abilities. Our shared passion lies in ensuring that higher education and workplace training foster the development of CDF thinking skills and dispositions, combining intellectual growth with the cultivation of essential human life skills. We take a humanity-centred approach to solving the world's most complex problems. AI should enhance our humanity, not diminish it, necessitating a focus on skills and dispositions that machines cannot replicate. Our collaborative journey in editing this book has been incredibly enriching. Working closely with contributing authors and professionals from diverse disciplines, we have learned from one another, gaining insights that have deepened our understanding of AI's multifaceted impact. This collaboration reflects the very interdisciplinary approach we advocate for within these pages.

Many people assisted in this journey. First and foremost, we thank the individual chapter authors for their contributions and for working to our tight deadlines. Without their efforts, there would be no book! The peer reviews were done by Peachy Pacquing (Managing Director, Hyper Island), Leong Yeng Wai (Programme Director, Hyper Island), Paviter Singh (Head of Courses, Hyper Island) and Ted Kilian (Principal Consultant, Encouraging Behaviour Ltd.), and Echo Han (Co-Founder, Star Papaya Story). We thank them for their reviews to improve the chapters. The reviewers of the book proposal provided us with valuable comments that helped guide the project at its inception, and we are grateful to have been greenlighted by them. Finally, we give our heartfelt thanks to Randi Slack, our publisher at Taylor & Francis for understanding the need for rapid publication and for encouraging and supporting us throughout the process. Thanks, Randi, for having our back!

Nadya Shaznay Patel and Nitin Indurkhya
Singapore
Northern Summer, 2024

About the Editors

Nadya Shaznay Patel, EdD Nadya has over 20 years of experience as an educator, researcher, and trainer. With a passion for Critical Design Futures (CDF) thinking, she regularly facilitates workshops to develop learners' critical competence, design dexterity, and futures flexibility for innovation. As a chief consultant of a management firm, committed to bridging the research-practice gap, she engages industry partners in applied research projects, leveraging her expertise in learning and development and interest in GenAI applications. Nadya graduated with a Doctor in Education from University College London in 2015, and a Master's degree in Digital Management from Teesside University in 2024 is a testament to her unwavering commitment to lifelong learning. Her research interests include CDF thinking, GenAI in learning and development, transdisciplinarity, and empathetic, dialogic communication. Nadya endeavours to remain at the forefront of transformational pedagogical approaches that leverage emerging technologies and transdisciplinary learning in close industry-higher education collaborations.

Nitin Indurkhya, PhD Nitin founded Data-Miner Pty. Ltd in 1997 and has been focused on data mining, text mining, language technologies consulting, and education since then. In the corporate sector, he led the language technologies group at a Fortune-100 company on Wall Street after establishing the Cloud Research Lab at Samsung Research America. Before that, he was a Principal Research Scientist at eBay in Silicon Valley. He has also been a professor/researcher in various universities and industrial research labs in Australia, Bhutan, Brazil, China, India, Japan, Malaysia, Portugal, Singapore, Spain, the USA, and Vietnam. He has an undergraduate degree from IIT, Kanpur and a PhD from Rutgers University.

Contributors

Danielle Ho Foong Ling thrives on the intersection of strategy and creativity with her passion for forging meaningful connections, and utilising data to tell stories and craft solutions.

Rebecca Lim Siok Hoon is a communications professional and leader in media, finance, fintech, public sector, and government, with expertise and experience in innovation, transformation, leadership and culture, strategic communications, crisis management, integrated marketing, public relations, corporate communications, and editorial and content strategy and production, while pivoting to health and wellness coaching and health tech.

Ernest Tey Cheng Hwee is an advertising veteran Blessed by the Lord.

Cynthia Mak is a seasoned go-to-market leader who helps businesses navigate complexities and unlock potential in the APAC markets across various industries.

Jeffrey Lee is a course manager at the School of Design & Media in Nanyang Polytechnic, Singapore. He is also a senior lecturer, lead student mentor, and skills coach under the Diploma for Communication & Motion Design.

Siok Khoon Lim is a former lawyer and now digital project consultant at Advokatfirmaet BAHR in Oslo, Norway.

Kenny Low Heng Khuen is a social entrepreneur and the founding director of O School and executive director of City Harvest Community Services Association.

Kenny Png is an independent film-maker and media consultant. He is the co-author of *On Happiness* (Math Paper Press), *Requiem for The Factory* (Deleres Press) and contributed to the 2020 edition of *The Birthday Book* (The Birthday Collective).

Edwin Tan Yurong, founder of branding consultancy Bravo Creative, creates award-winning brand projects and educates future designers.

Alan Teo Hui Yeong is a design lead at M1 Limited, Singapore.

1 Transforming Advertising Relationships

Revolutionising Client-Agency Dynamics with Large Language Advertising Model Brain (LLAMB)

Ernest Tey

INTRODUCTION: THE BRIEF

BACKGROUND OF THE BRIEF

The creative advertising industry (CAI) has evolved remarkably from its inception in ancient civilisations to the digital age. Symbols in Egyptian papyrus, trademarks on Grecian pottery, Roman signs, and Chinese printing innovations marked the early days of advertising. The industry's modern form took shape with the advent of mass marketing during the Second Industrial Revolution, as agencies like JWT expanded globally, emphasising consumer needs and branding (Beard, 2017; Davis, 2012). However, in the latter half of the last century, the CAI was reshaped and redefined by technological advancements, evolving consumer behaviours, client-agency dynamics, and significant mergers and acquisitions (M&As), leading to conglomerates like WPP, Publicis, and Omnicom (Ghobadian & O'Regan, 2011). The CAI has now evolved beyond the stage of static posters to leverage innovative technologies like data-driven targeting, influencer marketing, and AI to deliver tailored campaigns (Lamberton & Stephen, 2016).

PAIN POINTS AND BARRIERS OF THE BRIEF

In today's rapidly evolving advertising landscape, technological advancements have given rise to savvy consumers who engage across diverse platforms. This shift has heightened the market's sophistication and increased the complexity of understanding consumer needs, encapsulated in the "jobs to be done" (JTBD) concept (Christensen et al., 2016). This means agencies and their clients must forge closer relationships, nurturing deeper and solid client-agency collaborations to meet these nuanced consumer demands effectively. At the same time, the CAI must also keep to current advertising ethics, ensuring collaborative efforts must balance authentic engagement, uphold ethical branding, and protect consumer privacy (Uncles, 2008).

The business models within the CAI have also shifted. On the one hand, traditional full-service agencies, although comprehensive, often need help with agility and integrating new technologies. Digital agencies, on the other hand, excel in leveraging consumer data for targeted advertising but may require a broader service scope (Norris, 2017). M&As have surfaced as strategies to bridge these gaps, yet they are not without criticism; many industry leaders question their impact on creativity and overall value (Beard, 1999; Ducoffe & Smith, 1994). With the high rate of personal movement due to M&As, role ambiguity and miscommunications can erode trust, undermining the partnership's efficacy (Beard, 1999).

DOI: 10.1201/9781003469551-1

SIGNIFICANCE OF THE BRIEF

Integrating advanced technologies, particularly artificial intelligence (AI) and large language models (LLM), within the CAI is critical to revitalising client-agency relationships. This chapter posits that by embedding these innovative tools into agency operations, a new dimension of collaboration and trust can be established, fundamentally enhancing the effectiveness of the industry. Focusing on this aspect, the research explores how the strategic use of these technologies can mend the fraying bonds between clients and agencies. Adopting AI and LLM technologies is more than a mere upgrade in efficiency; it is a step towards rebuilding confidence and faith in the Ad World (Libai et al., 2020). By leveraging these innovations, two benefits stand out. More profound, prosperous, and meaningful client collaboration can be cultivated, leading to better operational efficiencies. This enhanced partnership is anticipated to improve the quality and impact of advertising work and re-establish the CAI as a beacon of innovation and trust. Successful integration of these technologies could mark a significant turning point for the CAI to reinvigorate the industry where advanced tools not only automate processes but also play a pivotal role in building and maintaining strong, trusting client-agency relationships, restoring the industry's prominence and influence in the modern Ad World.

HOW THE BRIEF IS TO BE ARTICULATED

This research focuses on pivotal questions affecting the CAI, mainly how agencies can stay relevant using creative future-forward strategies to lead in this ever-changing landscape, restructure business models, tackle business relationships, meet consumers' demands, and solve internal work dynamics. It examines if the industry can apply its creative solutions internally, using AI technologies like LLAMB, to address client challenges. A deep review of relevant literature followed by a comprehensive methodology begins with a Google Forms online survey for broad insights into the industry's status, followed by in-depth, face-to-face interviews to delve deeper into finer details and insights. After analysis, these methods comprehensively reveal the CAI's current challenges and potential opportunities. After this, proof of concept frameworks will be employed to determine this theoretical framework. Then, it will be implemented and prototyped as a theoretical prototype concept to be tested by industry leaders. After this, a review of iteration follows, then a discussion on the broader scope of the prototype, and, finally, a conclusion. The goal is to steer the CAI towards a future where innovative strategies enhance its standing, ensuring resilience and sustainability in a dynamic landscape.

LITERATURE REVIEW: UNPACKING THE BRIEF

FROM BILLBOARDS TO BYTES

The advertising landscape has profoundly evolved, shifting from the past's vibrant billboards and print media to the dynamic realm of digital advertising. This evolution signifies a pivotal change in how advertising agencies engage with clients and audiences. The turn of the 21st century marked a significant shift towards a data-driven approach in advertising. Platforms like Google Ads catalysed a transition from broad-based campaigns to focused, individual messaging, leveraging data analytics for precise targeting and return on investment (ROI) measurement (Szymanski & Lininski, 2018). Social media further revolutionised brand communication into an interactive, two-way dialogue. Influencers and user-generated content emerged as powerful tools, blending brand narratives into daily life and gaining trust among global audiences (Alalwan et al., 2017).

Central to this transformation is the advancement of AI in advertising. John McCarthy's pioneering work laid the foundation for what we now see as integral to modern advertising strategies – using AI for targeted, efficient campaign execution (Rajaraman, 2014). AI's evolution from generative artificial intelligence (GAI), which encompasses the use of generative adversarial networks (GANs), has been revolutionary. It uses a prevalent technique in which two neural networks work together to generate and evaluate synthetic content, such as images or music, blurring the line between real and virtual artefacts (Jovanovic & Campbell, 2022). Another form of AI, particularly analytical AI and LLMs, has also been monumental. These technologies, exemplified by models like ChatGPT, have revolutionised how advertising campaigns are created and executed, enabling more natural, human-like interactions and decision-making based on vast data analysis (Haenlein & Kaplan, 2019; Lu et al., 2023).

To develop a pre-trained LLM, one must first compile a vast unlabeled text collection from diverse sources like books and websites. This text is broken down and tokenised to form the model's vocabulary. Next, a learning objective, such as masked language modelling (MLM), is employed to train the model on this data. The model's parameters are initialised by selecting an architecture like transformer-based BERT or GPT. The LLM is then pre-trained on the text, followed by fine-tuning with specific labelled data for targeted tasks. This process equips the LLM to grasp and apply language patterns to various natural language processing applications (Schick & Hinrich Schütze, 2020).

GAI, GAN, AAI, and LLMs are pushing the next frontier and creating new possibilities for content creation. These technologies are redefining creative processes in advertising, offering novel ways to engage with audiences (Jovanovic & Campbell, 2022). The advertising sector is evolving with the introduction of these innovations. These advanced algorithm-based machines, adopted by giants like Microsoft and Google, redefine content creation, campaign strategy, and marketing personalisation (Omneky, 2023). However, LLMs come with challenges, such as potential biases and a lack of emotional depth, highlighting the necessity for human interference in strategic and creative roles. A strategy is to allow LLMs to take over the heavy loads of routine tasks, freeing professionals to focus on the cognitive thinking of strategy and creativity. The role of advertising experts needs to evolve, and these new hybrids are crucial in ensuring ethical AI usage, maintaining brand authenticity, and connecting emotionally with audiences. This blend of technology and human expertise helps balance AI efficacies and efficiency with human creativity and strategic thinking to enhance advertising effectiveness and audience resonance (Yuki, 2023).

Early adoption of advanced technologies in the CAI necessitates integrating new specialists to upskill the workforce with essential knowledge and tools. These experts will develop and train machines, preparing data sets to operationalise privately owned LLMs. Concurrently, creative, strategy, media, and production teams should leverage this technology, transitioning from a production-centric mindset to a strategy-focused approach in campaign crafting (Windels & Stuhlfaut, 2018).

One of the forerunners who has dived into the LLM trend is Omneky, an AI-driven ad tech platform that recently launched its advertising LLM. This tool, unveiled in June 2023, enables marketing teams to generate personalised, brand-aligned content efficiently and creatively. Designed specifically for the advertising industry, Omneky's LLM leverages structured and unstructured data to create targeted marketing strategies and creative content, ensuring data security and brand consistency. This innovation marks an essential advancement in applying AI technology in advertising (Omneky, 2023).

Envision a future where personalised AI becomes a digital butler for organisations, tailored to specific needs and evolving with user interaction. This represents a groundbreaking direction for the CAI and its clients. This vision, combined with the CAI's strategic strengths, points towards a future where technology like the prototype proposed in this chapter could be pivotal in reshaping the advertising landscape.

From Mainstream to Mavericks

The CAI has historically been shaped by M&As, technological advancements, evolving client-agency dynamics, and consumer behaviour shifts (Ghobadian & O'Regan, 2011). In the 1960s, advertising agencies expanded into various industries, later turning to acquisitions for growth.

This led to the emergence of "superagencies" in the 1980s through M&As (Ducoffe & Smith, 1994). While M&As offer benefits like enhanced capabilities and global reach, they also raise concerns about reduced creativity and diversity in media conglomerates (Hendrickson & Subotin, 2021). Surveys among top advertisers reveal a scepticism towards M&As, citing issues like stifled creativity and increased bureaucracy. These shifts have led to higher personnel turnover, resulting in possible role ambiguity and exhibiting low client responsiveness through reduced interactions, thus affecting business relationships (Ducoffe & Smith, 1994).

Amidst these changes, boutique agencies have emerged as significant players, challenging traditional large-scale firms. These smaller agencies can offer personalised services to cultivate closer client relationships, often excelling in creative agility and market adaptability (Norris, 2017). Focusing on personalised strategies and creative innovation, smaller agencies are proving to be formidable contenders against larger firms, demonstrating that success in the advertising industry is not always a matter of scale but of creativity and agility. Success stories like Droga 5, which rose to prominence from a modest start, highlight the potential of smaller, innovative agencies in the CAI (Dan, 2021). In this ever-evolving landscape of the CAI, marked by frequent M&As, a crucial question arises: will M&As scale and profits-driven mantra indeed be beneficial for the CAI (Ghobadian & O'Regan, 2011) or does the industry's success rely more heavily on creativity, agility, and flexibility to manoeuvre the intricacies of client relationships and meeting the demands of the consumer (Norris, 2017)?

FROM BUSINESS TO RELATIONSHIPS

"Make your buyer the hero of your story", advises Chris Brogan, encapsulating a key shift in modern advertising. Historically, consumerism was a domain of the wealthy elite, symbolising status. Over time, it has evolved, reshaping lifestyles, work, and social relations worldwide. Today's consumer culture, fuelled by technology and multinational corporations, has led to complex consumer behaviours and expectations (Stearns, 2006).

The Demanding Consumer

Consumer behaviour has evolved in today's digital landscape, prompting brands and the CAI to channel genuine value with new technological means to engage with these prosumers. Today's consumers are well-informed and empowered, reflecting "Metis" – a blend of local knowledge, practical skills, and digital literacy that encourages collaboration and shared expertise. These "prosumers" actively participate in co-creation, seeking meaningful, personalised brand interactions. Therefore, marketing and advertising agencies must deliver personalised experiences and co-creation opportunities, recognising and utilising the savvy consumer's competencies and empowerment (Macdonald & Uncles, 2007). Agencies must understand advertisers' and consumers' needs and wants, or JTBD (Christensen et al., 2016), to build trust in the relationship. However, this rise in consumer awareness in the digital age also brings ethical challenges for brands. Brands must now learn to balance consumer engagement with value, respect data privacy, avoid targeting vulnerable groups, and avoid deceptive marketing. By prioritising ethics, brands can build trust and offer more personalised, engaging experiences to increase value (Uncles, 2008). By cultivating intimate collaborations and engaging in co-creation, agencies can unearth deep insights, which can then be manifested into captivating communications that resonate with the "demanding consumer". This process enriches the consumer experience, creates substantial value for clients and their brands, and builds trust in the relationship (Berenguer-Contrí et al., 2020).

The Needy Client

The client and agency partnership is seen as a business-to-business (B2B) relationship that is not measured just on a professional level but requires deeper human interaction, such as value co-creation (VCC) and collaboration, to build trust, commitment, and social satisfaction. By understanding the causal relationships between these variables, agencies can create more value for clients and brands to build stronger, more sustainable relationships (Berenguer-Contrí et al., 2020). A study reveals that Clients part ways with agencies for various reasons, including the failure of agencies to understand client-specific needs, poor agency performance in terms of missed deadlines and lack of creativity, coordination issues, poor account servicing, a loss of confidence and trust in the agency's capabilities, and concerns about the profitability of their brands (Arul, 2011; Ducoffe & Smith, 1994). However, another study shows that agencies with open communication lines typically receive more comprehensive client information, enhancing creative output and streamlined execution. Conversely, clients appreciate agencies that confidently offer alternative perspectives, resulting in more productive interactions and higher-quality work, while clients who are accessible and decisive provide clear guidance, boosting the effectiveness of the collaboration (LaBahn & Kohli, 1997). Another study discusses how integrated marketing communication (IMC) is crucial for effective client-agency collaboration. It requires coordination and cooperation between clients and agencies to ensure that all brand touchpoints are integrated, ensuring the customer journey is smooth and effective. The four critical components of IMC include media neutrality, consumer centricity, coordination and consistency across the customer experience, and strategic involvement at the board level. By adopting an IMC approach, clients and agencies can work together to create synergy and added value, resulting in a strong positive impact on brand and financial performance (Laurie & Mortimer, 2019). The CAI needs to restrategise and shift from solely an "idea generator" to a "solution facilitator", emphasising collaboration to co-create compelling work (Lee & Lau, 2018). At its core, the CAI needs to build a more collaborative and understanding ecosystem to weave together industries, brands, and consumers to create value that helps to make lasting connections and impressions to drive businesses forward. This partnership's strength, characterised by mutual commitment, is a key determinant to achieving superior results in the advertising realm (Lee & Lau, 2018).

Romancing the Machine

LLMs like GPT-3 are pivotal in revolutionising human-AI collaboration, promising a new era of cooperative problem-solving and enhanced knowledge sharing. These LLMs, trained on extensive textual and multimodal data, can learn from diverse sources and apply knowledge to new situations. Their capacity to understand and generate human-like languages streamlines information retrieval, creative content generation, and decision-making across various domains like education, healthcare, and business (Tamkin et al., 2021).

One of the more notable AI innovations is intelligent embodied agents (IEAs). These novel AI innovations embody LLMs' planning, communication, and collaboration strengths, efficiently accomplishing complex tasks with humans. When fine-tuned properly, IEAs can be tailored for specific goals; these agents are highly adaptable and are capable of evolving based on experience. Researchers used a step-by-step approach to research and conduct tests on these IEAs by first developing a novel framework that utilises LLMs to plan and communicate collaboratively without the need for fine-tuning or few-shot learning. They then designed experiments to evaluate the performance of IEAs in various multi-agent cooperation challenges, including the Communicative Watch-And-Help (C-WAH) scenario and the ThreeDWorld Multi-Agent Transport (TDW-MAT) scenario. The IEAs were trained and tested on the designed experiments using the developed framework, and then human experiments were conducted to evaluate the performance of IEAs in collaboration with humans. Finally, results were analysed, and evidence shows that the research and prototypes have demonstrated the capability of IEAs to collaborate with humans effectively.

IEAs show significant promise in assisting humans across various settings. Their effectiveness in human-agent collaborations was proven in experiments where LLM-based agents and humans worked together on tasks like the C-WAH scenario. These interactions help nurture trust in IEAs among human participants and enhance mutual efficiency. The implications are vast, with IEAs positioned as valuable collaborators in both professional and personal domains, improving task performance and achieving shared goals. Moreover, IEAs have a unique capacity to forge and enhance human relationships. Their adeptness in communication and cooperation helps build trust and rapport, making them assistants and partners in cultivating positive human interactions. IEAs can also serve as mediators to enhance teamwork, leveraging their understanding of human intentions to facilitate meaningful interactions. Their role extends beyond task completion to nurturing healthy professional and personal relationships (Zhang et al., 2023). The reinforcement learning method is another way of training LLMs to understand, react to, and relate more to humans. This method aligns IEAs outputs with human expectations, making the human-machine interaction even more cohesive and relatable, thus creating better work relationships. With all these innovative progressions between technology and humans, concerns about bias and harmful language generation remain (Ziegler et al., 2020). Therefore, nurturing beneficial human-AI relationships necessitates ethical adherence, transparency in capabilities, and interdisciplinary efforts to address LLMs' societal impacts.

In this context, an IEA-trained LLAMB is a critical innovation that bridges and strengthens human-machine connections and enhances collaborative and trust-based relationships in various sectors. The "Human Experience" is vital in uncovering truthful human insights, resolving organisational conflicts and refining processes. Can introducing technology, especially advanced AI with analytics and human-like communication, enhance the VCC process and create better human relationships (Ziegler et al., 2020)?

From Traditional to Radical

The CAI has transitioned from a closed approach, keeping clients at arm's length during the creative process, to a more open and inclusive model (Turnbull & Wheeler, 2015). Besides adopting this collaborative VCC approach, recently, WPP and Nvidia's collaboration to harness generative AI for ad production has been a notable innovation, allowing for the scalable creation of customised content (Ziady, 2023). However, concerns remain about maintaining the essence of the relationship between client-agency-consumer to co-create and deliver insightful work.

Nvidia's CEO, Jensen Huang, acknowledges that while technologies and innovations can generate content at scale, creativity remains a fundamental human pursuit. In his own words, he says: "You can do content generation at scale. However, infinite content does not imply infinite creativity. You still must do that job." He implies that humans are still at the heart of the creation process (www.warc.com, 2023). Similarly, Doyle Dane Bernbach's (DDB) exploration into AI with "The Uncreative Agency" suggests a future where AI assists rather than replaces human creativity, supported by Omnicom's investment in AI integration (Baar, 2023). Publics, another big conglomerate, is pushing out its own AI, called Marcel.ai, which empowers employees with information, tools, and peer-to-peer connections to facilitate work processes (marcel.ai, n.d.). However, this platform is more like a resource pool than a highly engaging and pre-trained smart LLM capable of co-creating ideas with humans.

Navigating the Grey

Navigating the ethical landscape amidst technological advances is a critical challenge for the CAI. As technology and innovations revolutionise the field, they also bring socio-ethical considerations concerning privacy, data security, and human-machine interaction (Tene & Polonetsky, 2013). The balance between personalisation and intrusion is delicate, requiring agencies to act with ethical foresight. To harness technologies like generative AI responsibly, industry professionals must define clear objectives, identify beneficiaries, and ensure that applications align ethically with user intentions. Involving stakeholders – from GAI scientists to the end audience – is essential in circumventing unethical product behaviours and biases (Jovanovic & Campbell, 2022). One way to mitigate security concerns is to integrate blockchain with AI and IoT, creating transparent, secure, and decentralised networks that bolster data integrity and protect against cyber threats (Alharbi et al., 2022).

While technological advancements will displace some roles, they also create new opportunities, especially in areas that leverage creativity and technological and social intelligence (Paesano, 2021). Reflecting on its historical adaptability, the CAI is poised to harness and invest in innovative technologies through highly trained, advertising-specific, and privately owned LLMs to evolve sustainably and maintain its prominence. As this research progresses, it becomes increasingly clear that integrating technologies like "LLAMB", as mentioned at the very start of this chapter, can be pivotal in navigating the current digital landscape to shape a future where technology can augment work processes between clients and agencies, so that collaboration and value co-creation can happen resulting in better strategic and creative work, and ultimately lead to building better relationships.

This journey of adaptation and innovation is not just about survival but about seizing the opportunity to redefine the essence of client-agency relationships and value creation in the CAI.

CRITICAL ANALYSIS: THE WAR ROOM

TECHNOLOGICAL EVOLUTION AND HUMAN CREATIVITY

The shift from traditional billboards and print to digital and AI-driven platforms marks a significant evolution in the CAI. This change, mirroring society's move towards digital interconnectedness and data reliance, profoundly impacts how advertising agencies interact with their audiences and clients. Integrating AI and LLMs into advertising, inspired by pioneers like John McCarthy and modern entities like Omneky, signifies a leap towards smarter and more effective campaign execution (Haenlein & Kaplan, 2019; Lu et al., 2023; Omneky, 2023; Rajaraman, 2014; Szymanski & Lininski, 2018). However, one must emphasise that technological evolution in advertising is not about replacing human creativity with machines but about creating a synergy, a partnership, a romance where technology enhances human capabilities. This synergy promises a more efficient, responsive, and effective advertising industry where human creativity, bolstered by technological advancements, can thrive in new and exciting ways (Davenport et al., 2020; Haenlein & Kaplan, 2019; Jovanovic & Campbell, 2022; Lu et al., 2023; Schick & Hinrich Schütze, 2020).

The introduction of LLM advancements, along with IEAs, whose strengths lie in planning, communication, collaboration, and efficiently accomplishing complex tasks with humans, provides remarkable abilities for examining consumer data, forecasting trends, and crafting inventive content.. They act as instruments of automation and catalysts for a more nuanced, discerning form of advertising, merging data-informed decisions with the art of creative intuition. These machines can form that machine-human connection and build trust and better relationships (Zhang et al., 2023). However, the deployment of AI in advertising has its challenges. The ethical implications of using AI, particularly regarding data privacy and the chance for algorithmic bias, necessitate a careful, considered approach (Alharbi et al., 2022; Jovanovic & Campbell, 2022; Tene & Polonetsky, 2013).

DYNAMICS OF THE CREATIVE ADVERTISING INDUSTRY

M&As in the CAI are often perceived as a pursuit of scale, presenting opportunities and challenges (Ghobadian & O'Regan, 2011; Hendrickson & Subotin, 2021). While they offer resources and a global reach, there is a growing concern about their impact on creativity and individual agency cultures (Ducoffe & Smith, 1994). In contrast, boutique agencies, with their lean structures and focus on specialisation, are redefining success in the industry (Norris, 2017). They demonstrate that agility and a deep understanding of client needs often trump sheer size, exemplified by agencies like Droga5 (Dan, 2021). This landscape suggests a need for a balance where agencies, regardless of size, leverage technology to enhance their creative capabilities and maintain strong client connections. The CAI must consider its resource integration and hire technology specialists to help develop, machine train, and build up all relevant data sets to get a privately owned LLM functioning, like the revolutionary set-up example from Omneky (Omneky, 2023; Windels & Stuhlfaut, 2018). While technology can be an enabler, the CAI must also look at internal challenges related to organisational politics, role ambiguity, and resource integration (Beard, 1996; Cabiddu et al., 2019) Addressing these issues is key to maintaining a healthy workplace culture and effective client relationships. The STAR Modelemerges as a valuable framework for navigating these challenges, providing a structured approach to aligning strategy, structure, and human resources (Galbraith, 2011). The focus on resolving role ambiguity and cultivating effective internal communication and collaboration is essential for enhancing agency performance and client satisfaction.

THE CLIENT-AGENCY-CONSUMER RELATIONSHIP

According to the literature, evidence points to research on creative mediocrity, coordination issues, poor account servicing, and loss of trust as primary reasons for clients leaving agencies. These factors highlight the importance of effective communication, reliable service, and role clarity in a positive relationship (Arul, 2011; Ducoffe & Smith, 1994). Changes in an agency's creative personnel also significantly impact client retention, which can be seen when agencies go through M&As (Hendrickson & Subotin, 2021). Other factors like internal and external conflicts and power struggles can also fuel a client's loss of trust, leading to failed relationships (Sasser et al., 2013). Therefore, it is essential to maintain harmony and retain valuable resources so that the cohesion between clients and agencies is not broken. Hence, the VCC concept within the service-dominant logic (SDL) framework advocates for a more collaborative approach, involving clients as active participants in the creative process to nurture cohesion, collaboration, understanding, and better relationships (Levin et al., 2016; Vargo & Lusch, 2014). This shift towards co-creation and partnership is crucial for creating value and nurturing long-term relationships (Laurie & Mortimer, 2019). In modern advertising, understanding and engaging with the consumer is paramount. The digital age demands a nuanced approach to consumer behaviour, focusing on personalised marketing and ethical considerations (Stearns, 2006; Shavitt & Barnes, 2020; Uncles, 2008). Agencies must navigate complex consumer cultures, requiring culturally sensitive and ethically sound strategies. This aligns with the JTBD theory, which posits that effective advertising should directly address consumers' needs and wants (Christensen et al., 2016).

ETHICAL CONSIDERATIONS AND TECHNOLOGY'S ROLE

Integrating AI and other emerging technologies in advertising brings ethical considerations around privacy and data security to the forefront (Tene & Polonetsky, 2013). A fine line exists between personalisation and intrusion; maintaining this balance is critical (Jovanovic & Campbell, 2022). Blockchain technology's potential integration offers a solution to some of these challenges, providing a transparent and secure framework for AI and IoT applications in advertising (Alharbi et al., 2022). This suggests a future where technology can be efficient and ethically responsible.

THE FUTURE OF THE CAI – EMBRACING LLAMB AND INNOVATION

The literature suggests a future for the CAI that synergistically combines human creativity with technological innovation (Smith et al., 2018). Tools like LLAMB present an opportunity to enhance the creative process, offering new ways to engage with audiences and create impactful campaigns. However, the human element remains irreplaceable, guiding these technologies towards meaningful applications. Integrating LLAMB and similar technologies could redefine client-agency interactions, nurturing deeper partnerships and more effective value co-creation.

CONCLUSION – STRATEGIC ADAPTATION AND CREATIVE RESILIENCE

In synthesising the literature, it becomes evident that the CAI's future success lies in its ability to strategically adapt to technological advancements while preserving the human essence of creativity. The challenge is to leverage technologies like LLAMB not as replacements but as enhancers of human capabilities. Besides technological integration, the CAI must also manage and transform its internal structure to meet the digital world's requirements. The STAR Model stands out as a good framework that the CAI can leverage to evolve its operations, as this can help solve not just operations but also the issues of role ambiguity and responsibilities, which are the ultimate factors in maintaining good and healthy client-agency relationships. The forthcoming methodology section aims to explore this synergy empirically, providing insights into how the CAI can navigate its evolution, enhance relationships and innovation, and sustain its creative prominence in an ever-evolving landscape.

METHODOLOGY: STRATEGISING THE BRIEF

This study's methodology transcends mere data analysis, drawing significantly on the diverse cognitive competencies of people. It employs a mixed-methods approach, integrating quantitative and qualitative insights to uncover the CAI's evolving strategies. The intent is to pierce through the veneer of technological innovations, M&As, client-agency relationships, role responsibilities and value creation through research adopting an exploratory design to navigate the uncharted waters of technological integration within the CAI. Selected for its flexibility, this approach allows for a deeper understanding of complex, dynamic phenomena – particularly how a novel technological system could reshape agency-client interactions and operational models. This design supports an open-ended inquiry, enabling the discovery of new perspectives that align with the study's theoretical framework and paving the way for future-forward strategies in the advertising sector (Hay et al., 2019). The survey section consists of Part 1 – a Google Forms survey, and Part 2, face-to-face interviews.

PART 1 – DIGITAL FOOTPRINTS

This survey, conducted via Google Forms, effectively gathers quantitative and qualitative data to evaluate the advertising industry's current state and innovation receptivity. Designed to elicit insights into existing practices and future trends, it engaged 116 professionals from advertising and client sectors, ensuring a diverse sample. The data was then manually analysed to identify key sentiments and industry trends. The following Figures 1.1–1.8 show graphs and pie charts highlighting respondents' quantitative feedback. A total of 73.3% of the respondents are from the CAI, 23.3% are from clients and other industries (Figure 1.1). These data ensure insights are directly relevant to the CAI ecosystem. The demographic composition of participants, age, years of experience, and gender indicate a diverse and representative sample for a thorough analysis (Figures 1.2, 1.3, and 1.4).

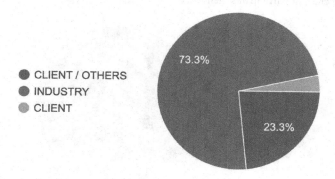

FIGURE 1.1 Pie chart representation of the sectors which participated.

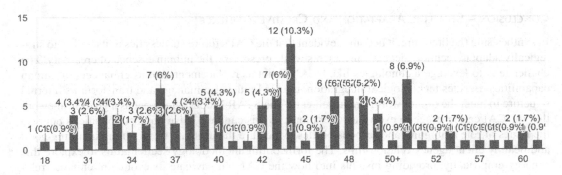

FIGURE 1.2   Graph chart of participants' age.

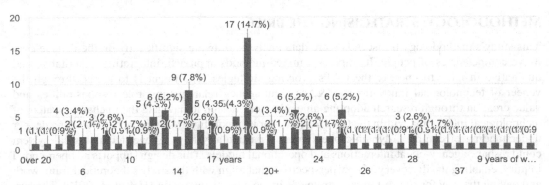

FIGURE 1.3   Graph chart of participants' work experience.

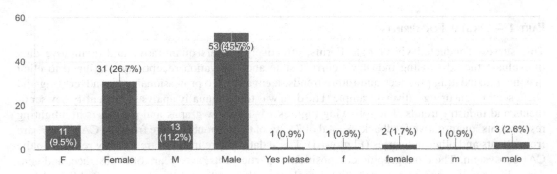

FIGURE 1.4   Graph chart of participants' gender.

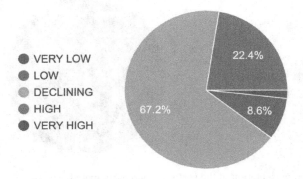

FIGURE 1.5   Trust and prominence in agencies.

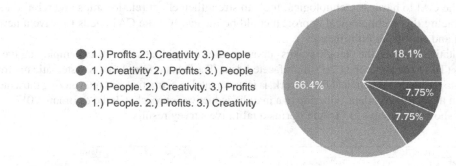

FIGURE 1.6   Preferred agency model culture.

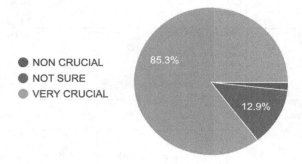

FIGURE 1.7   Technology influences.

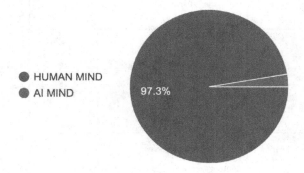

FIGURE 1.8   Balance between humans and AI.

A high of 67.2% reveals a noticeable decline in the trust and relationship connection between client-agency relationships (Figure 1.5). In all, 66.4% emphasised that people and talents are key in organisational structures (Figure 1.6). A staggering 85.3% agree that technology is crucial for the CAI (Figure 1.7) and 97.3% of respondents concur that the human mind is more important than technology's capabilities (Figure 1.8).

The data presented in Figure 1.5 highlight a notable decline in the CAI influence, a trend linked to eroding client trust and confidence. This issue is at the heart of the research, underlining the urgency for transformation within the industry. Figure 1.6 reinforces the idea that the intrinsic value of human talent holds more weight than profit margins, suggesting that the industry values creative human capital as the key to success and innovation.

Further insights from Figure 1.7 affirm the study's central argument: adopting technology is critical for revitalising the CAI. An impressive 85.3% of survey respondents recognise the indispensable role of technology in advancing the industry. This overwhelming consensus underscores the need for the CAI to harness technological tools to strengthen client relationships and rebuild its stature. Embracing this technological approach could be the catalyst the CAI needs to drive a new era of growth and restore its prominence.

Post-quantitative analysis, the Google Forms survey focuses on qualitative feedback, employing thematic analysis (TA) to decipher the CAI's multifaceted data. The TA, preferred for its systematic nature and relevance to the study's theoretical framework, is instrumental in extracting and interpreting intricate patterns within the qualitative data, adhering to a mainly deductive approach (Clarke & Braun, 2017).

Figure 1.9 shows the analysed Google Forms qualitative survey results.

QUESTION	CODES	FREQUENCY	THEMES
What factors influence the industry's prominence?	Technology/Platforms	37	Industry Dynamics and Challenges
	Savvy clients	19	
	Profits over Creativity	30	
	Consumer influence	26	Balancing Creativity with Commercial Goals
	Agency-Client Disconnect	8	
How should agencies evolve to stay relevant?	Embrace Technology	18	Agency Transformation
	Agile/Nimble	15	
	Creative Integrity	7	
	Agency-Client Relationship	15	Technological Integration and Agility
	Resource Integration	21	
	Incentivisation Change	7	
	Evolve Agency Model	4	
What adds more customer/client value: product or service?	Identifying the JTBD	27	User Goal Discovery
	Both are crucial	7	
What factors drive client choice in agency partnership?	Creativity	11	Value Dynamics
	Price	25	
	Managing relationships & needs	55	Brand Equity
	Reputation	20	
What are your views on co-creation partnerships?	Shared views, better outcomes	50	Collaborative Dynamics
	Trust and respect	7	
	Strengthens Partnerships	3	
	Can be messy	2	Operational Challenges
	Role Ambiguity	1	
What are your views on utilising AI for work?	Hasten Initial processes	27	Human-Centric Innovation
	Good to have only	5	
	Human first	9	
	Streamline Build Trust or Value	19	Agency Technological Evolution
	Against Machines	3	
	Hybrid	33	
	Reliability Issues	7	
	Free up time for thinking	13	
	New Job Roles	6	
	Evolution Of Agency Model	4	
What trends or innovations might shape the industry's future?	AI. GAI	50	Digital Experiences
	AR, VR, XR & Metaverses	17	
	Social Media	18	Technological Value Creation
	Content Creators/KOLs	3	
	WEB3	1	
	Innovative Mediums	1	
	Adding Value and Relationships	3	
	Hybrid	1	
	DATA	3	
How will technology and AI affect the industry and its workforce?	Job Displacement	24	Workforce Evolution
	Evolve to remain relevant	22	
	Efficiency	12	
	Emerging Jobs	36	Efficiency with Human Insight
	Free up time for thinking	4	
	New Opportunities	4	
	Over Reliance on Technology	10	
	Agency Model Restructure	20	
How do we integrate innovative strategies with human creativity to bring prominence back to the industry?	Technology as a tool only	25	Primacy of Human Creativity
	Technology is not solution	7	
	Technology lacks originality	3	
	Use Technology's efficiency	7	Technology as a Facilitator
	The Human mind is key	27	
	Educate, collaborate with Technology	26	
	Rules, Regulations, Mindfulness	3	

FIGURE 1.9   Analysis of qualitative survey data.

The questions presented have been succinctly reformulated for clarity while preserving their original purpose, achieving greater conciseness and precision. A coding method is employed manually by going through the data and examined to identify common themes, topics, ideas, and patterns of meaning that come up repeatedly. Themes will be generated based on the repeated codes and then reviewed against the data sets again to redefine and rename themes (Caulfield, 2019). Proof of concept tools will also be employed to support the evidence analysed from the data. Finally, a short conclusion explaining the main takeaways reveals how the analysis has answered the theoretical framework research question.

From the Google Forms qualitative survey analysis, 17 themes were conceived from the data sets shown in Figure 1.9. From the table, one can make out that "technology", "consumer influence", "human relationships", "value creation", "resource integration", and "transformation" registered top frequencies. Evidence points to Figure 1.5's pie chart that relationships need to be cultivated and strengthened, and in Figure 1.7, the influence of technology is high.

Part 2 – Voices from the Field

Face-to-face interviews provide in-depth insights into the advertising industry, focusing on the experiences and foresight of key leaders. Selected for their diverse and pioneering perspectives, these interviews offer unbiased, individualised feedback.

Figure 1.10 reflects the interviewee's answers, key excerpts, and insights.

The analysis in Figure 1.10 shows that human creativity remains central (1), and technology augments work, resulting in better relationships between agency and client (2).

To further scope the analysis and evidence to this research's theoretical framework, 12 key themes from the analysis in Part 1, Figure 1.9, will be correlated with the analysis from Part 2, Figure 1.10, to be illustrated in a causal loop diagram (CLD).

A hypothesis of an internal system structure of the CAI is mapped by linking causal relationships between variables to the model and developing a qualitative theoretical, conceptual CLD model. Since the structure of an agency is complex, the cause-and-effect relationships between variables can be multi-faceted. By employing the analysed data into a causal loop, one can distinguish complex interdependencies and feedback structures within the system, helping to identify potential reinforcing or balancing feedback loops that can drive internal system behaviour for possible transformation (Dhirasasna & Sahin, 2019). Figure 1.11 depicts the causal diagram.

The advertising industry's CLD encapsulates a dynamic interplay between human creativity, technological advancement, and the evolving nature of workforces. The primary loop suggests that the "primacy of human creativity" is central to generating "technological value creation". this relationship indicates that while technology catalyses innovation, human creativity ultimately drives value creation, supported by the literature identified (Davenport et al., 2020; Haenlein & Kaplan, 2019; Lu et al., 2023; Schick & Hinrich Schütze, 2020). Furthermore, the diagram highlights "Technology as a Facilitator", underpinning the notion that technology should augment human capabilities rather than replace them, as mentioned in Table 10's (1) face-to-face key excerpts:

> (1) Respondent acknowledges the central role of AI in the industry's future. However, he underscores that AI is based on past data and cannot predict the future, hence the need for human creativity to leverage AI in creating the future of advertising. This perspective highlights the synergistic relationship between AI and human creativity, essential for advancing the industry.

Hence, a highly trained machine with human creativity could become a central catalyst to bridge client-agency relationships with technological speed and accuracy. This theme leads to "workforce evolution", suggesting that as technology becomes more integrated into the industry, the skills and roles of the workforce must adapt accordingly, ensuring "efficiency with human insight" (Windels & Stuhlfaut, 2018). This is where the STAR Model framework can be applied to

---

Respondent 1-Industry practitioner of 30 years
Excerpts and key insights

Adaptation to Digital Technologies in Advertising:
The respondent highlights the critical shift from traditional to digital advertising, emphasising digital communication's interactive and two-way nature. He notes that those who have adapted to new technologies are thriving, suggesting that embracing digital transformation, including AI, is key to success in the advertising industry.

Role of AI and Human Creativity:
(1)   Respondent acknowledges the central role of AI in the industry's future. However, he underscores that AI is based on past data and cannot predict the future, hence the need for human creativity to leverage AI in creating the future of advertising. This perspective highlights the synergistic relationship between AI and human creativity, essential for advancing the industry.

Technology as a Catalyst, not a Replacement:
Emphasizing the advertising industry as a "people industry," respondent argues that technology is a tool or catalyst to enhance efficiency and the creative process but cannot replace the human element and interpersonal relationships. He suggests that technology can prompt desired behaviors in creativity and improve efficiencies, leading to cost savings and value enhancement. This view aligns with the idea of using technology like LLAMB not as a replacement for human touch but to enhance the creative and relational aspects of advertising.

Respondent 2-Industry practitioner of 24 years
Excerpts and key insights

Changing Dynamics in the Advertising Industry:
Respondent highlights how the advertising industry is no longer solely dictated by agencies. With the introduction of AI and the creation of in-house creative departments, clients are challenging traditional agency roles and seeking deeper business understanding from their agency partners. This shift signifies a need for agencies to adapt and provide more tailored solutions that align with client business goals.

AI's Role in Evolving Agency Operations:
Discussing the current trends in advertising, the respondent points out the growing use of AI tools like ChatGPT in strategy formulation. This evolution is changing the traditional agency structure, with some agencies operating without copywriters and relying on AI for creative tasks. This demonstrates AI's potential to redefine roles within agencies and offer innovative solutions, albeit with the risk of commoditisation.

AI as a Catalyst for Efficient Delivery:
(2)   Respondent observes that AI's incorporation into advertising processes can accelerate project completion and enhance trust in delivery timelines and budgets.
This efficiency underscores AI's role in nurturing stronger client-agency relationships by ensuring reliability and faster turnaround times.

FIGURE 1.10   Analysis of qualitative responses.

ensure organisational efficiency and prevent role ambiguity during resource integration. (Galbraith, 2011). "Human-centric innovation" is depicted as a core theme, reinforcing the idea that despite technological advances, the focus should remain on creating solutions prioritising human needs and experiences. This leads to the evolution of agency technology, where "agency technological evolution" pushes agencies to embrace new tools and platforms, for example, LLAMB, to enhance "collaborative dynamics" among team members, clients, and stakeholders. The "value dynamics" theme emerges from this collaboration, where the collective input of diverse individuals and

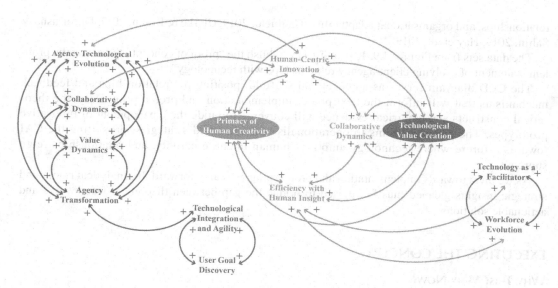

FIGURE 1.11 Causal diagram.

technology leads to a more profound creation of value. According to the literature, this is where the SDL theory fits in to encourage collaborations and build relationships so that VCC can happen to lead to a better understanding of the JTBD of users (Vargo & Lusch, 2014). This ties into Figure 1.10's (2) key excerpt on:

> (2) Respondent observes that AI's incorporation into advertising processes can accelerate project completion and enhance trust in delivery timelines and budgets. This efficiency underscores AI's role in nurturing stronger client-agency relationships by ensuring reliability and faster turnaround times.

This, in turn, necessitates an "agency transformation", evolving traditional structures to be more agile and responsive to market demands. Integrating technology with agility "technological integration and agility" enables agencies to discover and respond to "user goal discovery" swiftly, ensuring that advertising strategies align closely with user needs and market trends.

Overall, the CLD presents a holistic view of the advertising industry as a complex system where human creativity remains at the forefront, and technology serves as an enabler, nurturing strong relationship dynamics to evolve resources to achieve value to satisfy user goals and to adapt and drive the evolution of agencies constantly. It underscores the interconnectedness of these elements and their collective impact on maintaining the industry's prominence in an ever-changing digital landscape.

PROOF OF CONCEPT TO THE PROBLEM STATEMENT

This study's mixed-methods approach has provided a solid analytical foundation, merging quantitative data with qualitative insights to illuminate the CAI's intricacies. Survey findings and thematic analysis revealed 12 central themes in Figure 1.9. These themes have been juxtaposed with insights from face-to-face interviews in Figure 1.10 to construct the theoretical CLD in Figure 1.11, capturing the industry's essence. They reflect the industry's current state and suggest strategic directions for its evolution, highlighting the complex relationship between technology, human creativity, relationships, and organisational adaptability (Caulfield, 2019; Clarke & Braun, 2017; Dhirasasna & Sahin, 2019; Hay et al., 2019).

The data sets from Figures 1.9, 1.10, and 1.11 establish the "proof of concept" to answer the problem statement of "solving client-agency relationships with technology".

The CLD diagram serves as a conceptual scaffold, positing the relationships and feedback mechanisms that will inform the next phase: implementation and prototyping. Here, the theoretical constructs and empirical evidence will converge to guide the development of innovative prototypes. The prototype aims to operationalise the identified strategies, propelling the CAI towards a future where technology amplifies human talent, creativity, and relationships reign supreme.

As we pivot towards implementation and prototyping, we carry forward the analytical rigour and strategic insights gleaned thus far, poised to bridge the gap between theoretical frameworks and actionable solutions.

EXECUTING THE CONCEPT

WHY THIS? WHY NOW?

At a time when swift adaptation and meticulous precision are vital, the CAI stands on the cusp of a significant model shift. Highlighted and backed by empirical evidence and theoretical underpinnings through research and a push for digital innovation, the industry is urged to advance beyond old rules. Singapore's investment in a pioneering LLM programme epitomises this move towards targeted AI solutions (Goh, 2023). These emerging LLMs in the CAI mark a leap towards addressing complex, dynamic challenges and enhancing client-agency collaboration. They offer a harmonious blend of tech-driven efficiency and the irreplaceable spark of human creativity (1), crucial for strategic endeavours. LLMs can enhance collaboration and co-creation, resulting in richer idea generation, paving the way for impactful strategies and brand development, building trust, and closer client-agency relationships (2).

Privately owned LLMs can signify a shift in the CAI towards a more advanced business model. They go beyond mere ad campaign production to develop highly specialised "brains" tailored to individual brand stories and operational needs. According to the literature, this role evolution requires CAI professionals to become guardians of these digital intellects, feeding them with extensive historical and current data to stay ahead of market trends (Omneky, 2023; Windels & Stuhlfaut, 2018). Emphasising the need for an educational shift towards enhanced creativity and technological perception, as highlighted by Bill Gates, this approach prepares for a future where AI's complexity complements human skills, necessitating ongoing learning to stay relevant in a rapidly changing environment (Plante, 2023).

Imagine clients and brands equipped with their own fine-tuned LLMs, acting as digital stewards, instantly ready with extensive historical knowledge to tackle tasks, support, co-create, and spark innovation. These nimble, precise AI tools can overcome routine challenges, boosting client contentment. Freed from daily problem-solving, agencies can dive into deeper creative pursuits and co-create value, and in return, fuel LLMs with rich, human-driven insights. This creates a virtuous cycle where LLMs and human creativity enhance each other, transforming the advertising industry's landscape.

While the results can be exponential, this technological revolution brings many ethical and logistical difficulties. Considering the literature's critical review, while M&As can offer scale, they need more agility and response to facilitate connectedness between client and agency (Ghobadian & O'Regan, 2011). Another factor with size is that operational structures and resource integration can arise as pain points, resulting in role ambiguity and organisational failure (Ducoffe & Smith, 1994; Galbraith, 2011). In light of the concerns, a tighter, more agile setup proposal would seem more suitable and manageable for this research's theoretical concept (Norris, 2017). This revolutionary leap and vision promises to reinstate the CAI as a bastion of creative excellence, a reclamation of its erstwhile glory and innovative spirit.

TURNING VISION INTO REALITY

With research from the literature and a detailed methodology to unveil findings supported by a robust proof of concept CLD for this theoretical framework, this research now introduces the theoretical prototype "large language advertising model brain" (LLAMB), a groundbreaking tool designed to revolutionise the advertising industry.

This theoretical prototype presumes that the construction and programming of LLAMB, including the strategic resourcing of roles, have been carefully mapped out to ensure its effective operation. Future research must address the details of LLAMB's development and staffing.

LLAMB represents a pioneering blend of LLMs infused with IEAs, GAN, and analytical AI capabilities, going beyond the typical digital data bank assistant to become a dynamic partner in the creative process. This theoretical prototype is crafted to automate tasks and synergise with human teams, driving collaborative innovation and strategic planning within the CAI. As an intelligent system, LLAMB is equipped with a wealth of data, including historical campaigns and branding insights, enabling it to offer creative suggestions, generate visuals, craft narratives, brainstorm and provide strategic directions. With this knowledge, LLAMB is poised to revolutionise the advertising landscape, build stronger agency-client relationships through enhanced analysis and creative input and propel the industry's prominence towards a future of agility, insightful innovation and success (Haenlein & Kaplan, 2019; Jovanovic & Campbell, 2022; Tamkin et al., 2021; Yuki, 2023; Zhang et al., 2023). Continuing from this innovative leap forward with LLAMB, agencies are equipped to chase unparalleled creative heights and operational efficiency. LLAMB shines as a beacon of innovation, heralding a digital revolution within the CAI. This evolution is a proactive response to the industry's immediate needs and a visionary step towards defining its future. Embracing LLAMB is a testament to the industry's resolve to fuse technological prowess with human creativity, crafting advertising narratives that resonate deeply and forge lasting, collaborative relationships. (2)

An agency structure and operational framework (ASOF) is essential for creating a tech-forward agency to integrate LLAMB effectively. The Star Model, which focuses on aligning human-centric strategies with technological processes, stands out as a viable framework. Its five policies – strategy, structure, processes, rewards, and human resources – direct organisations and inform employees of engagement, ensuring cohesive direction and clear communication. Refer to Figure 1.12 for the Star Model framework.

FIGURE 1.12 Star Model.

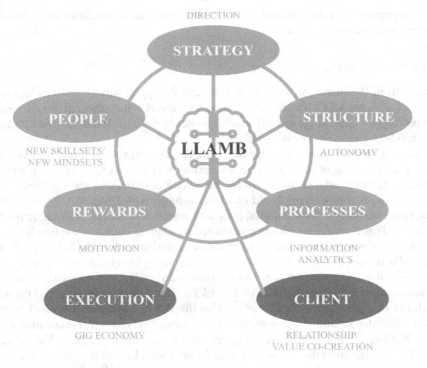

FIGURE 1.13   Star Model + LLAMB.

The integration of LLAMB as a central "brain" is depicted in Figure 1.13, modifying the Star Model to fit a modern agency. The ASOF adapts with slight changes to its actionable elements, namely the "people" policy with upgraded skills, the "structure" policy transitioning to autonomy, and the "processes" policy expanding to include analytics. Newly added execution and client policies focus on supplier relations and client engagement, respectively. This ASOF revision positions LLAMB at the core, bolstering policy efficacy and work processes and encouraging collaborative engagement within the agency ecosystem.

With the ASOF of the new technological agency structure complete, the next step is to build and test LLAMB's theoretical prototype functionalities. Graphical visual simulations will demonstrate LLAMB's role in enhancing data analysis, content creation, and decision-making in advertising workflow scenarios. These simulations, while theoretical, offer insight into LLAMB's potential impact on operational efficiency and collaboration within the CAI, considering both current technological limitations and future possibilities.

The prototype consists of two sets of simulated graphical visualisations. Figure 1.14 showcases LLAMB's ability to provide "data-driven insights" and "strategy optimisation". These visual simulations showcase LLAMB's capabilities in enabling efficient, responsive, and interactive meeting engagement. Its design streamlines data analysis and retrieval, enriching focused discussions.

Notably, LLAMB proactively assists in identifying media buy-in opportunities and cost-saving strategies, thereby enhancing client value and strengthening client-agency relations (Haenlein & Kaplan, 2019; Jovanovic & Campbell, 2022; Lu et al., 2023; Vargo & Lusch, 2014).

Figure 1.15 showcases LLAMB's ability to provide "collaborative workflow enhancement" and "generate creative content". The simulations show LLAMB's resourcefulness and process structure by demonstrating follow-ups in conversations with action. It also reflects how LLAMB integrates the most viable resources and links actors together to facilitate work progress internally and externally (as depicted in the ASOF Figure 1.13). LLAMB can also generate creative suggestions based on its deep

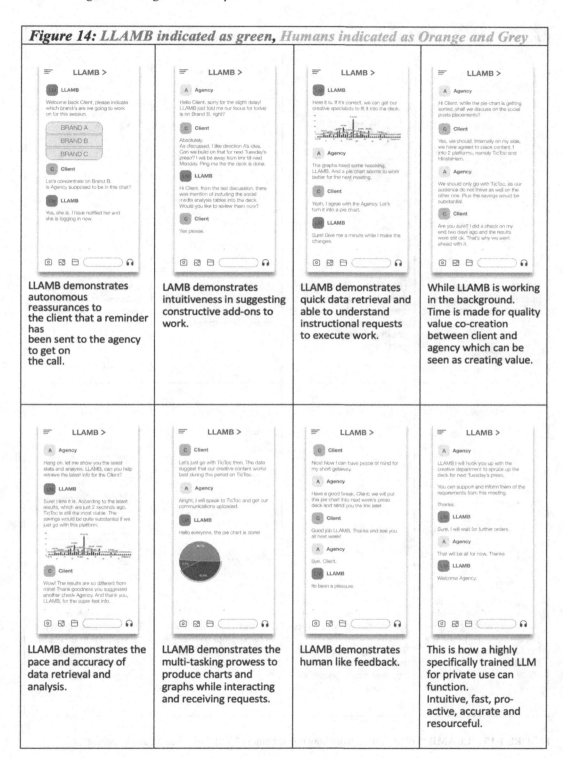

FIGURE 1.14 "Strategy optimisation".

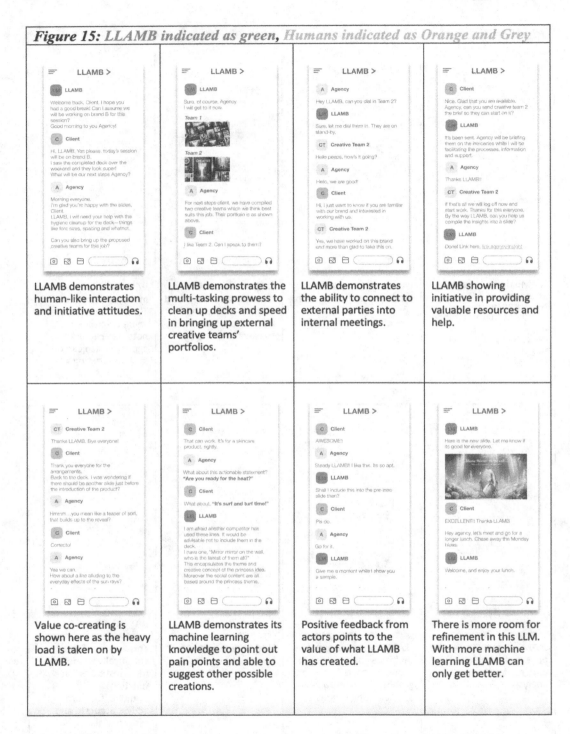

FIGURE 1.15   LLAMB "collaborative workflow enhancement" and "generate creative content".

industry knowledge. The demonstration shows how a specifically trained LLM model can support, facilitate and value co-create to produce value for clients and nurture better relationships (Haenlein & Kaplan, 2019; Jovanovic & Campbell, 2022; Lu et al., 2023; Vargo & Lusch, 2014; Zhang et al., 2023).

In summarising the prototyping phase of LLAMB, this theoretical model represents a significant stride in AI application within the advertising sector. LLAMB's development highlights the potential for AI to enrich the CAI with strategic and creative advancements simultaneously, bolstering the nurturing of trust in building better relationships. Its conceptual achievements include developing tailored AI knowledge and functions for advertising, an intuitive user interface for diverse professionals, and the ability to nurture collaborative environments for value co-creation, enhancing client engagement, campaign personalisation and, ultimately, better relationships.

TESTING AND FEEDBACK

In LLAMB's testing phase, theoretical validation and qualitative feedback will be gathered through expert reviews, peer evaluations, and hypothetical scenarios. This feedback, crucial for assessing the prototype's effectiveness and impact on client-agency relationships, will undergo Thematic Analysis to extract key themes and insights. Findings will be concisely presented, guiding LLAMB's future development in cultivating client relationships and value co-creation in the CAI. Figure 1.16 illustrates the

Response A	Response B	Response C	Response D	Response E	THEMES
Learning Curve, Collaboration Platform Suitability	API Access, User Interface Navigation	Practicality, Time Efficiency	Chat Functionality, Meeting Efficiency	Game-Changing, Easy Integration	(3) Efficient Integration
Creative Process Adaptation, AI Utility	Idea Generation, Human-Centric Approach	Statistical Tools, Shift from KPI to OKR	Knowledge Bank, Creative Ideation	Streamlining Processes, Alignment	(4) Creative Process Enhance-ment
Data Training	Data-Driven Planning	Efficiency, Information Retrieval	Advanced Analysis, Insight Generation	Knowledgeable Integrity, Unique Insights	(5) Data-Driven Strategic Insight
Human Element in Relations, AI Value Perception	Client Satisfaction, Remote Collaboration	Personalisation, Human-like Interaction	Chat-based Collaboration	Transparent Collaboration, Close Relationships	(6) Human-Centric Client Management
Ethical Risks	Privacy, Bias Prevention, Computing Requirements	Job Displace-ment		Security, Machine Learning Team, Cybersecurity	(7) Ethical Considerations in AI Deployment
ROI and Viability	Technological Evolution, Personalization		Real-time Feedback, Creative Material Collaboration	Versatility, Category Fine-Tuning, ROI Guarantee	(8) Future Potential and Industry Transformations

FIGURE 1.16 Prototype testing analysis.

LLAMB prototype testing analysis by five industry experts, focusing on its effectiveness and areas for improvement. The questions aim to gather in-depth insights, after which the feedback will be coded and themes derived.

After thoroughly analysing the feedback from the testing phase, results and findings are summarised and correlated to the CLD from Figure 1.11 to determine this research's theoretical framework further. The correlated results are as follows:

Theme (3) Efficient Integration:

Efficiency and user-friendliness were some feedback responses that acknowledged LLAMB's potential to improve workflow efficiency. Its intuitive response, fast analysis, and co-creation strengths were seen as factors that could facilitate smooth integration into existing systems. However, comments were on the learning curve, platform suitability, acceptance, and trust in using such technology.

Theme (4) Creative Process Enhancement:

LLAMB's impact on creative processes was perceived as a valuable tool for enhancing the creative ideation process. It could act as a knowledge bank, aiding in generating and refining ideas and strategies, especially by providing fast access to data and insights. It is suitable for streamlining work processes and collaborative alignment to prevent miscommunication or ambiguity.

Theme (5) Data-driven Strategic Insight:

The feedback highlighted LLAMB's ability to analyse consumer data effectively in data-driven strategic planning, thereby uncovering novel insights for campaign planning. Its predictive analytics and real-time monitoring capabilities were seen as key strengths. However, ethical implications were also mentioned in the form of data privacy and job displacements.

Theme (6) Human-centric Client Management:

LLAMB's potential to enhance client-agency relationships was a significant theme. Its capabilities in maintaining clear, transparent communication and collaboration were particularly emphasised, aligning with the research's focus on improving these relationships. The perception of value creation in using such technology is that it can improve trust and relationships between clients and agencies.

Theme (7) Ethical Considerations in AI Deployment:

Ethical and practical challenges, including data privacy, security, and potential job displacement concerns, were identified. The need for specialised teams to handle these aspects was also noted.

Theme (8) Future Potential and Industry Transformations:

Respondents envisioned LLAMB as a versatile tool with applications beyond advertising, suggesting its potential for cross-industry innovation. They also highlighted its capacity to evolve and adapt to future needs. One key mention was that the viability of LLAMB aiding in developing ROIs would be a game changer as this is a large part of running businesses.

Prototype test analysis in correlation with Figure 1.11 CLD:

- The findings reinforce the elements in the CLD Figure 1.11, particularly the loops involving 'technological value creation', 'collaborative dynamics', 'efficiency with human insight', and 'value dynamics'. These themes from the causal loop correlate integrally with (6) enforcing tighter client-agency relationship dynamics, (4) creative process efficiency, and (5) strategic decision-making.
- LLAMB's role in enhancing communication and collaboration from the testing feedback in (3), (4), (5), and (6) feeds into the positive feedback loop of client satisfaction and retention, as depicted in the CLD.
- LLAMB's efficient data analysis and strategic planning capabilities (5) align with the loops in the CLD as 'technology as a facilitator'.
- CLD Figure 1.11, 'Agency technological evolution', emphasises data-driven insights, leading to more effective campaigns and improved client outcomes. This aligns with (7) and (8).

Correlations from Figure 1.16 prototype testing analysis into Figure 1.11 causal loop diagram forms Figure 1.17.

In light of the testing phase, findings from Figure 1.17, the causal loop diagram, highlight LLAMB's potential as a transformative technological tool to be included in the internal system structure of the CAI. Together with Figure 1.13 of the Star Model ASOF, it resonates well with the chapter's theoretical framework and proof of concept, particularly its ability to streamline creative processes, enhance client-agency relationships, and provide strategic data-driven insights. However, the feedback also brings to light the need for careful consideration of ethical and security aspects, ensuring that LLAMB's integration is as beneficial as it is innovative. This comprehensive analysis, in conjunction with the two CLDs, solidifies the research and supports the envisioned role of LLAMB in shaping future advertising strategies.

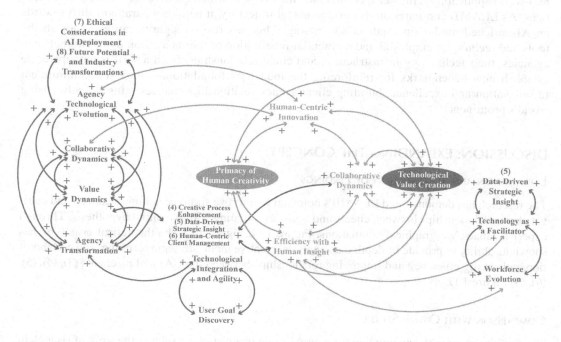

FIGURE 1.17 Causal loop on LLAMB analysis potential.

FUTURE AND IMPACT OF THE BRAIN

The Ascent of LLAMB in the Creative Advertising Industry

The testing phase of the LLAMB within the CAI has garnered optimistic feedback, signifying its capacity to bring about a transformative shift in advertising methodologies. LLAMB, by integrating IEAs and state-of-the-art Large Machine Models (LMMs), stands poised to fulfil the burgeoning industry need for synergised human-machine collaboration aimed at devising more effective and creative digital strategies and cultivating trusting relationships (Zhang et al., 2023). The effectiveness of LLAMB, however, hinges on a symbiotic relationship among AI developers, seasoned advertising professionals, and technology experts to forge a scalable and sufficiently versatile system that is ethically sound to adapt to the ever-evolving advertising landscape (5), (7).

Envisioning an initial rollout, a pilot programme within a smaller-scale agency might serve as the perfect incubator. Such an environment would offer a controlled setting for implementation while still promising a substantial impact. This approach aligns with the idea that agility and a deep understanding of client needs often trump sheer size (Dan, 2021). Starting small while thinking big can provide valuable insights into the operational intricacies of LLAMB in a real-world context. The specialised nature of LLAMB, with its unique model, is designed to revolutionise the utilisation of data, extracting nuanced consumer insights and thereby amplifying the creative output (4), (5), (8). Central to its design is the emphasis on value co-creation as this aspect is especially critical in nurturing deeper, more meaningful client-agency interactions, bolstering trust and fortifying relationships within the advertising ecosystem (2), (6) (Vargo & Lusch, 2017).

Despite its potential, it is imperative to recognise that LLAMB exists primarily as a theoretical prototype. While foreseeable, the bridge to its practical application remains a journey to be traversed. This recognition opens up vast avenues for further exploratory studies and technological refinements. Future research endeavours will be instrumental in ascertaining LLAMB's real-world applicability, impact spectrum, and the extent to which it can reshape advertising practices. As LLAMB continues on its developmental trajectory, it heralds a paradigm shift towards an AI-enriched modus operandi in advertising. This progression signifies an evolution in the tools and techniques employed and a potential redefinition of the interaction dynamics between agencies, their technology infrastructure, and client relationships. Such a shift is anticipated to establish new benchmarks for reinforcing the industry's foundational ethos and commitment to innovation and excellence, building client-agency relationships and restoring the advertising world's prominence.

DISCUSSION: EXPANDING THE CONCEPT

EXPLANATION AND JUSTIFICATION OF FINDINGS

The research has demonstrated LLAMB's potential to improve strategic communications and cultivate better relationships between clients and agencies, aligning with the initial hypothesis. Through expert feedback via graphical simulations, Figure 1.14, Figure 1.15, the theoretical prototype has shown its ability to provide in-depth data analysis, creative input, strategic planning, and collaboration, enhancing efficiency and improving relationships between the CAI and its clients (3), (4), (5), (6), (8) Figure 1.17.

COMPARISON WITH OTHER STUDIES

The findings align with existing literature on AI's role in advertising, such as the work of (Haenlein & Kaplan, 2019), who noted AI's potential to transform marketing strategies. Another study (Li,

2019) also mentions how AI is being integrated into the advertising process with various levels of impact and can be used to improve programmatic creatives and generate consumer insights in digital advertising. Further research by (Zhang et al., 2023) also points to evidence that IEAs can help cultivate trust and relationships between machines and humans. Unlike traditional agency models, LLAMB extends these capabilities, specifically focusing on the advertising sector, and offers a unique approach to nurturing client-agency relationships.

DISCUSSION OF LIMITATIONS

While the research offers promising insights, it acknowledges limitations. The prototype's evaluation was based on theoretical simulations rather than practical application, limiting empirical validation. Its current theoretical prototype concept still lacks the intricacies of actual programming and technicalities scope, presenting constraints and necessitating future development for real-world application. Furthermore, the feedback was gathered from a select group of industry experts, which has yet to exhaust the CAI's representation and diversity fully.

SUGGESTION OF FUTURE RESEARCH AREAS

Future research could explore the practical implementation of LLAMB, assessing its real-world efficacy and impact. Studies could also focus on a broader range of CAI professionals to better understand LLAMB's potential across different agency structures and client types. Moreover, exploring LLAMB's adaptability to other sectors could offer insights into its versatility and broader applicability (8). For instance, LLAMB can be leveraged to automate and personalise interactions at scale in sectors like customer service, public relations, and human resources. Beyond these, LLAMB's sophisticated analytical abilities, proven in understanding consumer behaviour and optimising campaign performance, could be invaluable in fields like healthcare for patient data analysis, finance for market trend prediction, and even the public sector for social issue analysis and public sentiment understanding (Ziegler et al., 2020). In the educational sphere, integrating technologies like LLAMB could prepare future professionals, including law enforcement officers, to work alongside AI, cultivating a workforce adept at utilising advanced tools for efficient problem-solving.

In the author's mind, one of the most compelling potential applications of LLAMB is in police investigative work, particularly in solving complex criminal cases like homicides. Here, LLAMB's advanced AI could be a game-changer, offering a new level of assistance in crime analysis. By accessing and analysing vast databases of criminal records, case histories, and forensic data, LLAMB could identify patterns, correlations, and anomalies that might elude human investigators. For instance, its ability to process and compare details from various cases could reveal a serial offender's signature methods or link seemingly unrelated incidents based on subtle similarities in methods, victim profiles, or geographical locations. The implementation of LLAMB in investigative teams could significantly accelerate the process of narrowing down suspects, understanding criminal behaviour, and even predicting potential future crimes based on historical data trends. This would expedite investigations and allow human officers to focus on critical aspects of the case, such as interviewing suspects, gathering physical evidence, and piecing together the crime narrative.

However, exploring LLAMB's role in crime-solving opens new research questions regarding the scalability of AI models, their ethical implications, and the balance between AI assistance and human decision-making in sensitive areas like criminal justice. These considerations are crucial to developing responsible AI policies and ensuring the ethical deployment of such technologies in sensitive sectors (Fischer et al., 2024).

More importantly, how can the CAI leverage LLAMB to make a change? As mentioned earlier, the CAI needs to restrategise and shift from solely an "idea generator" to a "solution facilitator", emphasising collaboration to co-create compelling work (Lee & Lau, 2018). A novel and creative business transformation would allow the industry to shift from mere creative solutions to providing strategic data and information services. The CAI and agencies can be providers of LLAMBs. Brains fine-tuned with a wealth of knowledge catered to specific brands and clients. These brains, which focus only on their specifics, are highly trained models capable of human-like interactions but with the speed and accuracy of technology. These brains can also be envisioned to be monetised in subscriptive forms, which means when a specific client needs to have a discussion or gather information, the charges could be time-based or clicked-based. The possibilities are beyond expected for now. New research and studies would have to address the further exponential possibilities.

INTERPRETING THE PRACTICAL IMPLICATIONS OF YOUR FINDINGS

The research into LLAMB reveals significant practical implications for the CAI. Using AI for in-depth data analysis and creative content generation, LLAMB presents a revolutionary approach to advertising strategies. Its capabilities in synthesising consumer insights and market trends could lead to more targeted and resonant advertising campaigns, thereby enhancing ROI for agencies and clients alike (8) (Szymanski & Lininski, 2018). LLAMB's potential in streamlining workflows and cultivating efficient decision-making processes indicates a shift towards more agile and responsive advertising practices. These data-driven shifts and technological infusion differ from traditional practices and could redefine the dynamics of collaborations and value co-creation, thus nurturing better client-agency relationships (2), (6). This suggests that LLAMB could be a critical tool in addressing the CAI's contemporary challenges, especially in adapting to the fast-paced digital environment and meeting the increasing demand for personalised and impactful advertising.

In the discussion surrounding the practical applications of the LLAMB, it is essential to underscore the ethical dimensions of its deployment. The responsibility for consumer data sits at the heart of LLAMB's ethical considerations, necessitating robust protocols to safeguard privacy and consent. In this era of data-driven decision-making, the potential for bias in algorithmic processes must be noticed; thus, LLAMB must incorporate mechanisms for fairness and equity, ensuring all consumer segments are represented accurately and without prejudice. Information of all clients and consumers must be made transparent on how it is used and for what purposes. LLAMB's development must be accompanied by clear lines of accountability, delineating who is responsible for its outputs and their consequences (5), (7) (Jovanovic & Campbell, 2022; Tene & Polonetsky, 2013). Furthermore, contemplation of LLAMB's influence on the employment landscape calls for a proactive approach to workforce transition and upskilling, ensuring that technological advancement does not come at the cost of human capital but rather augments and elevates the collective expertise of the field (7) (Smith et al., 2018).

On another note, with LLAMB's inclusion, the CAI might also see a rise in the freelance economy as agencies become more strategic and focus on their "new" brain-selling model and services. The freelance creative economy can leverage and support agencies in the outputs and execution of communications with better and faster outputs to meet clients' and consumers' demands.

Addressing these ethical imperatives aligns with moral business conduct and fortifies the trust in enduring client-agency relationships. (2)

REFLECTING ON THEORETICAL IMPLICATIONS

This research reveals that LLAMB's theoretical foundation in the CAI might provide valuable insights into the ongoing academic discussion about using AI in creative fields. This research bridges the gap between theoretical AI models and their application in advertising, contributing to a deeper understanding of how AI can augment human creativity and strategic thinking. It challenges traditional notions of AI as merely a tool for automation, proposing a more nuanced role of AI as a collaborator in creative processes. Furthermore, the study provides insights into the Service-Dominant Logic in marketing, emphasising the role of AI in co-creating value with clients (Vargo & Lusch, 2014; Zhang et al., 2023). By proposing LLAMB to enhance client-agency relationships, the research aligns with contemporary marketing theories that advocate for deeper, more interactive client engagement. This study thus positions LLAMB as a technological advancement and a catalyst for evolving advertising practices in line with modern marketing theories.

CONCLUSION: REVIEW THE CAMPAIGN

This chapter has researched the revolutionary potential of LLAMB within the CAI, focusing on how this advanced AI technology could revitalise client-agency relationships and instil renewed trust and collaboration in the industry. From a comprehensive review of the literature and data, the research emphasises the crucial role of LLAMB in bridging the gap between technological advancements and human-centric advertising strategies. The integration of LLAMB represents a significant leap into the future to address the challenges faced by the CAI, particularly in enhancing strategic communication and nurturing trust between clients and agencies (Laurie & Mortimer, 2019; Koslow et al., 2021). The theoretical concepts of this research are grounded in the belief that advanced AI, when tailored to the specific needs of the advertising ecosystem, can offer more targeted, efficient, and creative approaches in this digital age (Zhang et al., 2023. This aligns with the broader societal shift towards digital interconnectedness and data reliance, positioning LLAMB as a pivotal tool for future advertising success (2), (4), (5), (8).

The mixed-method methodology employed in this study, involving both quantitative and qualitative analyses, has validated the theoretical prototype of LLAMB. The findings reveal LLAMB's potential to augment operational processes, enrich the creative ideation phase, enhance the overall quality of work and cultivate the bonds of relationships. This aligns with the emerging needs of the CAI, where agility, precision, and a deep understanding of consumer behaviour are paramount. Furthermore, the research has illuminated the broader applicability of LLAMB-like technologies across various sectors. Theoretical predictions suggest that such advanced AI models could significantly contribute to healthcare, finance, education, and public service, particularly in crime-solving areas requiring complex data analysis and cognitive problem-solving Figure 1.9, Figure 1.17, (8) (Fischer et al., 2024; Ziegler et al., 2020).

As we conclude, it is recognised that LLAMB, in its theoretical form, may only become a practical model with further advancements in research. This prototype now serves as a foundation for exploration and development to shape LLAMB's trajectory, potentially making it an indispensable tool in the CAI and beyond. This research aims to stimulate discussions about AI in the CAI, offering a strategic approach to enhancing client-agency relationships, nurturing creativity, and fine-tuning human-machine collaboration in this ever-evolving advertising landscape. The findings lay the foundations for future research to integrate AI innovations like LLAMB across various sectors deeply, promoting industry advancement and technological progress. This journey of integrating AI innovations like LLAMB into the fabric of our industries signals a momentous shift, marking the dawn of the "Rise of the Machines".

REFERENCES

Alalwan, A. A., Rana, N. P., Dwivedi, Y. K., & Algharabat, R. (2017). Social media in marketing: A review and analysis of the existing literature. *Telematics and Informatics*, *34*(7), 1177–1190. https://doi.org/10.1016/j.tele.2017.05.008.

Alharbi, S., Attiah, A., & Alghazzawi, D. (2022). Integrating blockchain with artificial intelligence to secure IoT networks: Future trends. *Sustainability*, *14*(23), 16002. https://doi.org/10.3390/su142316002.

Arul, P. G. (2011). An explorative study on client's turnover reasons cited by ad-agencies and their clients. *Asia Pacific Business Review*, *7*(3), 146–154. https://doi.org/10.1177/097324701100700313.

Baar, A. (2023). DDB Taps AI to Quickly Produce Creative Proposals. Marketing Dive. https://www.marketingdive.com/news/ddb-uncreative-agency-ai-marketing/642555/.

Beard, F. (1996). Marketing client role ambiguity as a source of dissatisfaction in client-ad agency relationships. *Journal of Advertising Research*, *36*(5), 9.

Beard, F. K. (1999). Client role ambiguity and satisfaction in client-ad agency relationships. *Journal of Advertising Research*, *39*(2), 69.

Beard, F. K. (2017). The ancient history of advertising: Insights and implications for practitioners. *Journal of Advertising Research*, *57*(3), 239–244. https://doi.org/10.2501/jar-2017-033.

Berenguer-Contrí, G., Gallarza, M. G., Ruiz-Molina, M.-E., & Gil-Saura, I. (2020). Value co-creation in B-to-B environments. *Journal of Business & Industrial Marketing*, *35*(7), 1251–1271. https://doi.org/10.1108/jbim-01-2019-0061.

Cabiddu, F., Moreno, F., & Sebastiano, L. (2019). Toxic Collaborations: Co-Destroying Value in the B2B Context. *Journal of Service Research*, *22*(3), 241–255. https://doi.org/10.1177/1094670519835311.

Caulfield, J. (2019). How to Do Thematic Analysis | Step-by-Step Guide & Examples. *Scribbr*. https://www.scribbr.com/methodology/thematic-analysis/.

Chang, Y., Wang, X., Wang, J., Wu, Y., Zhu, K., Chen, H., Yang, L., Yi, X., Wang, C., Wang, Y., Ye, W., Zhang, Y., Chang, Y., Yu, P. S., Yang, Q., & Xie, X. (2023). *A survey on evaluation of large language models*. arXiv. https://doi.org/10.48550/arXiv.2307.03109.

Christensen, C. M., Hall, T., Dillon, K., & Duncan, D. S. (2016). Know your customers' jobs to be done. *Harvard Business Review*, *94*(9), 54. https://www.relativimpact.com/downloads/Christensen-etal-Jobs.pdf.

Clarke, V., & Braun, V. (2017). Thematic analysis. *The Journal of Positive Psychology*, *12*(3), 297–298. https://doi.org/10.1080/17439760.2016.1262613.

Dan, A. (2021). David Droga, who started in the mailroom, is now the king of advertising. *Forbes*. https://www.forbes.com/sites/avidan/2021/08/23/david-droga-who-started-in-the-mailroom-is-now-the-king-of-advertising/?sh=560ab6f464ee.

Davenport, T., Guha, A., Grewal, D., & Bressgott, T. (2020). How artificial intelligence will change the future of marketing. *Journal of the Academy of Marketing Science*, *48*(1), 24–42.

Davis, R. (2012). Negotiating local and global knowledge and history: J. Walter Thompson around the Globe 1928–1960. *Journal of Australian Studies*, *36*(1), 81–97. https://doi.org/10.1080/14443058.2011.647770.

Dhirasasna, N., & Sahin, O. (2019). A multi-methodology approach to creating a causal loop diagram. *Systems*, *7*(3), 42. https://doi.org/10.3390/systems7030042.

Ducoffe, R. H., & Smith, S. J. (1994). Mergers and acquisitions and the structure of the advertising agency industry. *Journal of Current Issues & Research in Advertising*, *16*(1), 15–27. https://doi.org/10.1080/10641734.1994.10505010.

Fischer, M., Metz, Y., Joos, L., Miller, M., & Keim, D. (2024). *MULTI-CASE: A transformer-based ethics-aware multimodal investigative intelligence framework*. arXiv. https://arxiv.org/abs/2401.01955.

Galbraith, J. R. (2011). *The Star Model™*. www.jaygalbraith.com. Retrieved from https://jaygalbraith.com/wp-content/uploads/2024/03/StarModel.pdf.

Ghobadian, A., & O'Regan, N. (2011). Building from scratch a marketing services giant by acquisition. *Journal of Strategy and Management*, *4*(3), 289–300. https://doi.org/10.1108/17554251111152298.

Goh, Y. H. (2023). $70m S'pore AI initiative to develop first large language model with Southeast Asian context. *The Straits Times*. 4 Dec. https://www.straitstimes.com/singapore/70m-s-pore-ai-initiative-to-develop-first-large-language-model-with-south-east-asian-context.

Haenlein, M., & Kaplan, A. (2019). A brief history of artificial intelligence: on the past, present, and future of artificial intelligence. *California Management Review*, *61*(4), 5–14. https://doi.org/10.1177/0008125619864925.

Hay, L., Duffy, A. H. B., Grealy, M., Tahsiri, M., McTeague, C., & Vuletic, T. (2019). A novel systematic approach for analysing exploratory design ideation. *Journal of Engineering Design*, *31*(3), 127–149.

Hendrickson, E. M., & Subotin, A. (2021). Mergers, acquisitions and magazine media in 2021. *Journalism Practice*, *17*(5), 1–15. https://doi.org/10.1080/17512786.2021.1960588.

Jovanovic, M., & Campbell, M. (2022). Generative artificial intelligence: Trends and prospects. *Computer*, *55*(10), 107–112. https://doi.org/10.1109/mc.2022.3192720.

Koslow, S., Sameti, A., van Noort, G., Smit, E.G., & Sasser, S. L. (2021). When bad is good: do good relationships between marketing clients and their advertising agencies challenge creativity? *Journal of Advertising*, *51*(3), 385–405.

LaBahn, D. W., & Kohli, C. (1997). Maintaining client commitment in advertising agency–client relationships. *Industrial Marketing Management*, 26(6), 497–508. https://doi.org/10.1016/s0019-8501(97)00025-4.

Lamberton, C., & Stephen, A. T. (2016). A thematic exploration of digital, social media, and mobile marketing: Research evolution from 2000 to 2015 and an agenda for future inquiry. *Journal of Marketing*, 80(6), 146–172. https://doi.org/10.1509/jm.15.0415.

Laurie, S., & Mortimer, K. (2019). How to achieve true integration: the Impact of integrated marketing communication on the client/agency relationship. *Journal of Marketing Management*, 35(3–4), 231–252. https://doi-org.ezproxy.tees.ac.uk/10.1080/0267257X.2019.1576755.

Lee, P. Y., & Lau, K. W. (2018). From an 'idea generator' to a 'solution facilitator'. *Career Development International*, 24(1), 2–17. https://doi.org/10.1108/cdi-03-2018-0080.

Levin, E., Lobo, A., & Thaichon, P. (2016). Enhancing client loyalty of advertising agencies: the influence of creativity and inter-firm relationships. *Journal of Contemporary Issues in Business and Government*, 22(1), 6. https://doi.org/10.7790/cibg.v22i1.30.

Li, H. (2019). Special section introduction: Artificial intelligence and advertising. *Journal of Advertising*, 48(4), 333–337.

Libai, B., Bart, Y., Gensler, S., Hofacker, C.F., Kaplan, A., Kötterheinrich, K., & Kroll, E. B. (2020). Brave new world? On AI and the management of customer relationships. *Journal of Interactive Marketing*, 51(1), 44–56.

Lu, Q., Qiu, B., Ding, L., Xie, L., & Tao, D. (2023). *Error analysis prompting enables human-like translation evaluation in large language models: a case study on ChatGPT*. ArXiv. https://doi.org/10.20944/preprints202303.0255.v1.

Macdonald, E. K., & Uncles, M. D. (2007). Consumer savvy: Conceptualisation and measurement. *Journal of Marketing Management*, 23(5–6), 497–517. https://doi.org/10.1362/026725707x212793.

marcel.ai. (n.d.). *Welcome | Marcel*. https://marcel.ai/public.

Norris, S. (2017). *Shifting advertising agency structure and evolving technology*. University Honors Theses. Paper 403. https://doi.org/10.15760/honors.399.

Omneky. (2023). *Omneky launches new 'Advertising LLM' capability*. https://www.prnewswire.com/news-releases/omneky-launches-new-advertising-llm-capability-301844472.html.

Paesano, A. (2021). Artificial intelligence and creative activities inside organizational behavior. *International Journal of Organizational Analysis*, 31(5), 1694–1723. https://doi.org/10.1108/ijoa-09-2020-2421.

Plante, J. (2023). Bill Gates on AI: People don't realize what's coming! *The Geopolitical Economist*. https://medium.com/the-geopolitical-economist/bill-gates-on-ai-people-dont-realize-what-s-coming-84dd9c248d35.

Rajaraman, V. (2014). John McCarthy — Father of artificial intelligence. *Resonance*, 19(3), 198–207. https://doi.org/10.1007/s12045-014-0027-9.

Sasser, S. L., Koslow, S., & Kilgour, M. (2013). Matching creative agencies with results-driven marketers. *Journal of Advertising Research*, 53(3), 297–312. https://doi.org/10.2501/jar-53-3-297-312.

Schick, T., & Schütze, H. (2020). *It's not just size that matters: Small language models are also few-shot learners*. ArXiv. https://doi.org/10.48550/arXiv.2009.07118.

Shavitt, S., & Barnes, A. J. (2020). Culture and the consumer journey. *Journal of Retailing*, 96(1), 40–54. https://doi.org/10.1016/j.jretai.2019.11.009.

Smith, N., Teerawanit, J., & Hamid, O. H. (2018). AI-driven automation in a human-centered cyber world. *2018 IEEE International Conference on Systems, Man, and Cybernetics (SMC), Miyazaki, Japan*, pp. 3255–3260. https://doi.org/10.1109/smc.2018.00551.

Stearns, P. N. (2006). *Consumerism in World History* (2nd ed). Routledge. https://doi.org/10.4324/9780203969885.

Szymanski, G., & Lininski, P. (2018). Model of the effectiveness of Google adwords advertising activities. 2018 IEEE 13th International Scientific and Technical Conference on Computer Sciences and Information Technologies (CSIT), Lviv, Ukraine, pp. 98–101. https://doi.org/10.1109/STC-CSIT.2018.8526633.

Tamkin, A., Brundage, M., Clark, J., & Ganguli, D. (2021). *Understanding the capabilities, limitations, and societal impact of large language models*. arXiv. doi https://doi.org/10.48550/arXiv.2102.02503.

Tene, O., & Polonetsky, J. (2013). Big data for all: Privacy and user control in the age of analytics. *Northwestern Journal of Technology and Intellectual Property*, 11(5), 239. https://scholarlycommons.law.northwestern.edu/njtip/vol11/iss5/1.

Turnbull, S., & Wheeler, C. (2015). The advertising creative process: A study of UK agencies. *Journal of Marketing Communications*, 23(2), 176–194.

Uncles, M. (2008). Know thy changing consumer. *Journal of Brand Management*, 15(4), 227–231. https://doi.org/10.1057/palgrave.bm.2550141.

Vargo, S. L., & Lusch, R. F. (2014). Inversions of service-dominant logic. *Marketing Theory, 14*(3), 239–248. https://doi.org/10.1177/1470593114534339.

Vargo, S. L., & Lusch, R. F. (2017). Service-dominant logic 2025. *International Journal of Research in Marketing, 34*(1), 46–67. https://doi.org/10.1016/j.ijresmar.2016.11.001.

Windels, K., & Stuhlfaut, M. (2018). New advertising agency roles in the ever-expanding media landscape. *Journal of Current Issues & Research in Advertising, 39*(3), 226–243.

www.warc.com. (2023). Will Agencies Become AI Factories? | *WARC | the Feed*. https://www.warc.com/content/feed/will-agencies-become-ai-factories/.

Yuki (2023). LLM in advertising. *Medium*. https://yukitaylor00.medium.com/llm-in-advertising-1093a61b4cde.

Zhang, H., Du, W., Shan, J., Zhou, Q., Du, Y., Tenenbaum, J. B., Shu, T., & Gan, C. (2023). *Building cooperative embodied agents modularly with large language models*. arXiv. https://doi.org/10.48550/arXiv.2307.02485.

Ziady, H. (2023). The world's biggest ad agency is going all in on AI with Nvidia's help. *CNN Business*. https://edition.cnn.com/2023/05/29/tech/nvidia-wpp-ai-advertising/index.html.

Ziegler, D. M., Stiennon, N., Wu, J., Brown, T. B., Radford, A., Amodei, D., Christiano, P., & Irving, G. (2020). *Fine-tuning language models from human preferences*. arXiv. https://doi.org/10.48550/arXiv.1909.08593.

2 Unlocking SME Potential
The Emergence of a Living Breathing Generative AI Knowledge Base

Edwin Tan Yurong

INTRODUCTION AND BACKGROUND

Established 12 years ago, Bravo has carved a niche as a boutique branding consultancy firm. Renowned for its creative and personalised approach, Bravo specialises in crafting distinct branding identities that resonate with a diverse clientele. Bravo's commitment to innovation and bespoke solutions has been a critical driver of its success in the ever-evolving landscape of branding and marketing.

Like many boutique firms, Bravo confronts operational challenges despite its achievements. These challenges encompass resource limitations and maintaining a competitive edge in a market characterised by rapid technological advancements and changing consumer preferences. Continuous evolution and adaptation are paramount for boutique firms to thrive.

This chapter delves into the operational hurdles Bravo faces and proposes a novel solution: integrating Gen AI technology into the firm's operations. The utilisation of Gen AI presents an opportunity to streamline workflows, enhance productivity, and optimise resource management. By harnessing the capabilities of Gen AI, Bravo can revolutionise its internal processes, making them more efficient and adaptable to the dynamic needs of the branding industry.

The focus of this investigation is twofold. First, it aims to explore how Gen AI can be integrated into Bravo's existing operational framework, identifying areas where AI-driven automation can be most beneficial. This includes examining the potential of Gen AI in automating mundane tasks, assisting with project planning, and providing data-driven insights for strategic decision-making. Secondly, the chapter seeks to evaluate the impact of Gen AI integration on team productivity and overall operational efficiency within Bravo. The goal is to assess whether Gen AI can streamline existing processes and foster innovation, ultimately leading to enhanced operations within Bravo.

In summary, this chapter aims to provide a comprehensive analysis of Gen AI's role in transforming the operations of a boutique branding consultancy firm. By exploring the potential benefits and challenges of Gen AI integration, it seeks to offer insights into how Bravo can leverage technology to maintain its competitive edge and streamline processes in a rapidly changing industry.

As we delve deeper into the challenges faced by boutique firms, it becomes evident that traditional operational models need to be fully equipped to handle the dynamic demands of today's market. This realisation leads us to a critical juncture where identifying specific operational challenges becomes imperative. The need to explore novel solutions that can enhance efficiency and productivity while managing resources effectively becomes a primary concern.

PROBLEM STATEMENT

In the ever-evolving landscape of branding consultancy, Bravo faces a significant challenge emblematic of many boutique-sized firms: balancing the need for specialised expertise against the constraints of a limited workforce. This challenge is particularly pertinent in today's fast-paced

DOI: 10.1201/9781003469551-2

and technologically driven business environment, where agility and innovation are crucial for success. Despite Bravo's proficiency in delivering highly specialised and creative branding solutions, its boutique nature limits scalability. This limitation is further compounded by the need for more specialised managerial roles, placing an undue burden on team members who must navigate many roles, from creative to administrative. Such a distribution of responsibilities leads to operational inefficiencies and risks employee burnout, potentially impeding creative processes and diminishing overall productivity.

This research chapter presents a novel approach to addressing these challenges, focusing on the potential integration of Gen AI as a transformative solution. Gen AI represents a cutting-edge development in artificial intelligence, offering unprecedented capabilities in automating routine tasks and enhancing decision-making processes. The integration of Gen AI in Bravo's operations could streamline administrative tasks, thereby allowing team members to concentrate more effectively on their core creative roles. Furthermore, Gen AI's ability to analyse complex data and provide insights could significantly enhance Bravo's operational efficiency.

This research is particularly relevant as it explores a unique intersection between technology and creativity within the niche field of branding consultancy. By investigating how Gen AI can be strategically leveraged to reinforce Bravo's unique strengths and overcome its operational challenges, this chapter aims to contribute valuable insights to branding consultancy and AI integration. Exploring Gen AI's potential in this context is not just about adopting new technology but strategically employing technology to foster innovation and competitive advantage. Thus, this chapter seeks to provide a comprehensive analysis of how Gen AI could revolutionise Bravo's approach to project management, team dynamics, and overall business strategy, potentially establishing a new paradigm in the competitive world of branding consultancy.

STATE OF THE ART: A LITERATURE REVIEW

The following literature review delves into the latest developments in branding consultancy, specifically focusing on integrating innovative technologies like Gen AI. This review covers studies, reports, and case studies that provide insights into how similar challenges have been addressed in other organisations, the impact of Gen AI on business models, workforce dynamics, and how it can be leveraged to improve operational efficiency and creative processes. Understanding these aspects is crucial for identifying how Gen AI can be tailored to meet Bravo's unique needs and overcome its specific challenges.

REAL-WORLD APPLICATIONS OF AI IN DIVERSE INDUSTRIES

The transformative impact of Gen AI across a range of industries has been well documented. Kanbach et al. (2023) present case studies that illustrate this impact, demonstrating how Gen AI revolutionises practices in software engineering, healthcare, and the financial sector.

In software engineering, AI-generated code now makes up a significant portion (41%) of new GitHub submissions, showcasing AI's potential to transform software development practices. In healthcare, AI aids in diagnostics and personalised treatments, particularly in mental health support. AI enhances investment strategies in the financial sector through advanced data analysis and risk assessment.

Furthermore, Soni (2023) explores the influence of Gen AI on revenue growth in small and medium-sized enterprises (SMEs), emphasising the interplay between human capital and technological infrastructure. The study reveals that while Gen AI can significantly boost revenues, its benefits are maximised when combined with skilled human resources and robust technological support. However, in highly competitive markets, the advantages of Gen AI may diminish as more competitors adopt AI technologies. Considering internal capabilities and market dynamics, this highlights the importance of a strategic approach to AI adoption.

Soni (2023) concludes that Gen AI's effectiveness in enhancing SME revenue growth varies depending on the competitive environment. The study underscores the need for a robust technological infrastructure to integrate AI effectively, suggesting that more than AI alone may be needed to guarantee improved financial performance in highly competitive environments. This calls for a balanced and strategic approach to AI adoption, considering internal capabilities and external market dynamics.

ECONOMICS AND PRODUCTIVITY

Current research is highly interested in the economic potential of Gen AI, as highlighted in McKinsey's report "The Economic Potential of Generative AI" (Chui et al., 2023). This report emphasises the vast economic impact Gen AI could have, with estimated contributions ranging from US$2.6 to US$4.4 trillion annually to the global economy. It identifies key areas such as customer operations and research and development where AI can significantly add value, transforming the way in which businesses operate and innovate.

One of the most notable insights from this report is the potential for Gen AI to change work dynamics, especially fundamentally in knowledge-based jobs. The report suggests that Gen AI and related technologies could automate up to 60–70% of activities currently occupying a substantial portion of employees' time. This shift would allow employees to dedicate more time to complex and creative tasks, enhancing productivity and fostering innovation in business processes and product development.

Furthermore, the report posits Gen AI as a pivotal force in the next wave of productivity growth. Rather than merely serving as a tool for task automation, Gen AI is seen as a fundamental driver of future productivity enhancements across various industries. This perspective underscores the transformative role of Gen AI in reshaping traditional business models and enabling new forms of creativity and innovation.

Eloundou et al. (2023) also contribute to this discussion by analysing large language models (LLMs) like generative pre-trained transformers (GPTs) and their potential impact on the US labour market. Their findings reveal that up to 80% of the workforce might see at least 10% of their work tasks influenced by LLMs. More strikingly, around 19% of workers could have over half of their functions affected by these technologies. This extensive influence of LLMs across all wage levels and industries highlights their potential as general-purpose technologies, capable of transforming a wide range of job roles and tasks.

ADAPTABILITY, VERSATILITY, AND BUSINESS MODEL INNOVATION

The adaptability and versatility of Gen AI are central to its transformative impact on business models and industry practices. As highlighted in McKinsey's report by Alex Singla and detailed by Chui et al. (2023), Gen AI's customisation capabilities are a key strength. This adaptability enables Gen AI to be tailored for specific tasks and industries, making it an exceptionally versatile tool capable of addressing unique business challenges. The ability to customise AI applications allows for a more precise and effective response to various business needs, ranging from routine operational tasks to complex strategic initiatives.

McKinsey's report further emphasises the broad applicability of Gen AI across different business functions and industries. Gen AI's influence extends to a wide range of business activities, from enhancing customer service and content creation to more complex tasks like product design and decision-making processes. This wide-ranging impact of Gen AI highlights its potential to reshape how businesses operate and innovate, offering new ways to address challenges and capitalise on opportunities.

Kanbach et al. (2023) delve into the role of Gen AI in business model innovation. Their analysis discusses Gen AI's impact across industries, highlighting its capacity to transform business operations, democratise knowledge, and create new revenue models. The integration of Gen AI in business strategies leads to a redefinition of skill sets, emphasising the need for workforce adaptation to these evolving technologies. This comprehensive analysis underscores Gen AI's transformative potential in reshaping not only business models but also workforce dynamics.

Additionally, Kanbach et al. (2023) discuss the evolution of workforce dynamics in the context of Gen AI. They highlight a significant shift from traditional roles to more AI-augmented functions, where humans transition from being creators to becoming curators and editors. This change in job roles and skill requirements underscores the need for new skills and approaches in the AI-driven business landscape. Their study reinforces the idea that Gen AI not only transforms business models but also reshapes job roles and skill requirements, necessitating a re-evaluation of traditional workforce structures and training methods.

KNOWLEDGE MANAGEMENT

In the realm of operational efficiency and decision-making, the role of Gen AI in revolutionising internal knowledge management systems (KMSs) is gaining increasing attention. McKinsey's report, as detailed by Chui et al. (2023), alongside earlier research by Tsui et al. (2000), provides valuable insights into this area. These reports suggest that AI can significantly enhance the way organisations manage and utilise their internal knowledge, leading to improved decision-making processes and increased operational efficiencies.

Sanzogni et al. (2017) address the limitations of AI in managing tacit knowledge, which is particularly critical in fields like branding consultancy. While AI is effective in processing explicit knowledge, it struggles with tacit knowledge, which includes the unspoken, intuitive insights gained from experience. This limitation is significant for industries that rely heavily on such knowledge, suggesting that AI's role in knowledge management is nuanced, with potential gaps in addressing the complexities of human expertise and intuition.

Jarrahi et al. (2023) explore AI's emerging role in knowledge creation, emphasising its capacity to recognise patterns and generate new insights. This is particularly relevant for organisations like Bravo, where AI's ability to analyse large data sets can lead to innovative branding solutions. The integration of AI into knowledge creation processes signifies a transformative shift in the way in which businesses harness and interpret vast data sets for strategic advantage.

Furthermore, Jarrahi et al. (2023) provide practical examples of AI in knowledge management. They highlight AI's application in legal sectors, where it is used to analyse and summarise legal precedents to determine case relevance. Another example is AI's role in marketing, where it facilitates feedback and smart sharing between marketing channels and sales pipelines. These instances demonstrate AI's practical utility in enhancing knowledge management processes, offering tangible benefits in various professional contexts.

Sanzogni et al. (2017) also delve into the role of AI in enhancing organisational learning and knowledge dissemination. They suggest that AI can facilitate the sharing and distribution of explicit knowledge across an organisation. However, this chapter also emphasises the crucial role of the human element in interpreting, understanding, and applying this knowledge effectively.

HUMAN-AI COLLABORATION

Human-AI collaboration is a critical aspect of integrating Gen AI into business processes, as explored in various studies. Sanzogni et al. (2017) emphasise the importance of contextual knowledge in AI applications, highlighting that AI systems, despite their advanced capabilities, may lack the context-driven understanding inherent to humans. This is particularly relevant in complex fields like branding consultancy, where contextual knowledge is key to strategic decision-making and creative processes. The chapter suggests that AI systems should be complemented by human expertise to navigate these nuanced domains effectively.

Prasad (2023) elaborates on the integration of Gen AI in business processes, highlighting the need to balance technological readiness with organisational dynamics. This balance is crucial for effectively adopting AI, especially for companies like Bravo, which are at the forefront of integrating AI into branding consultancy. The study underscores the complex interplay between technology and organisational strategy, advocating for a holistic approach to AI integration.

Sanzogni et al. (2017) also address the issue of AI dependency, warning against an over-reliance on AI systems for knowledge management and decision-making. They advocate for a balanced approach where AI complements rather than replaces human judgement. This perspective is vital to ensure that AI enhances rather than diminishes the role of human expertise in critical business functions.

Jarrahi et al. (2023) emphasise the symbiotic relationship between humans and AI in knowledge management. They argue that AI augments human capabilities, particularly in creative industries such as branding consultancy. This human-AI collaboration is pivotal as it enhances the creative and strategic processes, complementing rather than replacing human ingenuity.

Furthermore, Jarrahi et al. (2023) highlight the critical role of AI literacy among employees. They argue for the importance of training and skill development to enable effective utilisation of AI in knowledge management. This focus on human factors is essential for ensuring that AI tools are used to their full potential within an organisational context.

In the context of Bravo's integration of Gen AI, Dillion et al. (2023) offer crucial insights. While acknowledging AI's proficiency in mimicking human cognitive functions, they also highlight significant limitations, such as demographic representation and challenges in interpreting AI-generated data. This suggests that while Gen AI can enhance operational efficiency at Bravo by automating repetitive tasks and aiding in decision-making processes, it cannot fully replace the unique creative insights and nuanced understanding that human employees provide. Dillion et al. (2023) underline the need for a balanced approach in deploying AI within business environments, leveraging its capabilities while recognising the irreplaceable value of human contribution.

Bankins and Formosa (2023) explore ethical considerations in the integration of AI in the workplace. They highlight how AI influences the concept of meaningful work across various dimensions, such as task integrity, skill development, and autonomy. Importantly, they argue that while AI can enhance work experiences by automating mundane tasks and fostering skill growth, it also poses risks to workers' sense of autonomy and belonging. For example, in scenarios where AI systems make significant decisions, employees might feel their expertise and judgement are undervalued, leading to a reduced sense of autonomy. Additionally, if AI technologies lead to significant changes in team structures or reduce the need for collaborative tasks, this could impact the sense of belonging and camaraderie among team members. This underscores the need for a balanced approach in AI implementation that respects and enhances human roles and relationships in the workplace.

The ethical framework suggested by Bankins and Formosa (2023) emphasises beneficence, non-maleficence, and justice, advocating for a balanced approach in AI deployment to preserve the human-centric aspects of work. This comprehensive ethical framework ensures that AI contributes positively to the workplace, prevents inadvertent harm, and mandates fair and equitable distribution of AI's advantages and responsibilities. Dillon et al. (2023) emphasise that such an ethical approach is crucial not only for maintaining the integrity and meaningfulness of work but also for fostering a sense of belonging and autonomy among employees. This approach advocates for a balanced, human-centred AI integration, ensuring that technological advancements complement rather than compromise the human aspects of work environments.

IMPLEMENTATION OF AI

The implementation of AI in business processes, especially in boutique firms like Bravo, requires a careful consideration of various factors. Del Giudice et al. (2023) highlight the significant role

AI can play in enhancing human capabilities within the workplace. Their research emphasises the synergy between AI and human resource management, underscoring the potential for AI to augment rather than replace human skills. This is particularly relevant for boutique firms, where the integration of AI could lead to more efficient operations and improved employee productivity. The model proposed by Del Giudice et al. stresses the importance of a human-centred approach in AI adoption, ensuring that technology acts as a complement to human expertise, fostering a more dynamic and effective work environment.

Jarrahi et al. (2023) discuss the operational necessities for integrating AI into knowledge management, highlighting the importance of preparing data adequately, developing infrastructure for AI integration, and redesigning processes to facilitate AI-human collaboration. Prasad (2023) examines the factors influencing the adoption of generative AI in organisations, including technological readiness, organisational culture, and environmental influences. This analysis is crucial for understanding the multifaceted approach required for effective AI adoption in companies like Bravo.

Additionally, Weber et al. (2023) provide guidance for small-scale projects, advising that strategic AI integration involves tailored project planning and effective data management. This approach ensures that AI initiatives align closely with the boutique firm's specific goals and resource constraints. By doing so, Bravo can efficiently implement AI solutions, maximising the benefits of AI technologies while maintaining manageability and sustainability within the firm's operational scope. This strategy is pivotal for leveraging AI to enhance productivity and innovation in a resource-sensitive environment.

Low/No-Code

The low/no-code development movement is transforming the software industry, empowering individuals without formal coding skills to create applications and software solutions. This approach is beneficial for non-IT professionals, enabling them to build simple business applications rapidly. Yan (2021) discusses low/no-code development's role in digital transformation, noting its significant impact on future software development and digital transformation. Vera et al. (2022) examine the impact on software developers, highlighting the implications for software professionals and the evolution of their roles in the near future. Moskal (2021) discusses no-code technology in business digitalisation, viewing it as a solution for businesses needing customised IT solutions but facing budget constraints.

Wang and Wang (2021) explore educational applications, showing how no-code/low-code app development is transforming traditional information system development paradigms and being integrated into educational modules with positive outcomes in student learning. Kulkarni (2021) studies low-code/no-code platforms, common platform features, and their suitability for rapid application development. ElBatanony and Succi (2021) envision a future in software development where coding is not necessary, and applications can be created through principles like configuration-driven development and cloud computing.

Sundberg and Holmström (2023) discuss the advent of no-code AI platforms, offering a transformative approach for boutique firms like Bravo. These platforms democratise technology, enabling businesses without extensive technical expertise to leverage AI and machine learning capabilities effectively. This shift allows for rapid, user-friendly development and deployment of AI solutions, streamlining processes and enhancing efficiency. Such platforms could be instrumental in Bravo's pursuit of integrating AI into its operations, offering a cost-effective and accessible route to harnessing AI's potential.

However, Sundberg and Holmström (2023) emphasise the need for careful assessment of no-code AI platforms due to the unique challenges each business faces. While these platforms are user-friendly, they may not always align perfectly with specific business objectives or data requirements.

Therefore, firms like Bravo need to evaluate whether a particular no-code AI solution can effectively address their unique needs and enhance their specific business processes, ensuring that the AI integration is not only accessible but also strategically beneficial.

In summary, the literature review has highlighted the diverse applications of Gen AI in various industries, emphasising its potential in enhancing creativity, improving efficiency, and facilitating knowledge management. These insights have revealed key areas where Gen AI can significantly impact the operations of a boutique branding consultancy firm like Bravo.

Drawing on these insights, the methodology for this research has been carefully designed to address the specific challenges identified at Bravo. The literature underscores the importance of a tailored approach to AI integration, considering the unique needs and dynamics of the organisation. Accordingly, the following methodology section outlines a structured approach that not only aligns with the theoretical underpinnings identified in the literature but also caters to the practical realities faced by Bravo.

METHODOLOGY

The methodology for integrating Gen AI at Bravo adopts a comprehensive, multi-phased approach, beginning with an initial qualitative survey. This survey, utilising in-depth interviews and detailed questionnaires, aims to gather insights from various roles within the organisation to understand their current needs, challenges and expectations regarding Gen AI integration. The focus is to identify specific areas where Gen AI, particularly GPT, can be most beneficial, aligning with the findings of Soni (2023), who emphasises the importance of understanding internal capabilities and market dynamics when adopting AI technologies.

Based on the insights from this survey, we will develop hypotheses about Gen AI's potential impact at Bravo. The next phase involves the development and training of GPT models, selected for their cost-effectiveness and versatility. These models will be tailored to meet the identified needs, ensuring they address Bravo's diverse requirements without incurring high costs.

The integration of these GPT models into Bravo's operations will be conducted through hands-on experiments, focusing on areas identified as most beneficial for the team. This phased implementation will be guided by continuous feedback, allowing for the refinement of GPT applications. A final follow-up qualitative survey will conclude the project, evaluating the effectiveness of the integration and guiding future decisions regarding the continuation of Gen AI use at Bravo.

INITIAL QUALITATIVE SURVEY

Following our methodology, we conducted in-depth interviews with key Bravo team members Jane, Fangling, and Fred. These interviews were instrumental in understanding the specific challenges faced in their daily roles and how Gen AI could be effectively integrated to address these challenges. The insights gathered offer a nuanced view of the potential applications of Gen AI in a boutique branding consultancy setting. The detailed perspectives of each team member highlight their unique expectations and visions for Gen AI at Bravo. This approach ensures our Gen AI integration strategy is deeply informed by the actual needs and experiences of our team.

Jane

Jane holds a pivotal role at Bravo, frequently approached with a wide range of inquiries that reflect her involvement in both administrative and strategic domains. Her responsibilities include addressing logistical concerns such as office relocations and hiring initiatives, managing administrative processes like leave requests and onboarding procedures, and handling financial queries. Additionally, Jane is involved in more nuanced, judgement-based issues such as problem-solving strategies, client communication tactics, and project workflows. Beyond her professional responsibilities, Jane is also seen as a central figure for team support and morale, playing a role that extends to emotional support and team cohesion.

In her capacity at Bravo, Jane anticipates leveraging AI to enhance her operational capabilities. She expects AI to assist in identifying optimal workflows, managing resources efficiently, and making informed decisions about task delegation, such as selecting the most suitable designer for a project. Jane envisions AI as an analytical tool that augments decision-making and streamlines procedural tasks.

For the potential AI personal assistant, Jane envisions a comprehensive role that includes drafting quotations and invoices, planning project timelines, coordinating production schedules, composing detailed emails, ensuring systematic client feedback, and overseeing project management facets like planning, calendar organisation, and time management, as well as providing reminders. She also values AI's utility in transcribing verbal communications from platforms like Zoom, highlighting the importance of accurately capturing and utilising conversational data.

Looking to the future, Jane's aspirations for AI are twofold: to facilitate the team's adherence to their responsibilities through proactive reminders and to contribute to solving "human problems" such as team morale. This indicates an aspirational desire for AI to transcend its current bounds and engage with the more complex and subtle aspects of human dynamics within the workplace.

Fangling

As a new member of the Bravo team, Fangling, stepping into the role of project manager, naturally seeks guidance to navigate her new environment. Her inquiries to Jane cover a range of topics, from administrative issues such as leave policies and post-probation procedures to operational questions like the existence of templates or protocols for project actions and standards for evaluating performance.

Fangling's responsibilities as project manager involve coordinating schedules, assessing the work of creatives, and effective communication through email. She expresses interest in a generative AI tool that can support these tasks by identifying free time slots for meetings, offering a critical eye on creative work, and aiding in email composition. Fangling also values AI that can adapt to her personal work habits, styles, and preferences, thereby creating a more personalised and efficient workflow.

In addition to her immediate responsibilities, Fangling is curious about broader organisational objectives, seeking insight into the company's priorities, overarching goals, and strategies for individual career development. She is eager to understand how to enhance her performance and seeks advice for professional improvement.

Fangling's aspirations for AI include assisting in planning project timelines, gauging the time required by creatives for specific tasks, and optimising project planning and execution. Her vision for AI encompasses not just administrative efficiency but also refinement and enhancement of the strategic aspects of project management.

For daily use, Fangling sees AI as a tool to assist in writing emails, inspire creativity, aid in copywriting tasks, and serve as a digital aide for setting reminders and timing tasks. These expectations reflect a desire for AI to be integrated into the fabric of her daily work, providing both practical support and creative stimulation.

Fred

Fred, a senior strategist at Bravo, interacts closely with Jane to acquire updates and information crucial for strategic planning and execution. Her inquiries reflect a deep engagement with various facets of the company's operations, encompassing new client inquiries, the availability of team members like Satish, project timelines, client-provided information, and internal resource locations. Fred's requests also indicate a need for reminders about procedural matters, such as managing social responsibilities and strategic advice for improving project approaches.

In her daily activities, Fred desires AI to function as an extended cognitive apparatus – a "second brain" – that can remind her of tasks, expedite the search for information and documents, aid in research, foster creativity, and assist in note-taking. She values AI's ability to digest and synthesise data, which can lead to optimised day-to-day efficiency.

Fred's vision for AI extends beyond her immediate role, harbouring aspirations for it to augment disabilities and capabilities, improve the quality of life, and enhance educational prospects, particularly in developing countries. She hopes AI will become an inclusive tool that helps individuals from all walks of life, such as the elderly, to learn and adapt, ultimately contributing to the greater good without leaving anyone behind due to AI illiteracy.

Her utilisation of AI in daily tasks includes leveraging tools like ChatGPT for copywriting, editing, and brainstorming. Fred's perspective on AI is resoundingly optimistic; she appreciates its utility at work, especially its capacity to serve as an always-available editor. She acknowledges the need to develop proficiency in using AI effectively and does not perceive it as intimidating but as a beneficial addition to her professional toolkit.

ANTICIPATED OUTCOMES

Building on the valuable insights gleaned from our initial qualitative surveys with Jane, Fangling, and Fred, we are now poised to translate these findings into practical Gen AI applications at Bravo. The primary objective of this research is to explore and implement Gen AI solutions that are intricately tailored to address the specific challenges and aspirations identified by our team members.

Each team member's unique perspective on how Gen AI could enhance her role at Bravo has informed our approach to integrating this technology. Jane's desire for an AI tool to assist with operational and strategic tasks, Fangling's interest in AI support for project management, and Fred's vision for AI as an extended cognitive apparatus have collectively shaped the direction of our Gen AI strategy. We aim to develop AI solutions that optimise working hours, automate repetitive tasks, enhance decision-making processes, and provide advanced cognitive assistance, all while being attuned to the diverse needs of our team.

The anticipated outcomes of this initiative include increased overall productivity and profitability, streamlined operations, and the potential transformation of traditional roles. This endeavour aligns with Bravo's strategic goal of leveraging cutting-edge technology to maintain a competitive edge in the boutique branding consultancy sector. We anticipate that Gen AI will simplify current operations and open new avenues for creativity and innovation.

With the insights from our team members as our guide, we are now moving into the experimentation phase. This phase will involve hands-on trials of Gen AI applications, directly stemming from the needs and challenges highlighted in our surveys. These experiments are crucial for evaluating the practical effectiveness of Gen AI in a real-world setting and will provide a comprehensive understanding of its impact on Bravo's operations.

EXPERIMENT

To address the operational challenges at Bravo, we embarked on a strategic initiative to integrate Gen AI using a no-code approach. This decision was driven by two primary considerations: the limited budget for technological advancements and the need for specialised AI expertise within the firm. Our objective was to harness the potential of Gen AI in a way that was both cost-effective and accessible to all team members, regardless of their technical background.

The no-code strategy was particularly appealing as it allows for the quick and flexible implementation of AI tools without extensive programming knowledge. This approach empowers employees to experiment with AI tools, fostering an environment of innovation and creativity. By enabling team members to engage directly with AI, we aim to demystify the technology and encourage a culture of exploration and learning.

Our overarching hypothesis for this initiative is that Gen AI can significantly reduce workforce requirements, enhance productivity, and increase the firm's profitability. To test this hypothesis, we planned a series of experiments, each carefully designed to evaluate the impact of Gen AI in various aspects of Bravo's operations. These experiments are critical to understanding how AI can be practically applied in a boutique branding consultancy setting and what tangible benefits it can bring.

Each experiment is assessed based on three critical criteria: desirability, feasibility, and viability. Desirability refers to the degree to which the AI implementation meets the team members' needs and expectations. Feasibility considers the technical and operational aspects of integrating AI into existing processes. Viability evaluates the long-term sustainability and economic impact of AI solutions. This iterative process of experimentation and reflection aims to uncover the most beneficial applications of AI for Bravo, thereby optimising our approach to integrating these technologies into our business model.

The experiments, detailed in the following sections, represent a practical exploration of Gen AI's capabilities and limitations. They provide a real-world testing ground for our theories and assumptions about AI, offering valuable insights that will shape the future of technology integration at Bravo.

Experiment 1: Evaluating a GPT Chatbot as an Employee Handbook Knowledge Base

OBJECTIVE

The primary goal of Experiment 1 was to address a specific challenge at Bravo: the absence of a centralised knowledge base (KB) for administrative queries. This lack of a centralised resource led to an over-reliance on Jane, who became the go-to person for various inquiries, from administrative tasks to strategic decisions. This scenario could have been more efficient and placed undue pressure on Jane. A GPT chatbot could serve as a distributed knowledge system, easing Jane's burden and potentially reducing the need to hire additional managerial staff, such as a general manager. This experiment draws on the insights from McKinsey's report (Chui et al., 2023), which suggests that Gen AI can automate up to 60–70% of activities in knowledge-based jobs, potentially enhancing productivity at Bravo.

IMPLEMENTATION

In a proactive move, we rapidly utilised ChatGPT to develop an essential KB during a lunch break. The AI tool generated a content list for an employee handbook (see Figure 2.1), covering key topics relevant to Bravo's operations. We then expanded on these points to provide more detailed information, focusing on the most critical aspects for employees. The content was organised and compiled into a coherent KB using Notion, a versatile note-taking and project-management tool.

Our initial plan was to use Notion's web publishing feature to allow the GPT chatbot to access and update the KB in real time. However, we encountered technical challenges, as GPT could not access the published URL. As a workaround, we exported the KB from Notion as a PDF and uploaded it to ChatGPT, successfully facilitating access to the content. A group of employees was then selected to interact with the GPT-powered KB, and the group's queries and interactions were closely monitored and recorded. This data served as a basis for refining the GPT's responses and enhancing its ability to provide accurate and relevant answers to the commonly raised queries by Bravo's employees.

DESIRABILITY

Assessment: To evaluate the GPT chatbot's desirability, we conducted surveys and interviews with Bravo employees who interacted with the tool. We focused on user experience, relevance of information, and overall satisfaction with the chatbot as a knowledge resource.

Results and Analysis: The feedback revealed a positive reception to the chatbot's ease of use and ability to answer common queries quickly. However, some employees expressed concerns about the depth and comprehensiveness of the information provided, highlighting a preference

Employees' Handbook

By community builder A

Ask me about any admin queries you might have while working in Bravo.

Can I claim cab?	When do we start work after CNY?
When will I get paid?	Can I bring my pet to work?

📎 Message Employees' Handbook

FIGURE 2.1 Bravo employee handbook GPT

for a more structured and complete handbook alongside the chatbot. This indicates a need for a balanced approach, combining AI-driven efficiency with comprehensive human-driven resources.

FEASIBILITY

Technical Evaluation: Technically, integrating the GPT chatbot presented challenges, notably the inability to access the Notion-published KB directly. The workaround using a PDF export proved successful, but it highlighted the need for more seamless integration solutions.

Operational Assessment: Operationally, the chatbot required minimal employee training and fitted easily into their daily workflow. However, ongoing maintenance and updates to the KB content remain necessary to ensure its relevance and accuracy.

Results and Analysis: Overall, the chatbot was feasible to integrate from both a technical and an operational standpoint. The initial challenges were surmountable, and the tool has shown a potential to become a regular part of Bravo's workflow.

VIABILITY

Economic Impact: The development and maintenance costs of the chatbot were relatively low, especially considering the time saved in reducing the workload on key personnel like Jane. The potential for reducing the need for additional managerial staff could lead to significant cost savings.

Strategic Implications: Strategically, the chatbot aligns with Bravo's goals of innovation and efficiency. Its adaptability to evolving company needs and potential scalability make it a viable long-term solution.

Results and Analysis: The chatbot's economic and strategic viability appears promising. While it cannot replace human expertise in strategic decision-making, it is an effective tool for enhancing operational efficiency and reducing dependency on specific employees.

INSIGHTS AND REFLECTIONS

The experiment yielded several important insights. First, while the GPT chatbot effectively responded to specific queries from employees, it became apparent that it might only partially replace the comprehensive nature of a traditional employee handbook. Employees tend to ask only about topics they thought they needed to know, potentially missing critical information they needed to be made aware of. This limitation highlighted the potential challenges of relying solely on an AI-powered chatbot for knowledge dissemination. It underscored the importance of a structured and complete handbook that comprehensively conveys all necessary information.

Furthermore, the experiment shed light on Gen AI's potential in reducing dependency on key personnel like Jane. The chatbot successfully eased some of her administrative workload but also revealed the nuanced and strategic aspects of a general manager's role that AI could not replicate, such as high-level decision-making, strategic planning, and handling human management elements.

One of the most significant takeaways from this experiment was the rapid development of the KB using ChatGPT. The process demonstrated that what was previously thought to require months of work could be accomplished quickly with AI assistance. This revelation was a testament to ChatGPT's efficiency in expediting content creation and highlighted the potential for AI to streamline complex tasks and achieve significant time savings.

CONCLUSION

The findings from Experiment 1 suggest that, while AI can significantly streamline information creation and reduce dependency on specific employees, more is needed to replace the need for comprehensive human-driven resources, such as a traditional employee handbook or the strategic role of a general manager. The experiment highlighted the value of a hybrid approach that leverages both AI for efficiency and human expertise for comprehensive management and strategic guidance. This approach could be a more practical solution for Bravo, combining the speed and scalability of AI with the nuanced understanding and decision-making capabilities of human managers.

During the implementation of Experiment 1, our initial objective was to develop a rudimentary KB to test GPT's capabilities. Creating a comprehensive KB at Bravo had been postponed for a long time, mainly due to the perception that compiling all the necessary information would be overwhelming and time-consuming. However, as we began working with ChatGPT, we experienced a significant revelation: creating a fully-fledged, detailed KB was far less daunting than anticipated. It turned out to be remarkably straightforward and efficient.

This experience fundamentally altered our approach to task management at Bravo. It underscored the importance of re-evaluating long-held assumptions about the complexity of specific tasks and opened up new possibilities for leveraging AI to enhance productivity and operational efficiency. The experiment demonstrated that AI, when used creatively and in tandem with human expertise, can transform seemingly formidable tasks into manageable and efficient processes.

Experiment 2: Assisting in Bespoke Project Planning with GPT

OBJECTIVE AND BACKGROUND

The primary goal of Experiment 2 was to explore how GPT could assist in the project planning process for bespoke projects at Bravo. Given the unique nature of each project, standardising the planning process was challenging. Jane, a key employee frequently overwhelmed with managers' questions on planning these unique projects, was identified as a primary beneficiary of this AI assistance. The experiment aimed to use GPT to support project planning, reduce Jane's workload and enhance managerial efficiency.

METHODOLOGY

Recognising Bravo's diversity of project formats, we took an innovative approach by inputting a variety of past project proposals into GPT without any data preprocessing (See Figure 2.2 for Bravo's project planner GPT). The objective was to test whether GPT could comprehend the nuances of Bravo's bespoke projects and offer valuable planning insights. This methodology assessed GPT's learning capabilities from Bravo's distinct project styles and its ability to provide relevant and actionable project planning guidance.

DESIRABILITY

Assessment: The assessment focused on determining how well GPT met Bravo's managers' project planning needs, particularly regarding ease of use, relevance, and usefulness of the planning insights provided.

Results and Analysis: Feedback from managers, especially from Fangling, indicated high satisfaction with GPT's assistance in project planning. The tool was praised for its ability to understand the unique nature of Bravo's projects and provide actionable guidance. However, some limitations were noted in GPT's ability to grasp the intricacies of highly specialised projects fully, suggesting a need for human oversight.

Project Planner ∨

Project Planner

By community builder

I can work with you on plan project scope.

| Create a proposal for a new branding project. | Draft a proposal outline for a social media project. |
| Generate a scope table for a marketing campaign. | Show me an example of a detailed project timeline. |

Message Project Planner

FIGURE 2.2 Bravo's project planner GPT

Project Planner ⌄

(A) **Project Planner**
Based on the detailed proposal formats in the documents provided, I've crafted a comprehensive scope table for a hypothetical marketing campaign. This table incorporates the structure and details typically included in Bravo Creative Pte Ltd's proposals, ensuring clarity and thoroughness.

Marketing Campaign Scope Table

Phase	Description	Duration	Deliverables	Fees
Phase 1: Market Analysis	**Objective:** Understand market dynamics and consumer behavior to tailor the marketing strategy.	2 weeks	- Market research report - Consumer behavior analysis - Competitive landscape analysis	$12,000
Phase 2: Strategy Development	**Objective:** Develop a strategic marketing framework that aligns with business goals.	3 weeks	- Marketing strategy document - Campaign goals and KPIs - Channel and content strategy	$18,000
Phase 3: Creative Concepting	**Objective:** Design creative concepts that resonate with target audiences and embody brand values.	4 weeks	- Creative brief - Concept presentations (3 concepts) - Selection of final concept	$24,000
Phase 4: Campaign Execution	**Objective:** Implement the marketing strategy across chosen platforms.	6 weeks	- Campaign materials for all platforms - Execution plan - Monitoring and adjustment plan	$30,000
Phase 5: Monitoring & Optimization	**Objective:** Track campaign performance and optimize based on real-time data.	Ongoing	- Weekly performance reports - Recommendations for optimization - Post-campaign analysis and report	$10,000
Optional Add-Ons	- Advanced analytics setup - Influencer collaboration package - Extended campaign duration	Varied	- Custom analytics dashboard - Influencer contracts and management - Additional campaign content and management	$5,000+

🔗 Message Project Planner

FIGURE 2.3   Sample marketing campaign scope table.

FEASIBILITY

Technical Evaluation: The experiment tested GPT's learning capabilities with minimal data preprocessing, evaluating its effectiveness in comprehending and providing planning insights for diverse project formats.

Operational Assessment: The integration of GPT into Bravo's project planning process was relatively smooth. The ability to export project scopes to comma-separated values format for use in project management tools like Asana or Notion was particularly beneficial, streamlining the project initiation process.

Results and Analysis: GPT proved technically and operationally feasible for assisting in project planning. The ease of integration into Bravo's existing workflow and its adaptability to various project types demonstrated its potential as a valuable planning tool (See Figure 2.3 for a sample marketing campaign scope table).

VIABILITY

Economic Impact: Using GPT in project planning could lead to significant time savings and reduced dependence on crucial personnel like Jane, thus optimising managerial resources and decreasing the need for additional staffing.

Strategic Implications: Strategically, GPT aligns with Bravo's goals of innovation and bespoke solutions. Its ability to provide unique insights and aid in creative problem-solving enhances Bravo's competitive edge in project planning and management.

Results and Analysis: The economic and strategic viability of using GPT for project planning appears promising. The tool's ability to adapt to the unique demands of Bravo's projects and contribute to strategic thinking makes it a valuable asset for the firm's long-term success.

CONCLUSION

Experiment 2 showcased the potential of GPT in assisting with planning bespoke projects, offering a novel solution to alleviate the workload on key employees like Jane and enhance Bravo's project management efficiency. The success of this experiment opens up opportunities for further leveraging AI in project planning and management, particularly in contexts where projects require a unique and tailored approach.

As the experiment unfolded, we gained crucial insights into the expectations and capabilities of LLMs like GPT. Initially, there was a tendency to direct GPT towards generating outputs in a specific, structured manner, akin to traditional data processing tools. However, it became evident that the true strength of GPT lies in its ability to interpret and make sense of vast amounts of data more fluidly and expansively. This understanding shifted our perception of LLMs, recognising that GPT's value extends beyond simple data structuring to offering broader, more nuanced insights and thought expansion.

This realisation had important implications for utilising AI tools like GPT at Bravo. It highlighted the need for flexibility in our approach and openness to these models' unique strengths. GPT's capacity to process and synthesise information in innovative ways enhances creativity and aids in developing a comprehensive understanding of complex and diverse datasets. Instead of constraining GPT to a narrowly defined output, we learned to leverage its ability to provide a broader range of insights, thereby enriching our project planning processes.

Experiment 2 ultimately reinforced the idea that AI can be a powerful partner in creative problem-solving and strategic thinking when used thoughtfully. The success of this experiment suggests that AI can not only assist in administrative efficiency but also refine and enhance the strategic aspects of project management. This has opened up new possibilities for Bravo regarding AI integration, particularly in project planning and management where bespoke solutions are needed.

Experiment 3: Creating a Self-updating Knowledge Management System

BACKGROUND AND INITIATION

Building on the insights from Experiment 2, which demonstrated the utility of feeding large amounts of data to GPT for interpretation, I began to consider other sources of unprocessed data within Bravo that could be similarly leveraged. One particular type of data immediately came to mind: emails. As Bravo's creative director, I receive a plethora of project-related emails. While often overlooked, these emails are a treasure trove of information, meticulously documenting project details for record-keeping and protection. Typically, I rely on updates from managers who are client-facing individuals. However, I realised these email threads could be valuable data sources for enhancing our knowledge base.

OBJECTIVE AND METHODOLOGY

The objective was to use these extensive email communications to enrich Bravo's knowledge base and project planning resources on an ongoing basis. Specifically, the idea was to analyse project-specific email threads and synthesise frequently asked questions (FAQs) that could be integrated into the KB (from Experiment 1) and the project planner (from Experiment 2), thereby creating a self-updating knowledge loop.

To test this concept, I selected several long email threads from extensive projects and converted them into PDF format. These were then fed into ChatGPT with the request to synthesise the information and generate a list of FAQs reflecting Bravo's unique approach to project management, client interaction, and project delivery.

Example of synthesised FAQ from email thread: https://bybravo.notion.site/FAQ-from-email-thread-37ff19ab7bcd4058a0459b16ed996e08?pvs=4

DESIRABILITY

Assessment: The assessment focused on the usefulness and relevance of the FAQs generated by GPT from email threads and call transcripts. This also included evaluating how these FAQs enriched Bravo's existing KB and project planner.

Results and Analysis: Feedback indicated that the FAQs were highly valued for their practical insights into project management and client interaction. They provided a unique perspective by synthesising tacit knowledge from routine communications into explicit, accessible information. However, concerns were raised about the potential for information overload and the need to ensure relevance and accuracy in the evolving KMS.

FEASIBILITY

Technical Evaluation: The experiment tested GPT's ability to process unstructured data, such as email threads, and synthesise this information into valuable FAQs. The conversion process and integration of these FAQs into Bravo's existing systems were critical aspects of the feasibility assessment.

Operational Assessment: This AI-driven KMS was examined for its operational integration into Bravo's daily workflow. This included evaluating the ease of updating the system with new data and its adaptability to different types of communications.

Results and Analysis: The KMS proved technically feasible and operationally practical. Integrating AI into knowledge management processes demonstrated GPT's potential to handle diverse data formats and continuously evolve with the company's needs.

VIABILITY

Economic Impact: The economic viability was assessed by considering the time savings and efficiency gains from the AI-driven KMS and the potential reduction in the need for manual knowledge documentation and management.

Strategic Implications: Strategically, the self-updating KMS aligns with Bravo's goals of fostering a culture of continuous learning and innovation. The system's ability to adapt and grow with the company's experiences enhances Bravo's strategic capabilities in knowledge management.

Results and Analysis: The KMS shows solid economic and strategic viability. Its ability dynamically to synthesise and update information offers a transformative approach to knowledge management, potentially leading to significant long-term benefits for Bravo.

OUTCOMES AND SUCCESS

ChatGPT successfully captured the essence of how Bravo's managers handle projects and client relationships, translating these into easily understandable FAQs. These FAQs were specifically tailored to complement the existing content in our KB, effectively appending and enriching it with real-world, project-specific insights.

The result was a significant and unexpected success, marking a pivotal moment in this IRP. This experiment demonstrated the feasibility of creating a living, breathing KMS that can self-update and evolve based on ongoing project experiences and communications.

The outcomes of Experiment 3 are supported by the findings of Jarrahi et al. (2023), who explore AI's role in knowledge creation and management. They highlight practical examples of AI's utility in analysing and summarising complex data, as seen in legal and marketing sectors, demonstrating its ability to enhance knowledge management processes. This aligns with our experiment, where GPT successfully synthesised email data to enrich Bravo's knowledge base.

Building on the foundations laid in the first two experiments, Experiment 3 embarked on an innovative journey to harness the untapped potential of everyday communications as a source of valuable knowledge. Recognising the richness of interactions within the organisation, especially during remote work situations, I decided to experiment with converting a call transcript between a senior and junior designer into a knowledge-rich resource.

The transcript (Figure 2.4) generated automatically from the Zoom platform's speech-to-text function, was fed into ChatGPT. The aim was to explore the model's capability to generate FAQs solely based on this conversation. This approach addressed a common challenge in knowledge management: converting tacit, context-rich dialogue into explicit, accessible knowledge.

The results were strikingly effective. The GPT model identified key themes and queries from the conversation and articulated them into well-structured FAQs. These FAQs delved into specific aspects of design work, such as file preparation and naming conventions – topics critical for junior designers but often overlooked in formal training materials.

This outcome was enlightening in multiple ways. First, it demonstrated the feasibility of utilising AI to capture and codify practical knowledge from routine communications. This process, which once seemed daunting, was now achievable with remarkable efficiency. My business partner had long emphasised the need for a manual to guide new designers. The prospect had always appeared overwhelming, with doubts about where to start and whether it could be seen through to completion. However, this experiment shed light on a new pathway: using AI to build and maintain a manual.

The generated FAQs provided a starting point for a more comprehensive and continually evolving knowledge base. Each interaction, whether a Zoom call or a casual conversation, could now be transformed into an educational resource. This approach not only simplifies the process of knowledge documentation but also ensures that the information stays current and relevant.

GMT20231219-030345_Recording

Q ··· Built with 📷

GMT20231219-030345_Recording

GMT20231219-030345_Recording

[00:00:00] Ooh, recording in progress. So scary, right?

Hang on, I'm just trying to open it, not in my notes, cause my notes got a lot of trash at the side and I don't know how to You know how to hide the sidebar? Just move it lah, move it out, no? Okay, I guess. You open a new desktop ah? Oh Ah Oh, that's true. Oh no, my com My com when I do that right, it's like super Give me a second.

Let me just open like, what's that called? The, the rich tech, the text box thing, the, the paper. Uh, let me see how fast this is.

How are y'all? Um, [00:01:00] well,

it's the last week. Oh, right. That is true. Uh,

okay. 30 seconds. What have you been doing? What have I been doing? I mean. Um, taking my mental health break. Very good. Yes. You have packed your stuff, your room, whatever. Yes, I have. It is not, not all done, but, but I'm taking my time with it. Yes. Yes, okay. Okay. Okay. Um, okay. Let me share my notes. Oh my god. How do I use the share screen?

Okay. Okay. No, it's just that we [00:02:00] know it's too small. Okay. So, I have these notes um that I will send to Jane later. So, basically this is like Your Bible, I put it in the steps already, later I'll walk through all of this. Um, let me share my InDesign and my photos.

Okay, this one and photos.

Okay, okay, yeah. Okay, so, you are in your InDesign, like with your everything already. Basically, um, The first one, uh, is the SCTD. Can y'all see, is this very small? Yes, can see, can see. Okay, so basically SCTD doesn't want the GIF to look. So the only way that I found for this to really work, right, is that whatever you have exported your animation video, right, you put it into Photoshop and you export as GIF by the Save4Web.

And then you change [00:03:00] that, the default look from forever to one. So, Just in case, uh, you, so you know, you go to your, your over here, you say forward, right? After you say, yeah, over here, I changed the ones. So this one very important must always remember to do because I've tried other ways like online converters, but it doesn't work.

Only this way. I found that like, you will play once on me, then it will stop. Okay. So this one, okay. Then this is just like for STD. And after that, right. You in, in design, you put in your GIF, right? As per normal. So you will go to

You go to place when you place, right. Then you select your GIF from your, the pop up and then you put it in. Right. So basically this is a GIF file. You can see over here. Yeah. GIF file, right. You won't be able to, you won't be able to play it on in design, but you just put in and then your measurement of [00:04:00] everything, you fit your picture.

FIGURE 2.4 Sample transcript.

Synthesising Chat Transcript

Based on the transcript of the Zoom call you provided, here are some FAQs that can be
synthesized to add to Bravo's knowledge base, capturing unique practices and insights relevant
to project management and client collaboration:

FAQs from Zoom Call Transcript for Bravo's Knowledge Base

1. **How do we handle GIF animations in project deliverables?**

   - To properly incorporate GIF animations, export your animation video as a GIF using
     Photoshop's Save for Web feature. Remember to change the default loop setting from
     'forever' to 'one' to ensure the animation plays only once.

2. **What are the best practices for using GIFs in design projects?**

   - When adding a GIF to a design project, use the 'place' function in your design software.
     Note that while the GIF won't play within the software, it can be correctly positioned and
     sized.

3. **How should project files be named for web publishing?**

   - Name your project files according to the client's preferences, as the file name will be
     visible on the web page when published online. This helps ensure that the client's
     branding and naming conventions are accurately represented.

4. **What should we consider when publishing documents online?**

   - When publishing documents online, pay attention to the title settings, as they may differ
     from your design file's title. This title is crucial, especially when allowing viewers to
     download the document as a PDF.

5. **How do we handle revisions and updates to published documents?**

   - For minor revisions, you may update an existing document to retain the same link. For
     significant changes or multiple versions, consider publishing a new document to keep a
     record of all iterations.

6. **What are the tips for remote collaboration with clients?**

   - When collaborating remotely, prepare your files with correct naming and settings
     beforehand. This ensures a smooth process when you remotely control the client's
     computer for final adjustments or uploads.

7. **What additional steps are needed for covers in design projects?**

   - If your project includes a cover (especially for animated projects), use a separate JPEG
     file for the cover to ensure it appears as intended. This step is crucial to maintain the
     desired aesthetic and branding.

FIGURE 2.5   Sample synthesised FAQ from Zoom call.

RESULTS & RECOMMENDATIONS

The experiments conducted in this IRP revealed significant insights into Gen AI's potential to revolutionise boutique branding consultancy. The results provide a clear indication of how Gen AI can
be harnessed to enhance operational efficiency, creativity, and knowledge management.

SUMMARY OF RESULTS

- Experiment 1 established the foundation for an AI-powered KB, demonstrating Gen AI's
  ability to streamline information creation and reduce dependency on specific employees.
- Experiment 2 explored Gen AI's utility in bespoke project planning, revealing its capacity
  to reduce managerial workload and improve project management efficiency.
- Experiment 3 built upon these insights to create a self-updating KMS, showcasing the
  dynamic nature of AI in continuously enriching and evolving the company's KB through
  real-time data analysis.

RECOMMENDATIONS

To optimise operations and remain competitive, adopting a living breathing KB that evolves with
the company's needs is crucial, leveraging insights from internal communications for better decision-making. Gen AI should be utilised for tedious and complex tasks due to its unparalleled ability to analyse large datasets, synthesise information, and generate innovative solutions. SMEs, in

particular, can benefit from integrating such dynamic KBs, which provide real-time insights and guidance. Investing in Gen AI training and development is essential to maximise its potential, enabling employees to interact with AI tools effectively. Finally, continuous innovation with Gen AI should be pursued, and its applications should be regularly updated and refined based on new data, technologies, and business needs.

FUTURE POTENTIAL OF GEN AI IN BRAVO AND BEYOND

The successful implementation of Gen AI at Bravo serves as a model for other organisations in the branding consultancy sector and beyond. Gen AI's future potential includes its application in diverse areas such as customer relations, product development, and strategic planning. Businesses must embrace this technology as a tool for operational efficiency and as a driver of innovation and competitive advantage.

In conclusion, the IRP demonstrates that Gen AI can transform how boutique branding consultancies like Bravo operate, making them more efficient, productive, and adaptable to the changing market dynamics. Adopting a living, breathing KB and the strategic use of Gen AI can position SMEs at the forefront of innovation, enabling them to thrive in an increasingly competitive business landscape.

REVIEW AND FUTURE WORK

We have explored the hypothesis of a dynamic, living, breathing knowledge base (LBKB). This experiment builds upon the foundational KB established in Experiments 1 and 2, aiming to advance it by integrating diverse and real-time data sources. The idea is to use GPT algorithms to analyse emails exchanged within the organisation, extracting and synthesising relevant insights to enrich the KB continually.

This experiment hypothesis includes data from various platforms like Zoom, Asana, Notion, and emails. Text chats from Zoom meetings and interactions on project management and collaboration tools are proposed to be captured and analysed for proven integration into the KB. Moreover, transcribing and analysing spoken conversations from Zoom calls presents a treasure trove to diversify knowledge sources further.

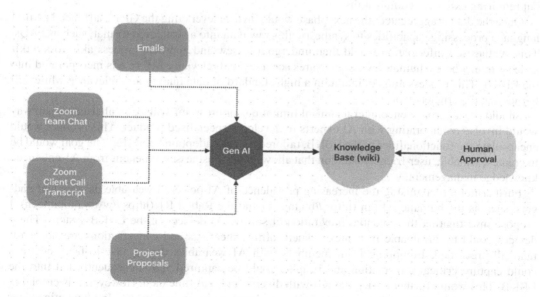

FIGURE 2.6 Living, breathing knowledge base (LBKB).

While we still need to gain the technical skills to implement this concept fully, we hypothesise that such a comprehensive and continuously updated KB could significantly enhance organisational efficiency and decision-making. This system could provide a more complete and current view of the organisation's knowledge and activities by effectively harnessing AI and machine learning.

The potential of this approach lies in its ability to adapt and evolve with the organisation, offering a living repository that grows and changes in response to the dynamic flow of information. This experiment aims to demonstrate the feasibility and benefits of such a system conceptually, setting the stage for future developments and practical implementations.

This experiment underscores the dynamic and adaptable nature of AI-powered knowledge management. It highlights the possibility of leveraging AI as a static tool for information retrieval and as an active participant in knowledge creation and curation.

The success of this experiment opens new avenues for Bravo regarding knowledge management. It provides a foundation for a more intelligent, responsive, and self-sustaining KMS that continuously adapts and grows with the company's experiences and project learnings. AI's ability to synthesise and update information dynamically offers a transformative approach to managing and disseminating knowledge within the organisation.

Furthermore, this experiment illustrates the potential of AI to enhance the efficiency of internal communication processes. By automating the extraction and synthesis of valuable insights from routine communications, AI can help streamline information-sharing and decision-making processes. This approach saves time and ensures that critical knowledge is captured and made accessible to all relevant team members.

Experiment 3 has revealed the extensive capabilities of AI in managing and enhancing organisational knowledge. It has opened the door to a more innovative and effective way of handling the wealth of information generated within a dynamic business environment like Bravo's. This experiment models how AI can be strategically implemented to foster a culture of continuous learning and improvement.

To enhance the LBKB while adhering to a no-code approach, I propose exploring the integration of automation tools like Zapier. This approach would enable the seamless feeding of data from various sources such as emails, Zoom, Asana, and Notion into the GPT. The objective would be to leverage Zapier's capabilities to automate data aggregation, thereby streamlining the process without requiring extensive coding skills.

Once the data is aggregated, the next phase would involve leveraging the GPT's advanced natural language processing capabilities to synthesise this raw data into a structured format, such as FAQs. This synthesised information would then undergo a review and approval process akin to a wiki-style system, where a human reviewer ensures accuracy and relevance before it is incorporated into the LBKB. This process aims to maintain a high standard of data quality and relevance while still harnessing the efficiency of AI.

Should this approach encounter technical limitations or prove infeasible, an alternative pathway would involve collaborating with AI experts to develop a specialised product. This product would encapsulate the functionalities of the LBKB, tailored for easy adoption by SMEs. The goal would be to create a scalable, user-friendly solution that allows many businesses to benefit from AI-enhanced knowledge management.

Furthermore, recognising the increasing prevalence of AI-powered wearable devices and gadgets, such as the humane AI pin (https://hu.ma.ne) and the Rabbit R1 (https://www.rabbit.tech), I propose investigating the potential integration of similar AI devices in the LBKB system. These devices could be invaluable in contexts where offline meetings and conversations are not automatically recorded. Equipping team members with AI wearables during face-to-face meetings could capture critical information and insights could be captured and subsequently fed into the LBKB. This would further enrich the KB with diverse and real-time data. However, it is crucial to acknowledge and address privacy and ethical considerations in such scenarios. Implementing clear disclaimers and obtaining consent for using these devices in meetings would be essential to ensure transparency and compliance with privacy standards.

An important consideration that emerged during this research is the potential impact of a highly reliable and comprehensive knowledge base, like the LBKB, on employees' critical thinking abilities and reliance. While the LBKB aims to enhance efficiency and decision-making by providing quick and easy access to a wealth of information, it raises a pertinent question: Could such a system inadvertently decrease critical thinking and problem-solving skills among employees?

As employees become accustomed to relying on the LBKB for information and solutions, there is a risk that they may become overly dependent on it, potentially diminishing their ability to think independently and critically analyse situations. This over-reliance could lead to a scenario where employees are less inclined to question or verify the information provided by the LBKB, potentially impacting their professional growth and the company's innovative capacity.

To mitigate this risk, it is crucial to balance leveraging the LBKB as a powerful resource and encouraging employees to engage in critical thinking and independent problem-solving. One approach could be integrating features within the LBKB that prompt users to reflect on the information provided, encouraging them to consider different perspectives and develop their insights. Additionally, regular training and development programmes can be implemented to foster critical thinking skills, ensuring that employees can effectively utilise the LBKB while maintaining their ability to think independently.

CONCLUSION

In conclusion, integrating Gen AI at Bravo has been a journey of discovery, innovation, and transformation. Our experiments have redefined operational processes and opened our eyes to AI's immense possibilities. The key takeaway from this experience is the paradigm shift from viewing AI as a mere tool for automating routine tasks to recognising it as a partner capable of achieving the extraordinary. By embracing AI, we have been able to envisage and approach tasks that were once considered daunting or even impossible.

Our experiments underscored the importance of harnessing the 'intelligence' in AI, leveraging it to accomplish tasks that are tedious or humanly impossible. This shift in perspective – from using AI for mundane tasks to embracing its potential for complex and innovative applications – opens new opportunities and efficiencies. For instance, AI's ability to automate information aggregation and analysis can maintain a comprehensive KB beyond human capabilities, transforming daunting tasks into achievable goals. This has empowered us to approach ambitious projects with newfound ease, levelling the playing field for SMEs by unlocking possibilities previously out of reach. As we harness the vast amounts of data generated daily, AI technologies like Google's Bard make the concept of a LBKB tangible, potentially revolutionising our operations. This underscores the need for continuous evolution and innovation, as AI's rapid advancement suggests a future that is both exciting and unpredictable. As we navigate this imagination age, embracing AI with a blend of excitement, caution, and responsibility will be crucial in unlocking its full potential.

This journey has also illuminated the importance of continuously adapting and evolving as individuals and organisations. As we step further into the imagination age, the potential for AI to reshape industries and societies is boundless. However, with this great power comes the responsibility to navigate the ethical and social implications thoughtfully. Brimming with opportunities, the future also challenges us to stay vigilant and proactive in harnessing AI responsibly and creatively. Ultimately, our experience with Gen AI at Bravo is a testament to technology's transformative power when aligned with human insight and creativity. As we look to the future, it is clear that AI will be a driver of business innovation and a catalyst for reimagining the fabric of our professional and personal lives. By continuing to explore the intersection of human ingenuity and AI capabilities, we can unlock new horizons and boldly shape the future we envision.

REFERENCES

Bankins, S., & Formosa, P. (2023). The ethical implications of artificial intelligence (AI) for meaningful work. *Journal of Business Ethics, 185*, 725–740.

Chui, M., Hazan, E., Roberts, R., Singla, A., & Smaje, K. (2023). The economic potential of generative AI: The next productivity frontier. McKinsey Digital. https://www.mckinsey.com/capabilities/mckinsey-digital/our-insights/the-economic-potential-of-generative-ai-the-next-productivity-frontier [accessed 2024-01-16]

Del Giudice, M., Scuotto, V., Orlando, B., & Mustilli, M. (2023). Toward the human–centered approach. A revised model of individual acceptance of AI. *Human Resource Management Review, 33*(1), 100856.

Dillion, D., Tandon, N., Gu, Y., & Gray, K. (2023). Can AI language models replace human participants?. *Trends in Cognitive Sciences, 27*(7), 597–600.

ElBatanony, A., & Succi, G. (2021, October). Towards the no-code era: a vision and plan for the future of software development. *Proceedings of the 1st ACM SIGPLAN International Workshop on Beyond Code: No Code, 1*(1), 29–35. https://doi.org/10.1145/3486949.3486965

Eloundou, T., Manning, S., Mishkin, P., & Rock, D., 2023. Gpts are gpts: An early look at the labor market impact potential of large language models. *arXiv preprint*, arXiv:2303.10130.

Jarrahi, M. H., Askay, D., Eshraghi, A., & Smith, P. (2023). Artificial intelligence and knowledge management: A partnership between human and AI. *Business Horizons, 66*(1), 87–99.

Kanbach, D. K., Heiduk, L., Blueher, G., Schreiter, M., & Lahmann, A. (2023). The GenAI is out of the bottle: Generative artificial intelligence from a business model innovation perspective. *Review of Managerial Science, 18*, 1189–1220.

Kulkarni, M. (2021). Deciphering low-code/No-code hype-study of trends overview of platforms and rapid application development suitability. *International Journal of Science and Research Publications, 11*(7), 536–540.

Moskal, M. (2021). No-code application development on the example of logotec app studio platform. *Informatyka, Automatyka, Pomiary w Gospodarce i Ochronie Środowiska, 11*(1), 54–57.

Prasad Agrawal, K. (2023). Towards adoption of generative AI in organizational settings. *Journal of Computer Information Systems*, 1–16. https://doi.org/10.1080/08874417.2023.2240744

Sanzogni, L., Guzman, G., & Busch, P. (2017). Artificial intelligence and knowledge management: Questioning the tacit dimension. *Prometheus, 35*(1), 37–56.

Soni, V. (2023). Impact of generative AI on small and medium Enterprises' revenue growth: The moderating role of human, technological, and market factors. *Reviews of Contemporary Business Analytics, 6*(1), 133–153.

Sundberg, L., & Holmström, J. (2023). Democratizing artificial intelligence: How no-code AI can leverage machine learning operations. *Business Horizons, 66*(6), 777–788.

Tsui, E., Garner, B. J., & Staab, S. (2000). The role of artificial intelligence in knowledge management. *Knowledge Based Systems, 13*(5), 235–239.

Vera, C. G. M., Vicente, M. V. O., Vera, I. L. A., Alexander, M. V. A., & Vera, H. F. B. (2022). Low/No-code development platforms and the future of software developers. *Minerva, 1*(Special), 21–33.

Wang, S., & Wang, H. (2021). A teaching module of no-code business app development. *Journal of Information Systems Education, 32*(1), 1.

Weber, M., Engert, M., Schaffer, N., Weking, J., & Krcmar, H. (2023). Organizational capabilities for AI implementation—Coping with inscrutability and data dependency in AI. *Information Systems Frontiers, 25*(4), 1549–1569.

Yan, Z., (2021). The impacts of low/no-code development on digital transformation and software development. *arXiv preprint*, arXiv:2112.14073.

3 Transforming Creativity through Strategic Change Management

Integrating AI and DesignOps in Workflows

Alan Teo Hui Yeong

INTRODUCTION

The world is experiencing a new era of technological advancement and digital transformation, resulting in the rise of artificial intelligence (AI) as a dominant force in various industries. In the design field, driven by creativity, innovation, and human-centred solutions, AI can significantly improve workflow efficiency and effectiveness by leveraging its capacity to analyse vast amounts of data, detect patterns, and generate insights (Verganti et al., 2020). Indeed, AI-powered design tools have the potential to facilitate the development of more user-focused, innovative, and constantly evolving solutions, as outlined by Verganti et al. (2020).

However, integrating AI into existing design workflows requires more than adding new tools. It demands a transformative shift in the design processes. This shift brings challenges such as resistance to established processes, scepticism about AI's creative potential, and concerns about its impact on traditional design roles (Yusa et al., 2022). Additionally, apprehension is fuelled by the fear of AI replacing human jobs. The increasing integration of AI in various sectors has intensified worries about job displacement (Huang & Rust, 2018). Therefore, careful planning, a deep understanding of human factors, and strategic change management are necessary to foster acceptance and overcome resistance to AI integration.

To tackle these challenges, this chapter aims to explore the role of change management strategies in facilitating the integration of AI into design practices. It will focus on three theoretical frameworks: Lewin's change management model, Kotter's 8-step change model, and the ADKAR model. These models offer insight into the underlying mechanisms of change and the factors influencing technology adoption, which are essential for successfully integrating AI into design practices.

Lewin's change management model explains that change involves three stages: unfreezing, changing and refreezing (Mind Tools, 2023). On the one hand, the unfreezing stage fosters a sense of urgency to persuade stakeholders about the necessity of change. The changing stage involves implementing the change, and the refreezing stage involves stabilising the change and making it permanent. On the other hand, Kotter's 8-step change model emphasises the importance of creating a coalition of supportive stakeholders, developing a vision for change, and communicating the vision effectively (Kotter, 1996). Finally, the ADKAR model focuses on the individual level of change by identifying the five stages a person must go through to adopt a new technology: awareness, desire, knowledge, ability, and reinforcement (Hiatt, 2006).

In addition to exploring change management strategies, this chapter will address the unintended consequences and ethical concerns surrounding integrating AI into design workflows. As AI technology progresses, it is essential to consider the possible negative impacts of AI-driven design solutions.

DOI: 10.1201/9781003469551-3

For instance, they may perpetuate systemic biases or exclude certain groups of users. Therefore, it is important to address these concerns to ensure that AI-powered design solutions align with ethical principles.

In conclusion, integrating AI into design workflows marks a significant shift in design methodologies, and its impact will likely be transformative. However, it demands careful planning, a deep understanding of human factors, and strategic change management to foster acceptance. By exploring the theoretical frameworks and ethical considerations surrounding AI integration, design professionals can ensure they can leverage AI's capabilities to create user-focused, innovative, and constantly evolving design solutions.

ABOUT THE CLIENT

In 2019, a landmark event occurred in the Singapore telecommunications industry. A leading company within the industry embarked on a significant transformation journey following its acquisition by two major conglomerates. This important event marked the beginning of a new era for the company, with high aspirations to emerge as Singapore's premier digital network operator. However, this ambitious vision was more than just advancements in technology. It was about a revolution in how customers interact and engage with the services, bringing automation and advanced self-service technology to the forefront and delivering a seamlessness that had never been experienced before.

As a crucial part of the organisation, the user interface (UI) and user experience (UX) team assumed a pivotal role. The team's responsibilities span a wide range of tasks, including designing customer experiences and developing design solutions for the various channels that served as critical touchpoints for product sales and customer service. In the initial phase, the department depended heavily on external vendors, with a few key in-house designers supplementing their efforts. However, to improve work quality and align the team more closely with the overarching values and vision of the organisation, a strategic shift was executed in late 2022 towards adopting an in-house approach. This strategic pivot proved to be a wise decision from a financial perspective, as it turned out to be more cost-effective than the previous heavy reliance on external vendors.

By the second half of 2023, the department had grown significantly, with more designers brought in to handle the increased workload. This influx of new members underscored the department head's urgent need to optimise the design process. It became clear that for the team to operate with maximum efficiency, effectiveness, and scalability, there was a pressing need to implement structured processes, methodologies, and tools. Additionally, in recognition of generative AI's increasing potential and capabilities, the department head expressed an interest in exploring the feasibility of integrating AI into its workflow. Consequently, the task of investigating potential implementation strategies and guiding the team in transitioning successfully to this new structure was assigned to me, marking a new chapter in my role within the department.

THE DAWN OF AI IN CREATIVE DESIGN: AN OVERVIEW

Integrating AI into workflows has become a topic of substantial debate in various industries. Discussions predominantly revolve around potential job displacement, the ethical implications of AI decisions and the integrity and security of AI systems. Notwithstanding these concerns, many individuals recognise AI's potential to enhance efficiency, creativity, and decision-making across numerous fields.

As the dialogue transitions into the realm of design, AI's role becomes increasingly significant and potentially transformative. In this landscape, AI is utilised to automate routine tasks, introduce innovative tools for creativity, and provide predictive design solutions. This integration necessitates a shift in the designer's role from manual creation to a more directorial position, supervising and guiding the AI's creative process. Consequently, the discussion evolves to question the authenticity of AI-generated designs and the need for an evolving skill set among designers. The application of AI in design not only challenges traditional methodologies but also unveils a multitude of possibilities for innovation and experimentation. Hence, integrating AI into the design landscape is a complex process that requires careful navigation and thoughtful discourse.

EXPLORING THE LANDSCAPE: AI IN DESIGN

Integrating AI into the design industry has significantly transformed it, redefining creativity and problem-solving. AI encompasses machine learning, neural networks, natural language processing, and computer vision, extending beyond being a mere addition to a designer's toolkit. Instead, it represents a novel approach to design methodologies and innovation (Verganti et al., 2020). Notably, AI has revolutionised developing and evaluating new solutions, exemplified in drone design, enhancing speed and efficiency (Song et al., 2022).

The transformative potential of AI extends to other fields, such as engineering, architecture, graphic design, and UI and UX design. In engineering and architecture, AI can redefine the creative process and envision novel structures (Dai, 2023; Jang et al., 2022). For instance, AI has been utilised to simulate and visualise complex designs in the Beijing Daxing Airport, resulting in enhanced efficiency and a competitive edge (ZIGURAT, 2023). In graphic design, AI automates repetitive tasks, generates personalised designs and aids in strategic decision-making for branding and marketing (Lee & Cho, 2020; Mustafa, 2023).

Furthermore, the study by Zhou et al. (2020) employed a machine learning framework called FEELER to explore and evaluate UI design solutions. By utilising collective learning and predictive modelling based on user feedback, FEELER enabled designers swiftly and conveniently to adjust UI modules to align with user preferences.

AI also significantly contributes to sustainable and efficient design practices by directing designers to use resources in an eco-friendly manner, thereby helping combat climate change (Vinuesa et al., 2020). Additionally, applying natural language processing (NLP) ensures regulatory compliance in design, bridging the gap between creative design and practical implementation (Zhang & El-Gohary, 2016).

However, integrating AI into existing design workflows requires more than adding new tools. It necessitates a transformative shift in design processes, accompanied by challenges such as resistance to change, scepticism about AI's creative potential, and concerns about job displacement (Huang & Rust, 2018; Yusa et al., 2022). Careful planning, a deep understanding of human factors, and strategic change management are crucial (Yusa et al., 2022).

In conclusion, AI's impact has spanned various fields, from architecture to UI and UX design, guiding designers towards more sustainable practices and enabling data-driven decision-making. This adoption trend of new technology will continue to grow exponentially soon.

While AI brings significant advancements to the UI and UX design field, it is essential to address the ethical considerations surrounding its utilisation. Safeguarding user privacy, mitigating biases, promoting user autonomy, and evaluating the broader societal implications are crucial steps in ensuring that AI-driven designs align with ethical principles and contribute positively to the field.

AI Brings Automation and Efficiency

AI has significantly impacted the design industry, especially in automating repetitive tasks (Mustafa, 2023). This automation has enabled designers to devote more attention to their work's creative and strategic aspects, facilitating the conceptualisation of new projects with greater ease and efficiency. A prime example of using AI for repetitive tasks is Adobe Sensei. This platform demonstrates how AI and machine learning can take over tedious tasks. For instance, AI has revolutionised digital art creation and manipulation through Adobe Sensei, which integrates AI capabilities into Adobe Photoshop. With AI-driven features such as automatic background reconstruction when objects are removed, advanced image recognition is showcased. It eliminates many steps in the creation process that require much clicking rather than deeply creative work (Adobe Sensei: Machine Learning and Artificial Intelligence, n.d.).

Research has demonstrated that AI enhances the speed and quality of deliverables produced by professionals in the business domain. For instance, a prestigious consulting company experienced a 33% increase in productivity and a 40% improvement in the quality of their deliverables when employing AI (Nielsen & Moran, 2023). Additionally, Nielsen and Moran (2023) point out that AI fosters creativity through its generative design approach and its capacity to generate numerous ideas rapidly, often matching or surpassing human-generated ideas that require extensive effort. These findings underscore the transformative impact of AI on design methodologies and highlight its potential to revolutionise the efficiency of the creative process in the design industry.

Bertão and Joo (2021) have also highlighted the potential of current AI algorithms to improve UI and UX design efficiency. They emphasise AI's ability to streamline data processing and automate design aspects, facilitating effective communication through concept visualisation and solution generation.

The arrival of generative AI, capable of creating high-quality written and visual content, has opened up possibilities for automating certain creative jobs in the design industry. This development not only offers opportunities for innovation but also carries inherent risks, such as the potential for AI tools to produce inaccurate or misleading content (McKendrick & Thurai, 2022). Alongside this technological advancement, there is a noticeable shift in the skill sets required of creative professionals. It is becoming increasingly important for designers to develop new competencies to effectively utilise AI technology and enhance their employability, suggesting broadening skill sets within the design industry (Girling, 2017).

As designers leverage AI technologies like Dalle2, ChatGPT, and Midjourney to streamline the creation process, ethical considerations arise regarding the authenticity and originality of AI-generated designs (Askari, 2023). This necessitates careful scrutiny and evaluation of the creative process to ensure that AI-driven designs align with ethical principles and maintain the integrity of the design profession.

In conclusion, the integration of generative AI into the design industry has the potential to bring about transformative changes. However, it requires a thorough evaluation of potential risks and ethical consequences. Striking a balance between utilising AI for creativity efficiency and managing its potential pitfalls is crucial in navigating this new design era.

AI Impacting Jobs

The emergence of smart technology, artificial intelligence, robotics, and algorithms (STARA) has ignited a critical debate about the future of employment. Notable figures like Stephen Hawking and Bill Gates and AI experts such as Ben Goertzel have warned of substantial job displacement, projecting that up to 80% of current jobs may become obsolete (Brougham & Haar, 2018; The Business Times, 2023). These advancements in robotic capabilities and cost-effective autonomous units present a significant challenge to the workforce.

Research conducted by Frey and Osborne (2017) suggests that AI and automation have resulted in a shift where high-skilled workers assume tasks traditionally performed by low-skilled workers, thereby pushing the latter down the occupational ladder and, in some cases, out of the workforce entirely. This evidence indicates that AI and automation have impacted the job market by redistributing tasks and potentially displacing certain categories of workers.

Conversely, Einola, Khoreva & Tienari (2023) present a more nuanced perspective on integrating AI in the workplace. They argue that, rather than resulting in job loss, the integration of AI leads to a transformation of job roles. Instead of fearing replacement, employees witness the evolution of their roles, which become more specialised and focused on managing and collaborating with AI. Supporting this viewpoint, Ullal et al. (2022) contend that occupations that require human empathy and emotional intelligence are less susceptible to automation. This perspective underscores the invaluable contribution of the human touch in specific professions (Einola & Khoreva, 2023; Ullal et al., 2022).

As noted by Asan et al. (2020) and echoed by McKendrick and Thurai (2022), AI's limitations can also lead to a lack of empathy and ethical considerations, requiring human oversight.

Supporting this transformative viewpoint, Masriadi et al. (2023) advocate for enhancing human skills to adapt to and integrate with AI. The author argues that upskilling safeguards employment and leverages human intuition and empathy, aspects of work where AI falls short. While it is true that automation may render certain roles obsolete, it is also essential to consider the resilience of other jobs and the new opportunities AI can create. For example, the rise of AI in healthcare has not only automated data analysis but has also created a demand for AI specialists to work alongside medical professionals, illustrating the evolving cooperation between human skills and machine capabilities (Sun, 2022).

It is also important to consider historical precedents that demonstrate how technological innovation can actually increase labour demand. Yehiav (2023) supports this viewpoint, citing examples such as the Industrial Revolution and the growth of paper production alongside the emergence of personal computers in the 1990s (Janicki, 2000). In both cases, new technologies led to increased labour demand and the creation of new opportunities.

As a design leader, I question AI's implications for UI and UX professionals and their futures. Should we embrace AI and allow it to assume control over our jobs?

As Sinek (2014) aptly stated, leaders are the ones who run head first into the unknown. As a leader, I am responsible for seeking answers and guiding the team through this uncertain time. In line with this, Nielsen (2023) suggests that UX professionals must acknowledge the imperative of AI in their careers to remain relevant. Drawing on his experience as a seasoned UX practitioner who witnessed the transformative impact of the dot-com boom, Nielsen (2023) highlights the complacency of UX professionals in embracing AI design and draws a parallel to their hesitation during the dot-com revolution. The author underscores the significance of overcoming inertia and fully embracing AI design while reflecting on missed opportunities for early UX intervention in the development of the internet. As Mark Twain aptly put it, "History never repeats itself, but it rhymes" (Doyle, Mieder and Shapiro, 2012).

From a design perspective, human input remains critical in creating designs. Implementing AI in design processes should consider the unique aspects of human psychology and culture to increase the likelihood of success. It is not a matter of completely replacing human involvement but rather integrating change management strategies to facilitate a smooth transition and effective collaboration between humans and AI (Uren & Edwards, 2023). Therefore, design professionals need to leverage AI's unique capabilities to adapt and find new opportunities in collaboration with AI.

EXPLORING THE LANDSCAPE: CHANGE MANAGEMENT

In the rapidly evolving business landscape, technological advancements such as AI present both opportunities and challenges for organisations. This incorporation extends beyond mere technical implementation, as it incites a comprehensive transformation affecting multiple aspects of the organisation (Dwivedi et al., 2021). This is where the pivotal role of change management becomes evident.

Defined as a structured approach, change management guides individuals, teams, and organisations from their current state to a desired future state (Figure 3.1). Successful technology adoption, therefore, requires not merely compliance with a checklist but comprehensive preparation, support, and guidance for individuals to adapt to and accept changes in their work environment, leading to a transformative impact on organisational culture and processes.

For instance, Netflix's successful transition from a service that delivers DVDs by mail to a platform that streams content online offers a notable example of effective change management. Netflix's comprehensive change management process, characterised by thorough risk assessment

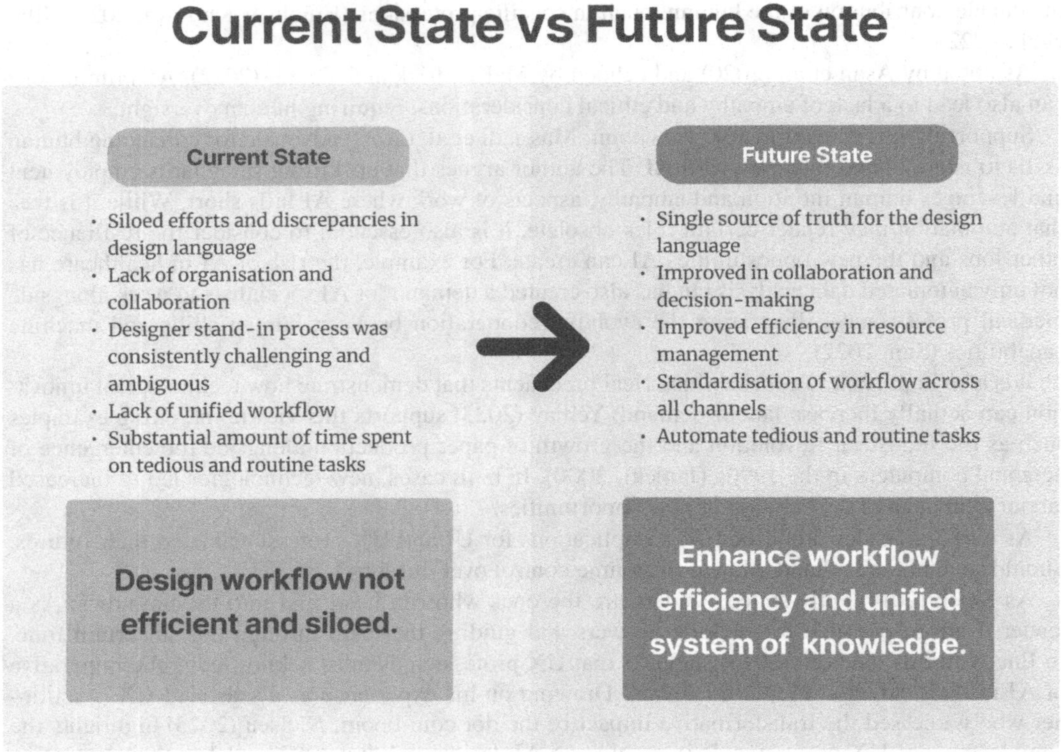

Current State vs Future State

Current State

- Siloed efforts and discrepancies in design language
- Lack of organisation and collaboration
- Designer stand-in process was consistently challenging and ambiguous
- Lack of unified workflow
- Substantial amount of time spent on tedious and routine tasks

Future State

- Single source of truth for the design language
- Improved in collaboration and decision-making
- Improved efficiency in resource management
- Standardisation of workflow across all channels
- Automate tedious and routine tasks

Design workflow not efficient and siloed.

Enhance workflow efficiency and unified system of knowledge.

FIGURE 3.1 Current state versus future state.

and post-change reviews, successfully minimised operational disruptions (Abbas, 2022). Their success was not just due to the technology itself but also how they managed the transition, including open communication, and proactively mitigated the impact of changes.

This transformation requires transparent communication, thorough training, and robust support for those impacted. The strategic vision and strong leadership are paramount in steering the organisation through this transition. Recognising and addressing these elements are critical steps in successfully navigating the complexities of technological change and harnessing its full potential to reach strategic goals and objectives.

For the context of this research, we will delve into three widely used change management models and strategies, elaborating on their practical application. Although numerous models are available that can aid organisations in implementing change and increasing the success rate of implementation, these three have been specifically chosen for their relevance. This comprehensive exploration underscores the pivotal role of change management in integrating new technology into organisations, highlighting its significance in an academic context.

KOTTER'S 8-STEP CHANGE MODEL

Developed by John Kotter, a Harvard Business School professor, the 8-step change model is a comprehensive method for managing and facilitating significant organisational change (Table 3.1). Introduced in his 1996 book, *Leading Change*, Kotter's model provides a step-by-step approach to navigating the complex change process, focusing on emotional and practical aspects. Each step is designed to build on the one before, encouraging participation and backing for the change.

TABLE 3.1

Kotter's (1996) Enhanced 8-Step Change Model

Leading Change 8-Step Process (1996)	Accelerate 8-Step Process (2014)
• Respond to or effect episodic change in finite and sequential ways.	• Run the steps concurrently and continuously.
• Drive change with a small, powerful core group.	• Form a large volunteer army from up, down, and across the organisation to be the change engine.
• Function within a traditional hierarchy.	• Function in a network flexibly and nimbly outside of, but in conjunction with, a traditional hierarchy.
• Focus on doing one thing very well in a linear fashion over time.	• Constantly seek opportunities, identify initiatives to capitalise on them, and complete them quickly.

In response to today's fast-paced changes, Dr John Kotter proposed an enhanced version of the 8-step change model in his book *Accelerate*. This updated model offers organisations the necessary capabilities to lead change effectively in today's fast-paced environment.

KEY PRINCIPLES

Kotter's 8-step change model provides a systematic approach to overcoming resistance to change and promoting lasting organisational transitions. This model is composed of a sequence of steps that include establishing a sense of urgency, forming a potent coalition, creating and communicating a vision for change, empowering broad-based action, generating short-term wins, consolidating gains to produce further change, and anchoring new approaches within the organisation's culture.

STRENGTHS AND WEAKNESSES

The effectiveness of Kotter's model in facilitating organisational change and enhancing productivity is well-documented, as demonstrated by the research of Laig and Abocejo (2021). Their study reported significant productivity increases following the application of this model, thereby attesting to its effectiveness in managing change. Furthermore, the versatility and adaptability of Kotter's model are evident in its successful application across various contexts, from libraries to medical education (Davis, 2022; Haas et al., 2019). However, it is crucial to acknowledge that this model's effectiveness can depend on specific contextual factors and the unique challenges an organisation faces. Considering the organisational context, this underscores the need for a nuanced approach when implementing change management models.

PRACTICAL APPLICATION

In terms of practical application, Kotter's model has proven instrumental in diverse scenarios, extending from the enhancement of residency didactics curricula in medical education to the management of departmental recruitment amidst the challenges posed by the Covid-19 pandemic (Haas et al., 2019; Miles et al., 2023). Such wide-ranging applications further underscore this model's broad applicability and utility in facilitating effective change management.

LEWIN'S CHANGE MANAGEMENT MODEL

Renowned psychologist and change management specialist Kurt Lewin proposed his seminal change management model in the 1940s (See Figure 3.2). This model, a cornerstone in change

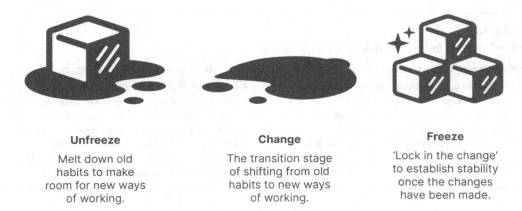

| **Unfreeze** | **Change** | **Freeze** |
| Melt down old habits to make room for new ways of working. | The transition stage of shifting from old habits to new ways of working. | 'Lock in the change' to establish stability once the changes have been made. |

FIGURE 3.2   Kurt Lewin's three-step model.

management, is lauded for its conciseness and effectiveness. It outlines organisational change as a three-part process, also known as changing as three steps (CATS), encompassing the stages of unfreezing, changing and refreezing. Each phase is vitally instrumental in facilitating a seamless transition while adopting novel technologies within an organisation (Mind Tools, 2023).

KEY PRINCIPLES

This model highlights the significance of preparing an organisation for change acceptance during the unfreezing phase and implementing the proposed modification during the change phase. Subsequently, the newly introduced state is solidified as the standard practice in the refreezing phase, ensuring sustainable transformation.

STRENGTHS AND WEAKNESSES

Lewin's change management model is widely acclaimed for its simplicity and comprehensibility, contributing to its popularity among diverse organisations. However, in today's dynamic and rapidly evolving business environments, the model's linear approach may present certain limitations on its effectiveness.

PRACTICAL APPLICATION

Despite the model's established utility, the recent literature does not provide explicit details on the contemporary applications of Lewin's change management model, indicating a research gap. Exploring its current utilisation in various contexts could be a beneficial addition to the existing body of knowledge.

PROSCI'S ADKAR MODEL

The ADKAR model, a goal-oriented instrument developed by Prosci founder Jeff Hiatt almost two decades ago, guides individual and organisational change. This model significantly emphasises the human dimension of change, outlining the steps individuals must undergo during a specific change process. Its versatility extends to its applicability, as it effectively accommodates both minor personal changes and major organisational transitions.

KEY PRINCIPLES

The ADKAR model, an acronym for awareness, desire, knowledge, ability, and reinforcement, focuses on individual change. It outlines a sequence individuals should follow for successful change management, emphasising the importance of personal transition within the broader organisational change framework.

STRENGTHS AND WEAKNESSES

The ADKAR model is distinct in that it primarily focuses on individual employees and their personal journey through the change process, a critical element for the overall success of organisational change. As Samosir and Jayadi (2023) point out, the model's strength is identifying potential gaps during the change process. Their research further demonstrates the model's relevance in modern, technology-driven environments, particularly in the context of digital banking transformations.

However, while its individual-centric approach presents certain advantages, it may also neglect broader organisational dynamics and systemic issues. This suggests a potential limitation of the model, emphasising the need for a balanced approach that considers both individual and organisational dynamics in the change management process.

PRACTICAL APPLICATION

The ADKAR model, with its unique focus on individual change, has been effectively deployed across varied sectors, encompassing organisational restructuring and adopting new technologies. The model's successful application in digital banking is a testament to its utility in managing organisational technological and structural changes (Samosir & Jayadi, 2023). This underscores the model's versatility and adaptability in addressing complex change management scenarios.

EFFECTIVE STRATEGIES FOR AI ADOPTION IN DESIGN

These strategies offer a comprehensive and structured framework to guide the transitional process, ensuring the effective incorporation of technological advancements into pre-existing organisational processes and the prevailing culture. They provide models that underscore the necessity of maintaining equilibrium between individual and organisational dynamics during change management.

While Kotter's and Lewin's models consider organisational strategies as the backbone of their proposed change management framework, the ADKAR model takes a slightly different approach by bringing the perspective of individual employees to the forefront. This nuanced approach provides a more inclusive perspective, helping to instigate and embrace technological changes across all hierarchical levels within the organisation. This, in turn, fosters an environment conducive to sustainable and effective integration of new technologies.

One common challenge organisations often face when implementing technological changes is employee resistance. This resistance can be significantly mitigated by adopting these models' strategic approaches. These include instilling a sense of urgency about the transition, facilitating clear and open communication channels, providing comprehensive training, and reinforcing new behaviours through positive reinforcement mechanisms. These strategies are instrumental in fostering a culture receptive to technology and innovation, smoothing the path to successfully implementing technological changes.

Moreover, change management models must demonstrate high adaptability in the rapidly evolving business landscape, particularly in AI and digital. These models must be flexible and dynamic to accommodate and keep pace with the fast-changing technological environment, ensuring their continued relevance and effectiveness.

These models also underscore the pivotal role of strategic vision and robust leadership in navigating organisations through the often complex process of technological transitions. Effective leadership, characterised by clear vision articulation, rallying support among employees, and guiding the organisation through the intricacies of adapting to new technologies, is brought to the forefront.

In conclusion, these change management models provide robust and invaluable insights for businesses embarking on technological transformations. They offer a roadmap for successfully navigating the intricate process of adopting and implementing new technologies, ensuring their technical implementation and cultural and operational integration within the organisation. This, in turn, significantly enhances organisational efficiency, adaptability, and competitiveness, paving the way for success in the contemporary technological era.

PRESENT CIRCUMSTANCES

DESIGNOPS AND AI

To address the current challenges in the team, the client and I researched and agreed upon implementing design operations (DesignOps) – a systematic approach to optimising people, processes, and design practices. Integrating AI tools within this DesignOps framework was also identified as a potential strategy to enhance design outcomes and enable automation within the team. Moreover, establishing a centralised design system, a crucial component of DesignOps, could potentially mitigate issues of work duplication and promote efficiency by providing a single reference for all design elements.

Adopting a DesignOps framework supplemented with AI tools emerged as a critical solution for the team to streamline the design process, enhance communication through clear guidelines and processes, and ensure consistent, high-quality designs.

Before implementing change, it was essential to assess our team's situation comprehensively, examine each team member's mindset, and gauge the prevailing sentiments among creative professionals.

As the design team expanded, the client faced substantial challenges in scaling operations due to an influx of projects and the task of integrating new designers with seasoned ones, leading to inconsistent workflows. The increased diversity within the team also made it challenging to maintain high-quality designs. Additionally, the team struggled with issues surrounding inherited legacy works and projects, primarily attributed to unstructured workflows, resulting in confusion, inefficiencies, and frequent miscommunications.

In order to conduct a comprehensive evaluation of the department's strengths and weaknesses, the business model canvas (Figure 3.3) was employed as an analytical tool to dissect the design team's operation. This visual framework offers a holistic depiction of the design team's critical aspects, explaining its role and contribution to the organisation's structure. It facilitated the identification of crucial components, such as the resources required by the design team, the key activities they enact, their primary partners, and the overall value they give to the organisation. This analytical process provides strategic insights by highlighting areas of strength and revealing potential gaps within the design team's role in the business model.

Upon analysis, it becomes evident that our department operated without a unified design language and tools system. This lack of standardisation hinders the team's ability to automate routine tasks, particularly those related to research and UX writing. Developing a standardised system will enhance our designs' consistency and free up time for ideation and creative thinking. Therefore, we must establish a cohesive system and explore avenues for automating repetitive tasks.

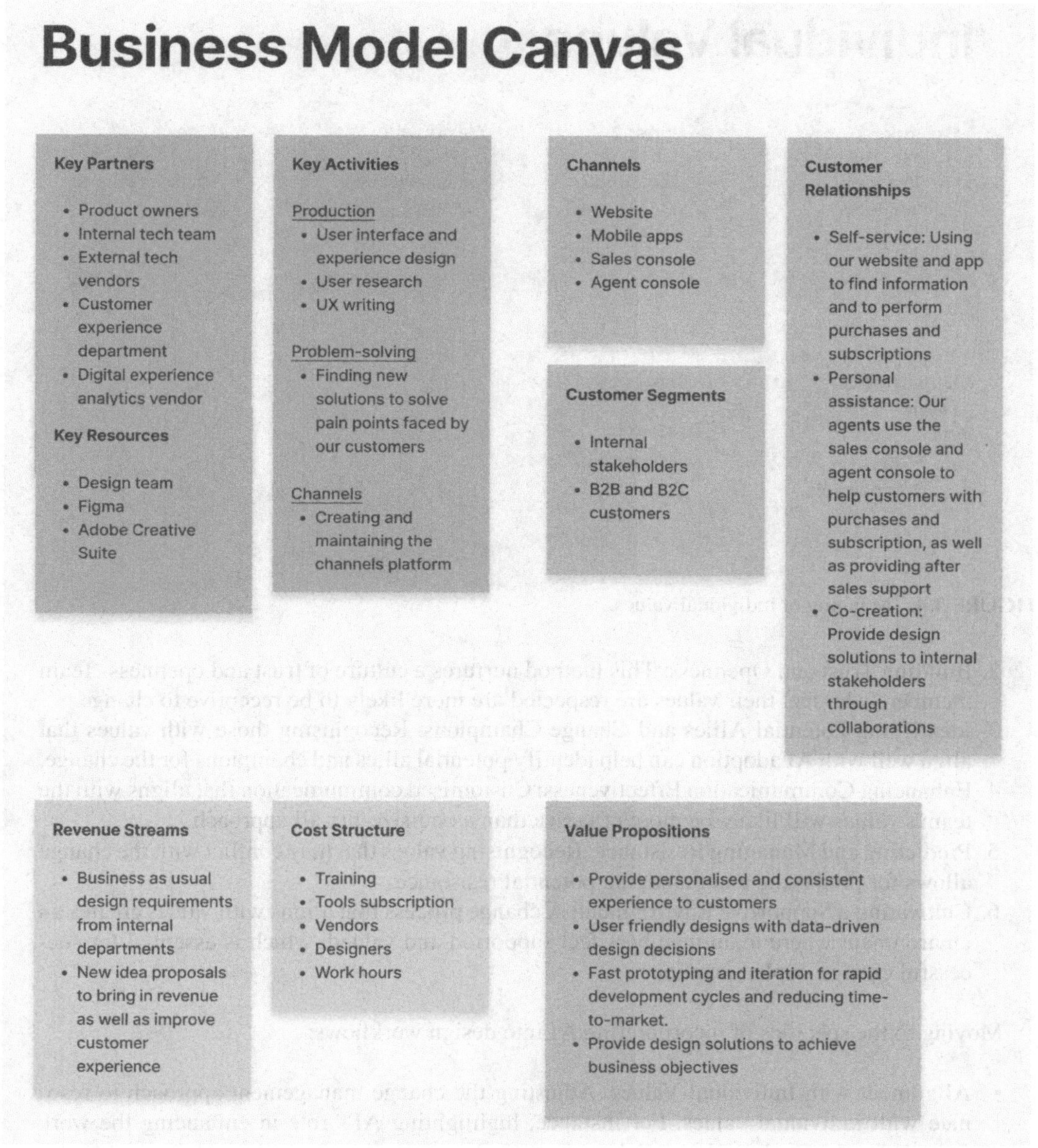

Business Model Canvas

Key Partners

- Product owners
- Internal tech team
- External tech vendors
- Customer experience department
- Digital experience analytics vendor

Key Resources

- Design team
- Figma
- Adobe Creative Suite

Key Activities

Production
- User interface and experience design
- User research
- UX writing

Problem-solving
- Finding new solutions to solve pain points faced by our customers

Channels
- Creating and maintaining the channels platform

Channels

- Website
- Mobile apps
- Sales console
- Agent console

Customer Segments

- Internal stakeholders
- B2B and B2C customers

Customer Relationships

- Self-service: Using our website and app to find information and to perform purchases and subscriptions
- Personal assistance: Our agents use the sales console and agent console to help customers with purchases and subscription, as well as providing after sales support
- Co-creation: Provide design solutions to internal stakeholders through collaborations

Revenue Streams

- Business as usual design requirements from internal departments
- New idea proposals to bring in revenue as well as improve customer experience

Cost Structure

- Training
- Tools subscription
- Vendors
- Designers
- Work hours

Value Propositions

- Provide personalised and consistent experience to customers
- User friendly designs with data-driven design decisions
- Fast prototyping and iteration for rapid development cycles and reducing time-to-market.
- Provide design solutions to achieve business objectives

FIGURE 3.3 The business model canvas.

LEVERAGING CORE VALUES FOR EFFECTIVE AI INTEGRATION IN DESIGN WORKFLOWS

To comprehend the design team's core values, a critical factor in incorporating AI into design workflows, I asked each designer, including myself and the client, to reflect on their core values using Brown's (2023) extensive list as a guide. We examined this list (Figure 3.4), selected values aligned with our own, and then shared two core values with me, explaining their selections.

This exercise is vital for several reasons:

1. Personalising Change Management: By understanding individual values, I can tailor change strategies to align with what team members deem essential.

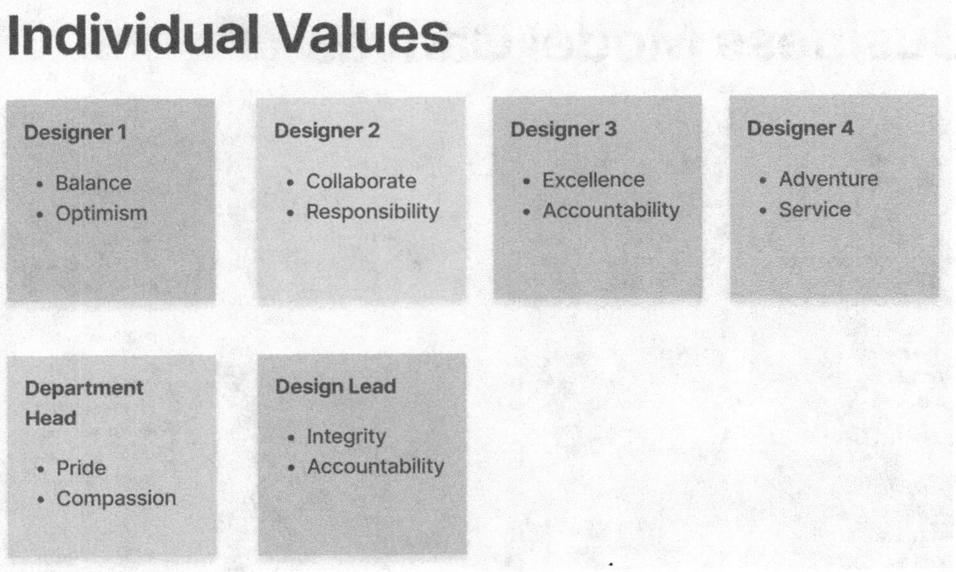

Individual Values

Designer 1	Designer 2	Designer 3	Designer 4
• Balance • Optimism	• Collaborate • Responsibility	• Excellence • Accountability	• Adventure • Service

Department Head	Design Lead
• Pride • Compassion	• Integrity • Accountability

FIGURE 3.4   Mapping of individual values.

2. Building Trust and Openness: This method nurtures a culture of trust and openness. Team members who feel their values are respected are more likely to be receptive to change.
3. Identifying Potential Allies and Change Champions: Recognising those with values that align well with AI adoption can help identify potential allies and champions for the change.
4. Enhancing Communication Effectiveness: Customised communication that aligns with the team's values will likely be more effective than a one-size-fits-all approach.
5. Predicting and Managing Resistance: Recognising values that may conflict with the change allows for predicting and managing potential resistance.
6. Cultivating a Supportive Environment: A change process that aligns with values creates an environment where team members feel supported and valued, which is essential for successful change implementation.

Moving to the specifics of incorporating AI into design workflows:

- Alignment with Individual Values: Adjusting the change management approach to resonate with individual values. For instance, highlighting AI's role in enhancing the work process for those who value collaboration.
- Addressing Concerns and Resistance: Concentrating on addressing concerns related to values like job security by demonstrating how AI can enhance, not replace, human skills.
- Facilitating Engagement and Buy-in: Ensuring the introduction of AI aligns with their core values, thereby increasing buy-in.
- Customised Training and Support: Creating training programmes that cater to the team's diverse values, such as focusing on efficiency or learning opportunities.
- Promoting a Positive Work Culture: Ensuring that the implementation of AI brings technological progress and aligns with and boosts individual value and morale.

In this way, understanding individual values becomes a key component in successfully managing the transition to AI-enhanced workflows in design, ensuring a smoother, more accepted, and effective integration.

INSIGHTS FROM CREATIVE PROFESSIONALS

Before embarking on implementation, it is also crucial to ascertain the general perspectives regarding using AI tools within creative workflows. Additionally, understanding individual views and concerns with AI tools is equally important. Such insights significantly aid in AI implementation at the individual level, fostering a more informed and practical integration process. This, in turn, reinforces the staying and acceptance of the changes.

From the various studies and literature research conducted, it is evident that creative professionals have diverse and rather complex views on implementing and using AI technology in their respective fields. These views span a wide spectrum, ranging from apprehension and concern to excitement and enthusiasm. On the one hand, some individuals within these professions have expressed worry about how AI could potentially transform or disrupt their work and the conventional ways they have been accustomed to operating (Inie et al., 2023; Yusa et al., 2022). On the other hand, some welcome this potential revolution in the creative process, seeing it as an opportunity to reshape and redefine their work in exciting and innovative ways.

Furthermore, the introduction of AI in the realm of art has sparked a heated and ongoing debate among professionals in the field. Some artists and critics have praised AI's potential to revolutionise creative processes, seeing it as a tool that can push the boundaries of what is possible in art. They believe AI could bring about novel art forms and unprecedented innovation (Yusa et al., 2022).

However, others question whether AI can indeed be creative like humans. They are sceptical of the impact of AI on human creativity and wonder whether it could lead to a loss of authenticity in art, as they believe that true creativity is inherently human and cannot be replicated by a machine (Yusa et al., 2022).

According to recent research, there is a strong suggestion that AI has the potential to open up new avenues for creativity (Lundman & Nordström, 2023). Regarding AI technology, the perspectives among creative professionals are incredibly diverse. These viewpoints often intersect with concerns about AI's impact on the creative industry, specifically on job security, originality, and the role of human creativity. In order to make an informed decision about how to implement changes in the design department, it was necessary to gather detailed insights from those who would be directly affected by these developments.

To do this, I surveyed various professionals, from UI and UX designers to creative directors (Figure 3.5). The purpose of this survey was to understand their views on AI technology and any concerns they might have regarding its integration into their work processes.

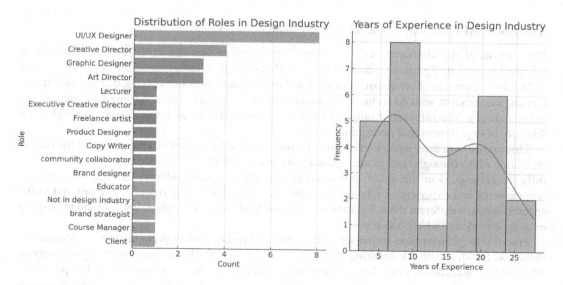

FIGURE 3.5 Demographics data from a survey with 30 creative professionals.

ANALYSIS OF ROLES AND EXPERIENCE IN THE DESIGN INDUSTRY

The exploration of the diverse roles and experience levels in the design industry, as evidenced by my dataset, offers significant insights into the potential impact and successful implementation of AI on workflow efficiency. The prevalence of roles such as UI/UX designers, creative directors, graphic designers, and lecturers indicates a multifaceted industry with varying needs and expectations from technological advancements.

AI's influence on workflow efficiency in design is multifaceted. AI presents a dual-edged sword for seasoned professionals who have witnessed the industry's evolution over 20 years or more. On the one hand, it promises enhanced efficiency and the ability to streamline complex processes. On the other hand, it challenges established methodologies, requiring a shift in mindset and approach. Thus, implementing AI in this context should involve technological integration and address the potential resistance and apprehension associated with changing long-standing practices.

For the newer generation of professionals, with less than ten years of experience, AI is often seen as a natural progression of the digital landscape. This group may be more open to embracing AI, viewing it as a tool to augment their capabilities and streamline workflows. However, they also require guidance to optimise AI's use effectively, ensuring that it complements rather than replaces human creativity and expertise.

Therefore, the successful implementation of AI in the design industry hinges on tailored change management strategies. These strategies must recognise the diverse perspectives and varying degrees of readiness for AI adoption across different roles and experience levels. For instance, change management for seasoned professionals could focus on demonstrating how AI can enhance rather than replace their expertise, coupled with training programmes that ease the transition. For the newer generation, the emphasis could be on integrating AI as a tool that complements their skills and innovative approaches.

The impact of AI on workflow efficiency in the design industry is undeniable. Adopting change management strategies as diverse and multifaceted as the industry itself is crucial to implement AI successfully. By doing so, we can ensure that AI is a powerful ally in the creative process, enhancing efficiency and innovation across all experience levels and expertise in the design industry. This approach facilitates a smoother transition into an AI-augmented future and ensures that the unique talents and perspectives within the industry are preserved and enhanced.

ANALYSIS OF AI FAMILIARITY BY ROLE AND EXPERIENCE

The analysis of AI familiarity within the design industry, as depicted in the box plot illustration (Figure 3.6), reveals significant variation across different roles, notably among UI and UX designers and creative directors. This disparity in AI familiarity underscores the varying degrees of interaction and engagement with AI technologies tied to specific roles rather than the length of one's career in the industry. The substantial range in familiarity scores within each role further highlights the diversity of experiences and perceptions of AI.

This insight is crucial when considering the impact of AI on workflow efficiency and the implementation of AI through change management strategies. The data suggests that the key to successfully integrating AI in the design industry lies in recognising and addressing each role's unique requirements and experiences. Change management strategies should, therefore, be tailored to the specific needs of different roles, focusing on increasing AI familiarity where it is low and leveraging existing knowledge where it is high.

Practical implementation strategies include targeted training and development programmes for those less familiar with AI, ensuring they have the necessary skills and understanding to integrate AI into their workflows. For roles with higher AI familiarity, the focus could shift to optimising

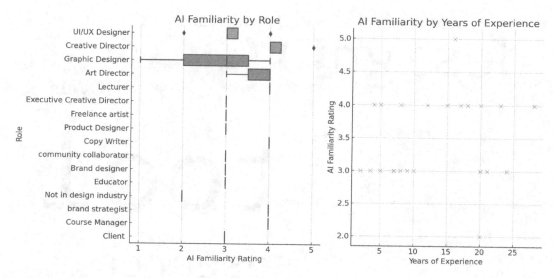

FIGURE 3.6 Survey data on AI familiarity.

the use of AI technologies to enhance workflow efficiency, encouraging innovation and advanced application of AI.

In summary, the role-specific variability in AI familiarity within the design industry calls for a nuanced approach to implementing AI. By adopting role-specific change management strategies that cater to the varying levels of AI familiarity, we can facilitate a more effective and seamless integration of AI, thereby maximising its impact on workflow efficiency in the design industry.

OPINION ANALYSIS ON AI INTEGRATION AND ETHICAL CONCERNS IN DESIGN

The comprehensive examination of attitudes towards AI integration in the design field and ethical considerations illuminates several key themes directly relevant to the research topic (Figure 3.7). The recognition of AI as a supportive and creative tool, as evidenced by terms like "tool", "help", and "creative" in the word cloud analysis, aligns well with its potential to augment creative capabilities and boost workflow efficiency. However, the successful implementation of AI is not just about its capabilities as a tool but also hinges on its seamless integration into existing workflows and team dynamics, a notion supported by the frequent appearance of terms such as "work", "process", and "team".

In addition to the operational aspects of AI integration, the ethical considerations, highlighted by concerns around intellectual property, authenticity, and artistic integrity, underscore the complexity of its incorporation into the design process. Terms like "copyright", "originality", and "use" reflect the legal implications, while "art", "creativity", and "replicate" point to the impact on artistic uniqueness.

A balanced and comprehensive approach is necessary to address these themes within the context of the research topic. This approach should focus on leveraging AI for efficiency gains and ensure that it is integrated in a way that respects and enhances the creative and collaborative aspects of design work. Additionally, it must navigate the ethical landscape by establishing guidelines and best practices that respect intellectual property rights and maintain the authenticity of designs.

Incorporating AI in the design industry should be approached with a dual focus: operational and ethical. Change management strategies should be developed to facilitate smooth technological

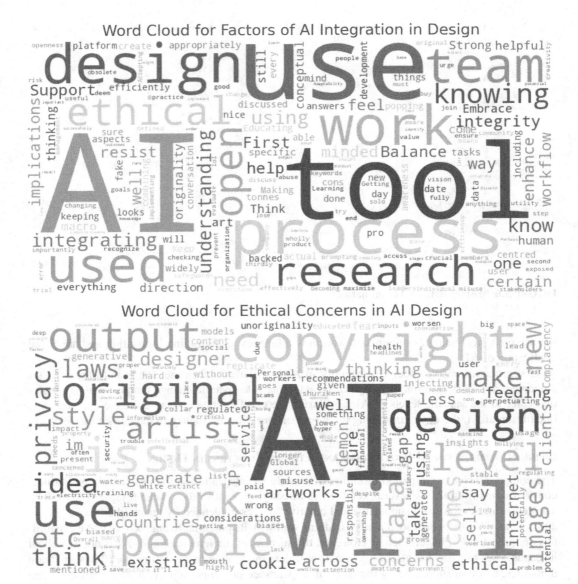

FIGURE 3.7   Qualitative data on AI integration factors and ethical concerns in AI design.

integration, foster team collaboration, and address ethical concerns. By doing so, AI can be successfully integrated as a tool that enhances workflow efficiency and respects the artistic and legal nuances of the design industry.

SENTIMENT ANALYSIS FINDINGS AND THEIR IMPLICATIONS FOR THE DESIGN INDUSTRY

The sentiment analysis from the dataset reveals a multifaceted perspective on integrating AI in the design industry (Figure 3.8). This perspective is crucial for understanding how AI can impact workflow efficiency and the strategies for successful implementation. The positive sentiments towards AI integration reflect optimism and readiness in the industry, suggesting a trend towards increased reliance on AI-assisted tools and platforms. This trend underscores the potential for AI to enhance workflow efficiency in design, aligning with the focus of this research.

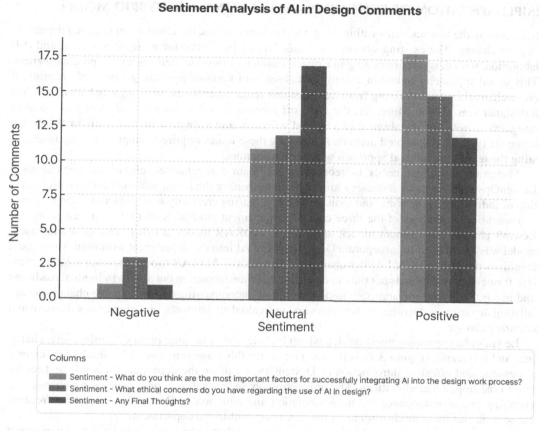

FIGURE 3.8   Sentiment analysis of AI in design comments.

However, negative sentiments, particularly concerning ethical issues like copyright, data privacy, and authenticity, highlight critical areas that must be addressed in AI implementation. These concerns have significant implications for policy-making and the establishment of ethical guidelines and best practices in AI usage. Thus, they should be essential to change management strategies, ensuring a responsible and conscientious approach to AI integration.

The neutral sentiments, indicating a possible lack of awareness or understanding of AI's role, point to a need for enhanced education and training within the design field. This suggests that change management strategies should include educational components, focusing on AI literacy and ethical considerations. Design firms and educational institutions might need to invest in workshops and training sessions to bridge this knowledge gap, facilitating a smoother transition to AI-enhanced workflows.

In conclusion, the varied sentiments expressed about AI in the design industry highlight the need for a comprehensive approach to integrating AI into design practices. This approach should leverage AI for enhanced workflow efficiency and address the ethical concerns and educational needs identified in the sentiment analysis. Recognising the optimism and addressing the concerns reflected in these sentiments is essential for guiding the industry towards an informed, ethical, and practical adoption of AI. Change management strategies, therefore, must be multifaceted, combining technological integration with ethical considerations and educational initiatives to ensure successful AI implementation in the design industry.

IMPLEMENTATION OF DESIGNOPS AND AI USING THE HYBRID MODEL

Recognising the inefficiencies within the current design process, the client acknowledged the necessity for change. The existing system was characterised by a need for more organisation and collaboration, with designers operating in isolated channels alongside their respective product owners. This siloed approach resulted in a significant absence of knowledge-sharing, obscured visibility of cross-channel impacts deriving from new business requirements, and no designated stand-in when a designer was absent. Moreover, the hand-off process was consistently challenging as stand-in designers varied, and the absence of a unified workflow and a single source of truth for the design language further complicated matters. Addressing these issues required comprehensively re-evaluating the team's operational approach within each channel.

Therefore, the client needs to reorganise the team for enhanced efficiency. Implementing DesignOps with AI tools, a strategy aimed at strengthening the team, will facilitate the orchestration of individuals, processes, and tools, thereby increasing creativity and impact on a larger scale.

Based on the analysis of the three change management models, Kotter's 8-step change model, Lewin's change management model, and Prosci's ADKAR model, a hybrid change management model was established to incorporate DesignOps and AI into the department workflow. This model combines the principles of Lewin's change model with the ADKAR model, creating a comprehensive framework that attends not only to departmental transformation but also to individual readiness and needs. This strategic approach underpins an all-encompassing and sustainable change, encapsulating departmental structure alterations and individual adjustments, ensuring comprehensive and enduring change.

Lewin's change management model, which includes the structured phases of unfreezing, changing, and refreezing (Figure 3.9), is the foundation for this transformation. This model facilitates a systematic and effective introduction of DesignOps, ensuring the department is ready and receptive to the impending alterations. The unfreezing phase instigates a consciousness for change, the changing phase underscores the implementation and adoption of DesignOps, and the refreezing phase solidifies these modifications as the new norm within the department.

Simultaneously, integrating AI tools into design workflows requires a focus on the human aspect of change, a concept that the ADKAR model will address. This model, which centres on the stages of awareness, desire, knowledge, ability, and reinforcement, understands that successful organisational change is dependent upon individual transitions. By ensuring each team member is aware of, motivated, and equipped to use AI tools in their workflow, the model considers personal transitions and responses to change, promoting a seamless integration of AI into design processes.

Combining these two models aims to devise a comprehensive change management strategy, ensuring every aspect of the department, be it structural or individual, is noticed. This dual approach facilitates the adoption of new systems and technologies. It fosters a culture of adaptability and continuous learning, which are crucial in the dynamic field of design and technology.

The decision to utilise a hybrid model is strategic, ensuring that as the department evolves structurally with DesignOps and technologically with AI, our team remains at the core of this transformation, adequately equipped and confident in navigating these changes. It will provide our team with the required tools and confidence for effective change management and will also provide a controlled testing ground for the hypothesis suggesting the enhanced effectiveness of change when implemented at both macro and micro levels through a hybrid model. Moreover, this approach paves the way for investigating potential unintended consequences that might surface from the concurrent

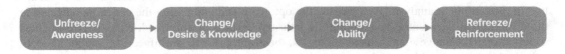

FIGURE 3.9   Lewin's change management model.

application of both models. Carefully examining these outcomes could yield insights into the intricacies of change management within our department, thereby fine-tuning our strategies for a more cohesive and successful transformation.

UNFREEZE/AWARENESS

The initial step in this reorganisation involved considering a team structure based on "design pods". This approach was informed by a detailed analysis of each designer's daily tasks, considering the four primary touch points for product sales and customer service (website, mobile app, sales console, and agent console).

The change was to divide responsibilities into two distinct design pods: the "iBuy" pod, which focused on the customer-facing website and in-store sales console, and the "iManage" pod, which was responsible for after-sales service touch points, including the mobile app and agent console. This reorganisation facilitated the cross-sharing of knowledge within each pod and enabled designers to concentrate on solving customer pain points specific to their assigned touch points.

Additionally, a responsibility matrix was developed to provide clarity in decision-making processes. This matrix, grounded in the DACI decision-making framework, assigns distinct roles to team members, thereby fostering a deeper understanding of the decision-making structure within the team with clarity (Figure 3.10).

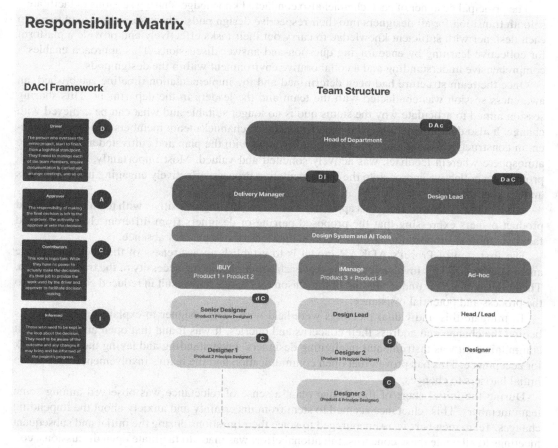

FIGURE 3.10 Responsibility matrix.

Implementation Timeline

FIGURE 3.11   Implementation timeline.

A comprehensive visual timeline has also been crafted to outline the timeline for implementing the new team structure, providing a clear understanding of its implementation schedule (Figure 3.11). In addition, a one-month trial period is strategically established in the timeline to monitor preliminary transformations and gather feedback from the product owners and designers, enabling possible changes to the team structure and operational processes if needed.

The principal designer of each channel also conducted knowledge-sharing sessions to facilitate a smooth transition for all designers into their respective design pods. These sessions aimed to equip each designer with sufficient knowledge to carry out their tasks effectively and provide a platform for collective learning by encouraging question-and-answer discussions. This approach enables a comprehensive understanding and a collaborative environment within the design pods.

Once the team structure had been determined and the implementation timeline established, an awareness session was conducted with the team and the leaders in the department. This sharing session aimed to articulate why the status quo is no longer suitable and what can be achieved with change. It also disseminated the plan and engaged each channel's team members and product owners in constructive discussion. This ensured alignment with the plan and cultivated a collaborative atmosphere wherein feedback was actively solicited and valued. Most importantly, this approach promoted a collective approach to the implementation process, effectively engaging team members and product owners.

The feedback received from this approach was overwhelmingly positive, with team members and product owners expressing that the proposed pairing of designers from different channels would facilitate mutual support during high workload periods or instances of absence.

The first step of Prosci's ADKAR model is to establish an awareness of the need for change among designers. This involves ensuring that each comprehends the necessity of the transformation. The absence of a clear understanding of the reason for change may result in reduced commitment to the process and potential resistance.

To mitigate this, individual meetings were held with each designer to explain the motivations behind the change and address their concerns and queries. It was found that open and transparent communication was instrumental in aligning designers' understanding and laying the groundwork for acceptance. This has proven that open communication and the team's involvement removed the initial barriers to changes.

During the initial stage of the change plan, a sense of reluctance was observed among some team members. This reluctance seemed to stem from uncertainty and anxiety about the impending changes. Team members were encouraged to voice their questions during the initial and subsequent meetings to alleviate these concerns. Intentional effort was made to facilitate open discussions, collectively and individually, to instil acceptance of the change within the team gradually.

CHANGE/DESIRE AND KNOWLEDGE

This stage involved introducing DesignOps methodologies, providing training sessions to the team members, and gradually incorporating DesignOps into the design process. The team leaders were essential in this stage, addressing any concerns and ensuring that everyone was on board with the new process.

During this phase of the change process, an essential step was re-emphasising the newly devised operational approach to the entire team. It included an in-depth explanation of the methodologies and rationale behind the proposed alterations, ensuring that all team members understood the new strategic direction well.

Concurrently, a culture of open communication was cultivated by inviting feedback and suggestions from all team members regarding the proposed structure. That facilitated the refinement of the new changes, ensuring that they were optimised for our context and that all potential issues were addressed. Moreover, because the team was actively involved in the change process, it fostered a sense of ownership and responsibility among the team members, which anchored the success of the change implementation. The contribution from the team to the change process also resulted in increased investment and commitment, thereby enhancing the likelihood of lasting and successful change. This stage showed how clear communication, teamwork, and inclusive decision-making can facilitate the smooth initiation of our new strategies.

THE VALUE OF A DESIGN SYSTEM

The rapid and widespread evolution of user interface (UI) design requires the development of effective systems and processes to boost innovation and improve user experience (Fessenden, 2021). That will enable the team to manage many design screens and UI components across the four channels and uphold consistency across all interfaces.

Hence, establishing a design system is an integral part of our DesignOps implementation, and building this system constitutes the second vital phase of change within the department (Figures 3.12 & 3.13).

Moreover, to instil a willingness to adapt and support change amongst the designers, the impact of a design system was exemplified through a project in which a design system was utilised to govern the design language. This demonstration addressed the personal implications of change and

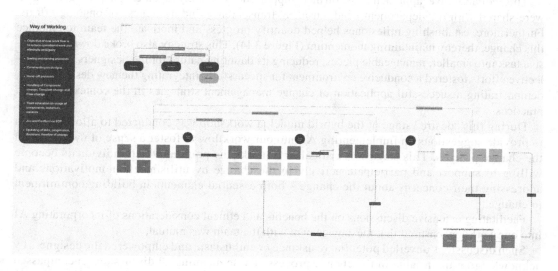

FIGURE 3.12 New ways of working (design system structure).

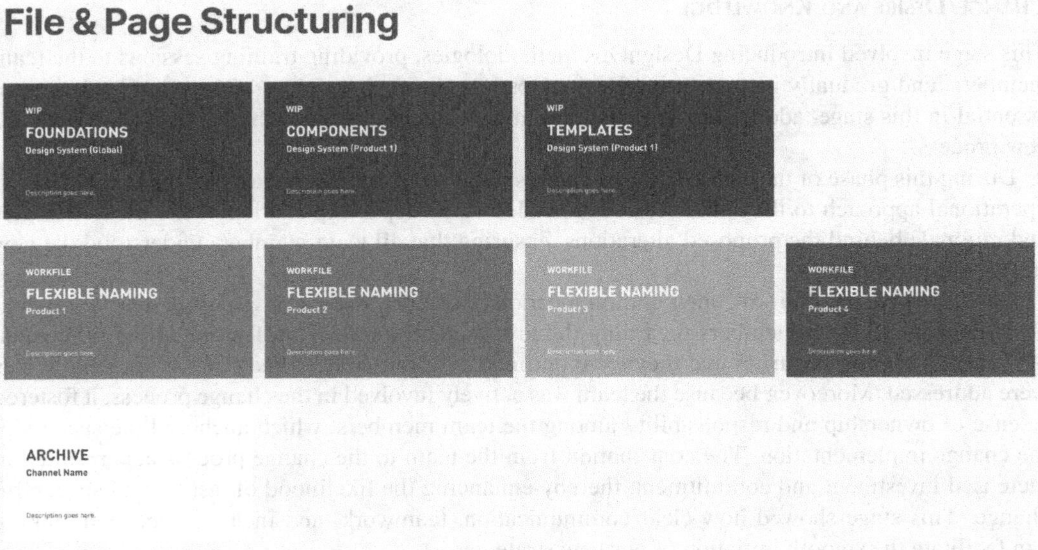

File & Page Structuring

FIGURE 3.13 File organisation.

aligned it with individual motivations and departmental objectives. The process of building the design system was also shared with and guided by the designers to enhance their understanding of the process involved in developing design systems for each of their channels.

Given the team's limited experience establishing a design system, one key designer was appointed to supervise its development with my guidance. This individual collaborated closely with me to steer this initiative. Monthly meetings with all the designers were organised to allow me to guide and support the team. In conjunction with this, I periodically shared educational resources in the form of online articles and tutorials on building design systems with the team to facilitate their learning.

This collaborative approach was further supported by weekly updates on progress, which were shared with the department head, ensuring transparency and awareness of ongoing efforts. Furthermore, establishing milestones helped quantify progress and motivate the team to embrace this change, thereby maintaining momentum (Figure 3.14). This strategy also broke down the extensive task into smaller, manageable pieces, reducing its daunting nature for the designers. These collective efforts fostered a conducive environment for successfully integrating the new design system, demonstrating a successful application of change management strategies in the context of design practices.

During this "desire" stage of the hybrid model, a workshop was conducted to allow the team to provide suggestions on implementing AI into our workflow to foster a sense of ownership in the change process. This stage was critical as it marks the point at which individuals become willing to support and participate actively in the change by utilising their motivations and addressing their concerns about the change – both essential elements in building commitment to change.

Facilitating extensive discussions on the benefits and ethical considerations of incorporating AI into our workflow ensured that the engagement with the team was mutual.

Such discussions unveiled potential resistance or enthusiasm and empowered the designers by acknowledging their value in the change process. These meaningful discussions encompassed

Task	Date	Agenda	Remarks	Completed
Designers' Sharing Session	20 Sep, Wed	1. Share design workflow using global design system library with designers 2. Designers to update team on design foundation progress	**To do by next session:** 1. Update channels' foundation using template provided	Done
Foundation Discussion 1	4 Oct, Wed	1. Designers to update team on design foundation progress 2. Discuss foundation alignment and standardisation		Done
Foundation Discussion 2	18 Oct, Wed	1. Designers to update team on design foundation progress 2. Standardisation of naming convention for foundation		Done
Foundation Discussion 3	1 Nov , Wed	1. Alignment on component timeline 2. Final run through of open points for foundation and making of necessary changes to complete global foundation		Done
Component Discussion 1	15 Nov, Wed	Designers to update team on design component progress and discuss potential challenges.	New proposed date: 20 Dec 23	Done
Component Discussion 2	29 Nov, Wed	Designers to update team on design component progress and discuss potential challenges.	New proposed date: 10 Jan 24	Done
Component Discussion 3	13 Dec, Wed	Designers to update team on design component progress and discuss potential challenges.	New proposed date: 24 Jan 24	
Closing	20 Dec, Wed		New proposed date: 31 Jan 24	

FIGURE 3.14 Design system milestones.

the types of AI and their applications and a thorough examination of the benefits and ethical considerations involved in incorporating AI into our workflow (Figure 3.15). The alignment between the benefits and concerns listed during the discussions and the findings from an earlier survey validated the insights from the survey as a reliable basis for planning the change implementation.

Subsequently, the team was tasked with identifying the areas where AI could potentially enhance our workflow (Figure 3.16). This exercise allowed the team to visualise how AI could augment the workflow and align with broader departmental goals, thus rendering the change more meaningful to them.

Upon identifying these opportunities, the team was asked to list potential AI tools available to help us capitalise on them (Figure 3.17). Seeing their ideas infused the team with a personal stake in the change's success, transforming them from passive observers to active participants.

This early involvement of the team also facilitated the identification and cultivation of potential change champions who could positively influence their peers, further instilling a sense of ownership and commitment. However, considering the size of our team, the impact of having a champion on the change process warrants further exploration.

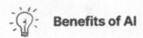

Benefits of AI

Increased Efficiency

Automate repetitive tasks, allowing team members to focus on more strategic and creative work.

Improved Accuracy

Analyse large amounts of data with precision, reducing the risk of human error.

Enhanced Decision-Making

Provide valuable insights and predictions based on data analysis, helping us make informed decisions.

Streamlined Processes

Optimise workflows by identifying bottlenecks, suggesting improvements, and automating manual processes.

Research

AI can provide initial data to kick start research tasks.

Ethical Considerations

Data Privacy

AI tools often require access to large amounts of data, raising concerns about privacy and security.

Algorithmic Bias

AI algorithms can be biased if trained on biased data, leading to unfair outcomes or discrimination.

Job Displacement

The integration of AI tools may result in job displacement or changes in job roles and responsibilities.

Training

Using AI tools can be a complex process that requires learning new skills to fully harness their potential.

Over-reliance

Over-reliance on AI tools could potentially stifle creativity and critical thinking, as well as lead to a lack of understanding and control over processes that are fully automated.

FIGURE 3.15   Benefits and ethical considerations of using AI.

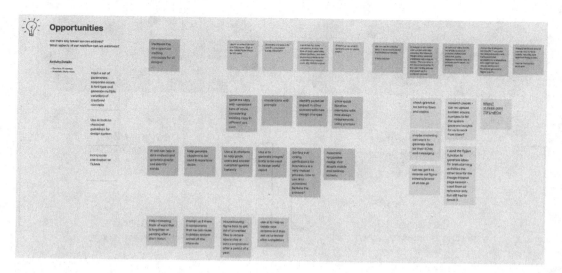

FIGURE 3.16 Identify areas of opportunities for AI.

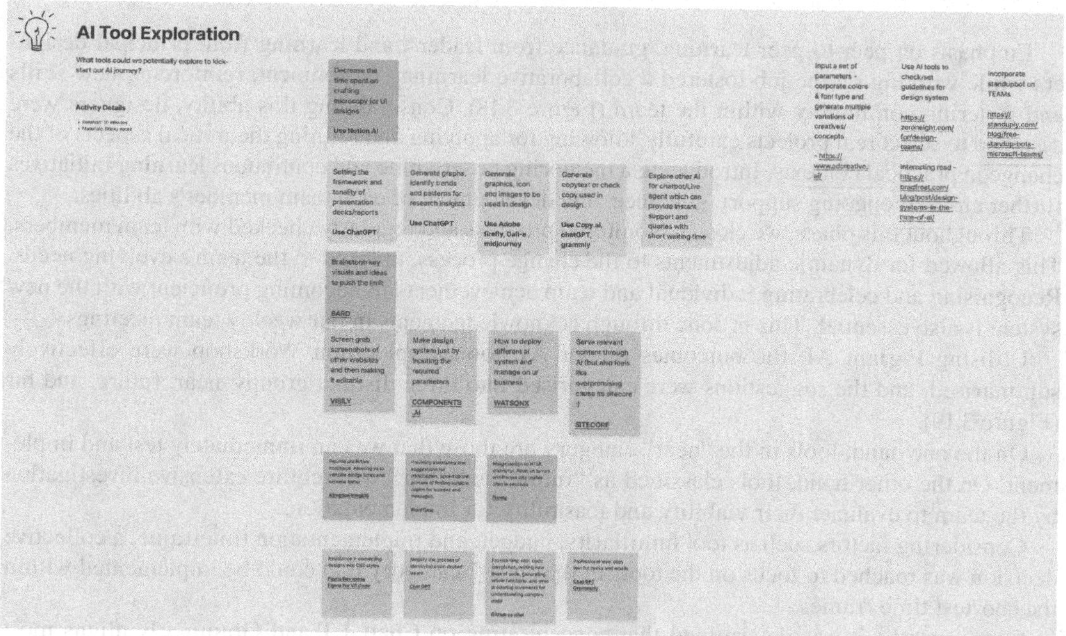

FIGURE 3.17 AI tools we could use.

CHANGE/ABILITY

Empowering Designers for Sustained Change

With the structure for the new DesignOps system in place and the team's understanding and acceptance of the design system, we transitioned into the change/ability phase. This stage focused on empowering the design team with the skills, tools, and confidence they need to work effectively within the new system and contribute to its ongoing evolution.

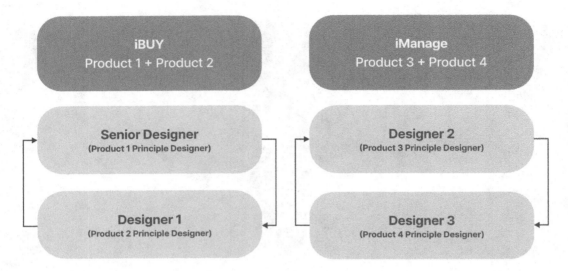

FIGURE 3.18 Peer-to-peer learning and support.

Emphasis on peer-to-peer learning, guidance from leaders, and learning from principal design-ers while working on the job fostered a collaborative learning environment, reinforcing new skills and fostering community within the team (Figure 3.18). Consolidating this ability, designers were assigned to select real projects carefully, allowing for applying and refining theoretical aspects of the change in practical contexts. Introducing a mentoring programme and continuous learning initiatives further ensured ongoing support, guidance, and development of each team member's abilities.

Throughout this phase, we closely monitored progress and regularly checked with team members. This allowed for dynamic adjustments to the change process, tailored to the team's evolving needs. Recognising and celebrating individual and team achievements in becoming proficient with the new system is also essential. This is done through acknowledgements in our weekly team meetings.

Utilising FigJam AI, the outcomes of our AI Tools Exploration Workshop were effectively summarised, and the suggestions were categorised into three distinct groups: near, future, and far (Figure 3.19).

On the one hand, tools in the "near" category are those that we can immediately test and imple-ment. On the other hand, tools classified as "future" and "far" will require extensive investigation by the team to evaluate their viability and feasibility for implementation.

Considering factors such as tool familiarity, budget, and implementation timeframe, a collective decision was reached to focus on the tools in the "near" category that could be implemented within the shortest time frame.

Consequently, it was determined that concentrating on ChatGPT and Grammarly aligns most appropriately with the objectives of the DesignOps implementation and the familiarity with the tools.

In conclusion, this phase, characterised by a hands-on, supportive, and collaborative approach, significantly enhanced the team's ability to adapt to and embrace the new operational structure. This laid the groundwork for a successful and sustainable transformation within the department.

REFREEZE/REINFORCEMENT

The "refreeze/reinforcement" phase must embed the new DesignOps methodologies and tools into the department's standard operating procedures to solidify the new practices and behaviours and ensure their stability. This process is more than just implementing the new changes; it is about

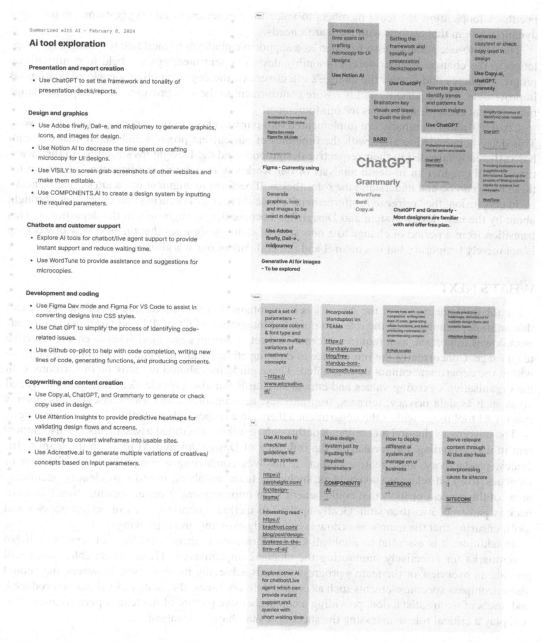

FIGURE 3.19 Summarisation of the workshop.

integrating them into policy documents and guidelines and setting specific targets aligning with the broader organisational goals.

A systematic approach is adopted to assure stability and consistency in the newly implemented practices, incorporating regular audits and reviews and maintaining feedback loops for ongoing refinement.

Monthly one-to-one sessions with individual designers further bolster this approach. These sessions serve dual purposes. First, they offer an opportunity for designers to provide feedback.

Feedback loops allow the team members to voice their experiences and suggestions, promoting a dynamic system that evolves with the team's needs.

The one-to-one sessions also function as a supportive platform to facilitate the designers' adaptation to the change. Furthermore, bi-monthly design system meetings are held to communicate progress updates and ensure the change's effectiveness, thereby promoting a cycle of continuous improvement. These periodic audits ensure commitment to the new processes and help the change management team identify areas for reinforcement.

ChatGPT and Grammarly are implemented into our workflow in the second quarter of the year. This delay is intended to align with the timelines of our current projects and ensure that the team has sufficiently adjusted to their respective design pods and the DesignOps structure. In the meantime, I will conduct an in-depth analysis of our workflow to determine where the integration for these tools can be and understand the roles they will perform to augment our team's effectiveness.

In conclusion, the "refreeze/reinforcement" phase is crucial in ensuring that the changes brought about by the new design system and DesignOps become deeply ingrained in the department. This transition from a period of change to a new state of stability signifies that the new way of working is not merely temporary but is a natural and integral part of our workflow.

WHAT'S NEXT

Since the implementation of the AI tools is being planned in the second quarter of the year, I feel that an in-depth analysis of the role and potential integration of ChatGPT and Grammarly into our workflow is important, as it requires a formalised set of writings and guidelines to ensure effective team usage. Concurrently, the importance of creating ethically aligned guidelines for AI tool usage within the department cannot be overstated. These guidelines should not only be in harmony with the organisation's existing values and ethical standards but also provide clear direction on crucial areas such as data privacy, fairness, transparency, accountability, and user consent. This ensures that all AI tool usage within the department adheres to a responsible and ethical way of working.

The ongoing evaluation and refinement through regularly scheduled assessments are also important in understanding and measuring the influence of DesignOps and AI tools on the team's efficiency, productivity, and creative output. It involves conducting a comprehensive collection of feedback from all relevant stakeholders, including team members, in order to identify accurately areas of the process or tools that could benefit from improvement or enhancement. Once this feedback is gathered, it is then strategically applied to facilitate improvements in both processes and tools, ensuring that the team's workflow is constantly evolving and improving.

In addition, it is essential to establish key performance indicators (KPIs) that serve as reliable benchmarks for effectively monitoring the changes implemented. These measurable values will provide an overview of the team's progress towards achieving its objectives. However, they could also encompass specific elements such as project delivery times, the quality of the output produced, and levels of team satisfaction, providing a comprehensive picture of the team's performance. This will play a critical role in assessing the success of our change management.

CONCLUSION

This research showed me that integrating new technology into design workflows goes beyond technological advancements. While it is natural to perceive breakthrough technology like AI as disruptive to our traditional ways of working, viewing it as an enhancement rather than as a disruption is essential. AI is not here to replace or disrupt our processes but to augment them, enabling us to work smarter and more efficiently. It allows us to automate repetitive tasks, freeing our time to focus on more creative and strategic aspects of our work.

As designers, we now have a unique opportunity to shape AI's evolution in our field. Our interaction with AI and feedback on its performance can guide its development and refinement. We can influence how AI evolves to meet our needs and enhance our craft. Therefore, it is important that we embrace AI as a tool that can assist us and actively participate in its evolution to ensure it develops in a way that truly supports and enhances our work.

In evaluating the complexities of transformation within my department, I conclude that it is also vital to have strategic planning, employee engagement, effective communication, and continuous learning to implement change successfully. The essence of change management resides in its capacity to align the objectives with individual values and capabilities, thereby cultivating a resilient, adaptable, and innovative environment.

My research has demonstrated that proficiency in managing change can help organisations attain enhanced operational efficiency, elevated employee satisfaction, and strengthened competitive advantage. However, the journey towards successful change is also full of challenges. A holistic approach is crucial to address them – one that considers the emotional and psychological dimensions of change with visionary, empathetic, and communicative leadership.

Looking ahead, I feel that the significance of change management is set to escalate further, propelled by rapid technological advancements and changing market demands. Organisations need to excel in change management to endure and prosper, transforming challenges into opportunities for growth and innovation.

REFERENCES

Abbas, T. (2022). Netflix change management case study. https://changemanagementinsight.com/netflix-change-management-case-study.

Adobe Sensei: Machine learning and Artificial Intelligence (no date). https://www.adobe.com/sg/sensei.html.

ZIGURAT Institute of Technology, (2023, November 27). AI in Architecture: The Key to Enhancing Design Efficiency and Gaining a Competitive Edge [2024 GUIDE]. Zigurat Institue of Technology. https://www.e-zigurat.com/en/blog/ai-in-architecture-guide.

Asan, O., Bayrak, A. E., & Choudhury, A. (2020). Artificial intelligence and human trust in healthcare: Focus on clinicians. *Journal of Medical Internet Research*, 22 (6), e15154. https://doi.org/10.2196/15154.

Askari, M. (2023). The ethical implications of AI in design | UX Collective. *Medium*, 15 February. https://uxdesign.cc/the-ethical-implications-of-ai-in-design-87ab731489fd.

Bertão, R. A., & Joo, J. (2021). Artificial intelligence in UX/UI design: A survey on current adoption and [future] practices. *Blucher Design Proceedings*, 9(5), 404–413. https://doi.org/10.5151/ead2021-123.

Brougham, D., & Haar, J. (2018). Smart technology, artificial intelligence, robotics, and algorithms (STARA): Employees' perceptions of our future workplace. *Journal of Management & Organization*, 24(2), 239–257. https://doi.org/10.1017/jmo.2016.55.

Brown, B. (2023) Dare to Lead | List of Values. https://brenebrown.com/resources/dare-to-lead-list-of-values.

Dai, A. (2023). Co-creation: space reconfiguration by architect and agent simulation based machine learning. In: Yuan, P.F., Chai, H., Yan, C., Li, K., Sun, T. (eds) *Hybrid Intelligence. CDRF 2022. Computational Design and Robotic Fabrication* (pp. 304–313). Springer. https://doi.org/10.1007/978-981-19-8637-6_27.

Davis, J. (2022). Dewey Goes corporate: Examining the suitability of Kotter's change management model for use in libraries. *Journal of Library Administration*, 62(3), 275–290. https://doi.org/10.1080/01930826.2022.2043687.

Dwivedi, Y. K., Hughes, L., Ismagilova, E., Aarts, G., Coombs, C., Crick, T., Duan, Y., Dwivedi, R., Edwards, J., Eirug, A., Galanos, V., Vigneswara Ilavarasan, P., Janssen, M., Jones, P., Kar, A. K., Kizgin, H., Kronemann, B., Lal, B., ... Williams, M. D. (2021). Artificial intelligence (AI): Multidisciplinary perspectives on emerging challenges, opportunities, and agenda for research, practice and policy. *International Journal of Information Management 57*, 101994. https://doi.org/10.1016/j.ijinfomgt.2019.08.002.

Einola, K., & Khoreva, V. (2023). Best friend or broken tool? Exploring the co-existence of humans and artificial intelligence in the workplace ecosystem. *Human Resource Management*, 62(1), 117–135. https://doi.org/10.1002/hrm.22147.

Einola, K., Khoreva, V., & Tienari, J. (2023). A colleague named Max: A critical inquiry into affects when an anthropomorphised AI (ro)bot enters the workplace. *Human Relations*, 62, 117–135. https://doi.org/10.1177/00187267231206328

Fessenden, T. (ed.) (2021). Design systems 101. Nielsen Norman Group. https://www.nngroup.com/articles/design-systems-101.

Frey, C. B., & Osborne, M. A. (2017). The future of employment: How susceptible are jobs to computerisation?. *Technological Forecasting and Social Change*, *114*, 254–280. https://doi.org/10.1016/j.techfore.2016.08.019.

Girling, R. (2017) AI and the future of design: What will the designer of 2025 look like? https://www.oreilly.com/radar/ai-and-the-future-of-design-what-will-the-designer-of-2025-look-like/.

Haas, M., Munzer, B., Santen, S., Hopson, L., Haas, N., Overbeek, D., Peterson, W., Cranford, J., & Huang, R. (2019). #didacticsRevolution: Applying Kotter's 8-step change management model to residency didactics. *Western Journal of Emergency Medicine*, *21*(1), 65–70. https://doi.org/10.5811/westjem.2019.11.44510.

Hiatt, J. M. (2006). *ADKAR: A model for change in business, government, and our community*. Prosci Research (US).

Huang, M.-H., & Rust, R. T. (2018). Artificial intelligence in service. *Journal of Service Research*, *21*(2), 155–172. https://doi.org/10.1177/1094670517752459.

Inie, N., Falk, J., & Tanimoto, S. (2023). Designing participatory AI: Creative professionals' worries and expectations about generative AI. *Extended Abstracts of the 2023 CHI Conference on Human Factors in Computing Systems* [Preprint], 1–8. https://doi.org/10.1145/3544549.3585657.

Jang, S., Yoo, S., & Kang, N. (2022). Generative design by reinforcement learning: Enhancing the diversity of topology optimization designs. *Computer-Aided Design*, *146*, 103225. https://doi.org/10.1016/j.cad.2022.103225.

Janicki, M. M. (2000). Paper Consumption Reduction. https://www.cga.ct.gov/2000/rpt/2000-r-1041.htm.

Kotter, J. P. (1996). *Leading change*. Harvard Business School Press.

Laig, R. B. D., & Abocejo, F. T. (2021). Change management process in a mining company: Kotter's 8-step change model. *Journal of Management, Economics, and Industrial Organization*, *5*(3), 31–50. https://doi.org/10.31039/jomeino.2021.5.3.3.

Lee, H., & Cho, C.-H. (2020). Digital advertising: Present and future prospects. *International Journal of Advertising*, *39*(3), 332–341. https://doi.org/10.1080/02650487.2019.1642015.

Lundman, R., & Nordström, P. (2023). Creative geographies in the age of AI: Co-creative spatiality and the emerging techno-material relations between artists and artificial intelligence. *Transactions of the Institute of British Geographers*, *48*(3), 650–664. https://doi.org/10.1111/tran.12608.

Masriadi, D., Ekaningrum, N. E., Hidayat, M. S., & Yuliaty, F. (2023). Exploring the future of work: Impact of automation and artificial intelligence on employment. *ENDLESS: International Journal of Future Studies*, *6*(1), 125–136. https://doi.org/10.54783/endlessjournal.v6i1.131.

McKendrick, J., & Thurai, A. (2022). AI isn't ready to make unsupervised decisions. *Harvard Business Reviews*. https://hbr.org/2022/09/ai-isnt-ready-to-make-unsupervised-decisions.

Miles, M. C., Richardson, K. M., Wolfe, R., Hairston, K., Cleveland, M., Kelly, C., Lippert, J., Mastandrea, N., & Pruitt, Z. (2023). Using Kotter's change management framework to redesign departmental GME recruitment. *Journal of Graduate Medical Education*, *15*(1), 98–104. https://doi.org/10.4300/JGME-D-22-00191.1.

Mind Tools. (2023). 'Lewin's Change Management Model': Understanding the three stages of change [Preprint]. https://www.mindtools.com/ajm9l1e/lewins-change-management-model.

Mustafa, B. (2023). The impact of artificial intelligence on the graphic design industry. *Arts and Design Studies*, *104*. https://doi.org/10.7176/ADS/104-01.

Nielsen, J. (2023). UX Needs a Sense of Urgency About AI. Jakob Nielsen on UX, 14 June. https://jakobnielsenphd.substack.com/p/ux-needs-a-sense-of-urgency-about.

Nielsen, J., & Moran, K. (2023). Getting Started with AI for UX. Jakob Nielsen on UX, 18 October. https://jakobnielsenphd.substack.com/p/get-started-ai-for-ux.

Samosir, P., & Jayadi, R. (2023). Change management for transformation of digital banking. *Jurnal Sistem Cerdas*, *6*(1), 29–43.

Sinek, S. (2014). *Leaders eat last: Why some teams pull together and others don't*. Penguin.

Song, B., Soria Zurita, N. F., Nolte, H., Singh, H., Cagan, J., & McComb, C. (2022). When faced with increasing complexity: The effectiveness of artificial intelligence assistance for drone design. *Journal of Mechanical Design*, *144*(2), 021701. https://doi.org/10.1115/1.4051871.

Sun, P. (2022). How AI helps physicians improve telehealth patient care in real-time. https://telemedicine.arizona.edu/blog/how-ai-helps-physicians-improve-telehealth-patient-care-real-time.

The Business Times. (2023). AI could replace 80% of jobs in next few years: AGI guru. *The Business Times*, 9 May. https://www.businesstimes.com.sg/international/ai-could-replace-80-jobs-next-few-years-agi-guru.

Doyle, C.C., Mieder, W. and Shapiro, F.R. (2012) 'The dictionary of modern proverbs,' *Choice Reviews Online*, 50(02), pp. 50–0613. https://doi.org/10.5860/choice.50-0613.

Ullal, M. S., Nayak, P. M., Dais, R. T., Spulbar, C., & Birau, R. (2022). Investigating the nexus between artificial intelligence and machine learning technologies in the case of Indian services industry. *Business: Theory and Practice*, 23 (2), 323–333. https://doi.org/10.3846/btp.2022.15366.

Uren, V., & Edwards, J. S. (2023). Technology readiness and the organizational journey towards AI adoption: An empirical study. *International Journal of Information Management*, 68, 102588. https://doi.org/10.1016/j.ijinfomgt.2022.102588.

Verganti, R., Vendraminelli, L., & Iansiti, M. (2020). Innovation and design in the age of artificial intelligence. *Journal of Product Innovation Management*, 37(3), 212–227. https://doi.org/10.1111/jpim.12523.

Vinuesa, R., Azizpour, H., Leite, I., Balaam, M., Dignum, V., Domisch, S., Felländer, A., Langhans, S. D., Tegmark, M., & Nerini, F. F. (2020). The role of artificial intelligence in achieving the sustainable development goals. *Nature Communications*, *11*(1), 233. https://doi.org/10.1038/s41467-019-14108-y.

Yehiav, G. (2023). Will AI augment or replace workers?, *Forbes*, 8 August. https://www.forbes.com/sites/forbestechcouncil/2023/08/08/will-ai-augment-or-replace-workers/?sh=37df90146cb5.

Yusa, I. M. M., Yu, Y., & Sovhyra, T. (2022). Reflections on the use of artificial intelligence in works of art. *Journal of Aesthetics, Design, and Art Management*, 2(2), 152–167. https://doi.org/10.58982/jadam.v2i2.334.

Zhang, J., & El-Gohary, N. M. (2016). Semantic NLP-based information extraction from construction regulatory documents for automated compliance checking. *Journal of Computing in Civil Engineering*, 30(2), 04015014. https://doi.org/10.1061/(ASCE)CP.1943-5487.0000346.

Zhou, J., Tang, Z., Zhao, M., Ge, X., Zhuang, F., Zhou, M., Zou, L., Yang, C., & Xiong, H. (2020). Intelligent exploration for user interface modules of mobile app with collective learning. In *Proceedings of the 26th ACM SIGKDD International Conference on Knowledge Discovery & Data Mining* (pp. 3346–3355). ACM. https://doi.org/10.1145/3394486.3403387.

4 Innovative Eldercare
Designing Strategies for Care Capacity Expansion in House of Joy Active Ageing Centres

Kenny Low Heng Khuen

INTRODUCTION

THE CONTEXT

Singapore faces a pivotal demographic challenge: the rapid ageing of its population, a trend shared globally but acutely felt in this island nation. A 2020 study by Vollset et al. highlighted a broader pattern of population decline projected in 23 countries, including Japan, Thailand, and Spain. In Singapore, this ageing trajectory is pronounced, with forecasts indicating that by 2030, at least a quarter of the population will be over 60 years old (Kwok et al., 2014). Driving this shift are two critical factors: a historically low total fertility rate (TFR) of 1.05 in 2022 (Ng, 2023) and an increased life expectancy from 82.1 in 2012 to 83 years in 2022 (Tan, 2023b). Consequently, the number of seniors aged 65 or older is expected to exceed 900,000 by 2030 (Koh, 2023). Moreover, an increase in single elderly households, from about 63,800 in 2020 to an anticipated 83,000 in 2030 (Prakash, 2023), marks a shift in family dynamics, intensifying the challenges in eldercare.

THE MACRO PLAYBOOK

Singapore has traditionally leaned on the "ageing in place" model, where family members assume primary caregiving responsibilities. However, demographic shifts challenge this model's sustainability. The emerging "middle group" of seniors, needing more support than families can provide but less than nursing homes offer, highlights a care gap (Chia et al., 2022). Tackling this issue is crucial, as older adults are significantly more likely to require hospitalisation (Tan Chorh Chuan, 2023). In response, Singapore has launched the Healthier SG and Age Well SG initiatives, emphasising preventive care. These programmes aim to transform the role of family doctors into health coaches and leverage community and technological resources for proactive health management. The Age Well SG initiative, aligning with Healthier SG, focuses on engaging seniors through active ageing centres (AACs) in preventive care activities. The government's commitment is evident in the substantial investment planned for AACs, aiming to increase their number and enhance their capabilities to reach more seniors effectively (Koh, 2023).

UNPACKING THE TASK AND IDENTIFYING THE CHALLENGES

With an ageing population of around 717,000 seniors in 2023 (Singapore Department of Statistics), AACs face the daunting task of effectively engaging an average of 2,607 seniors within their assigned region. House of Joy (HOJ), an AAC chain in Tampines and Mountbatten, is a case study in this research. HOJ, through its range of programmes and a team of five to six care staff, has engaged about 1,500 unique seniors annually. This is also with the support of contracted programme

DOI: 10.1201/9781003469551-4

instructors and volunteers. This study explores how HOJ, using the design thinking approach, can expand its senior engagement capacity without compromising care quality. It will examine the process stages of empathise, define, ideate, prototype, and test to uncover innovative solutions that can enhance HOJ's impact and efficacy in addressing the needs of Singapore's rapidly ageing population.

METHODOLOGY

The methodology for this chapter will employ the design thinking framework, which encompasses four key stages: empathise, define, ideate, and prototype.

1. Empathise and Define: The initial phase involves in-depth secondary research to understand the challenges faced comprehensively. This stage is crucial for empathising with stakeholders, particularly the state, elderly, and care providers. By immersing in the context and experiences of these groups, the research will identify and clearly define the specific problem statements to be addressed. This empathetic approach ensures that the solutions developed are grounded in the real needs and challenges of those directly impacted by the issues.
2. Ideate: A literature review will be conducted based on the insights gained from the empathise and define stages. The review will focus on collating and synthesising ideas from various sources, including successful case studies from other countries, innovative practices in eldercare, and emerging trends in healthcare and technology. This stage aims to generate diverse creative and feasible ideas that can be tailored to the specific context of Singapore and HOJ.
3. Prototype and Testing: The most appropriate and promising ideas from the ideation stage will be tested through surveys or deployed experiments.

Feedback gathered from the surveys and tests will then inform us of possible iterations required for the solutions.

EMPATHISE AND DEFINE

One of the key challenges in scaling the outreach capacity of active ageing centres (AACs) lies in fulfilling their multifaceted responsibilities amidst an expanding demographic of seniors. AACs are tasked with conducting outreach strategies to understand and meet the needs of seniors and to engage between 1,000 and 4,000 seniors. Within three years of operation, an AAC is expected to have reached or engaged at least 50% of its assigned population. Furthermore, 80% of these engaged seniors should participate in at least one in-person active ageing programme (AAP). These programmes, conveniently located near seniors' homes, must achieve a high satisfaction rate of 85%.

In addition to offering AAPs, AACs provide essential befriending services for lonely or vulnerable seniors, estimated to range from 150 to 600 individuals, depending on the service boundary. These services, too, must maintain an 85% satisfaction rate. AACs are also responsible for providing information and referral services for national and community schemes, acting as a social connector for social prescription interventions, conducting community health screenings, and partnering with regional health systems to improve community health outcomes.

While technology can assist in achieving some of these goals, the central role of care staff, programme coaches, and volunteers remains indispensable. The government encourages the recruitment and engagement of volunteers to meet these performance indicators. For instance, at HOJ, 50 volunteers assist with befriending efforts on weekends, and approximately 500 volunteers support project-based outreach programmes.

However, as AACs are projected to increase by over 40% by 2025, the demand for manpower is expected to rise exponentially. This surge comes from long-standing challenges in the sector, including manpower shortages and high turnover rates, as highlighted by Ng and Yang (2024) and Lindquist (2023). Thus, the primary challenge in scaling the engagement capacity of AACs is attracting and retaining talented individuals dedicated to this vital work. Given the increasing responsibilities and expanding reach of AACs, effectively addressing this manpower challenge is crucial for ensuring the continued delivery of high-quality services and fulfilling the needs of Singapore's rapidly ageing population.

Unpacking the Workforce Challenge

Navigating the manpower conundrum in AACs is complex. Each AAC is allocated an average annual budget of about S$400,000, and if 60% of this is dedicated to staffing costs, it equates to an average salary of around S$3,150 per staff member for a team of five. According to www.salary.sg, this salary level sits at the 37.9th percentile of all resident taxpayers, placing it below Singapore's national average wage.

Given this salary range, AACs may primarily attract candidates like fresh polytechnic graduates with limited experience. However, out of Singapore's five polytechnics, only Temasek Polytechnic offers a diploma in gerontology, with an annual intake of 50 students. This figure starkly contrasts with the demand for at least 330 new positions created by expanding AACs, outstripping the supply.

In the short term, AACs like HOJ have sourced staff from varied backgrounds, typically not formally trained in gerontology but with experience in service industries. Notably, of the 11 centre executives at HOJ, four are above 50 years old, indicating a segment of the workforce that may prioritise stability over high-income growth.

Singapore's approach to enhancing eldercare services through financial grants faces a sustainability issue. Unlike investments in education, which generate economic returns, funding for ageing-in-place strategies is limited, viewed as a cost-saving alternative to hospital stays. This perspective potentially caps the sector's funding and, by extension, salary growth potential for its workers.

Therefore, the workforce challenge in AACs is threefold: identifying and tapping into new viable recruitment sources, developing effective training programmes, and enhancing productivity to enable salary growth within a sector constrained by finite funding. Addressing these issues is critical for ensuring AACs' sustainable expansion and effectiveness in catering to Singapore's ageing population.

1. How might we enhance the productivity of present eldercare workers?
2. How might we identify and tap into new viable recruitment sources?
3. How might we develop effective training programmes to prepare these pools of new workers?

The responses to the third statement could depend greatly on the answers to the second. For this chapter, I will attempt to scope the ideation and solutions to the first and second questions.

IDEATION

In addressing the manpower challenges in eldercare, various countries have implemented diverse strategies, providing insights and potential solutions.

Hong Kong

Hong Kong has embraced "Ageing in Place" since 2007, while the "Action Plan for Successful Ageing in Singapore" was only published in 2016. Hong Kong policy-makers also encourage the elderly to remain in their homes and communities. Despite the increase in the number of elderly service providers, demand still significantly outstrips supply. Elderly service workers often face overwork, underpayment, and high stress levels. A study by Kwok et al. (2014) highlighted that poor remuneration (74.1%) and limited career prospects (71.6%) are major hurdles in hiring frontline staff for active ageing facilitation. Respondents to the research suggested that government funding and policies to trigger more provision of vocational training in this sector would be most helpful. These manpower challenges are already present in Singapore's healthcare sector, especially in employing nurses (Chua, 2020). So, while we do not want elderly patients to overrun hospital beds, the ramping up of eldercare facilities will also put pressure on the supply of care sector workers.

Canada

Canada has focused on improving the efficiency of eldercare workers. Fraser et al. (2019) identified that a substantial portion of caregivers' time is spent on non-direct care processes. Solutions proposed include enhancing technology and information system infrastructures in home care services, boosting nurses' technology literacy, balancing workload distribution, and streamlining data transmission to save time and improve accuracy. The study also underscored the diverse and often invisible aspects of caregivers' workload, including emotional labour and caseload management.

Denmark

In Denmark, research conducted by Clausen et al. (2011) found a significant correlation between job demands, job resources, and the risk of long-term sickness absence in eldercare services. They suggested interventions to improve the psychosocial work environment could help prevent long-term absenteeism. One of Singapore's private hospitals, Gleneagles, had recognised that the daily travelling time and routine borne by its pool of Malaysian nurses had added great fatigue to them apart from the actual work in the hospital. As such, the management initiated a shuttle bus service for these Malaysian nurses, easing their commute and providing more rest time, thereby helping to improve the work experience and reducing the rate of possible attritions (Min, 2023).

Japan

Japan, anticipating a care crisis due to its ageing population, is turning to technology and robotics. Initiatives include Sketter, a job-matching site for non-essential nursing home tasks, and data-based systems to monitor and provide personalised healthcare to the elderly remotely. The Japanese government also incentivises new technologies, like AI and robotics, in nursing (Eun-young, 2023; Matsuyama, 2023). The bigger play is the investment in robotic solutions to supplement or replace human caregivers, with machines like Hug, HAL, and Pepper designed to perform care tasks (Wright, 2023b).

These international examples reveal common challenges in eldercare manpower: a limited pool of trained labour, the difficulty of retaining staff due to demanding workloads and inadequate compensation, and the need for innovative solutions. These range from sourcing alternative labour pools (such as foreign workers or robots) to reimagining work processes, enabling care workers to concentrate on direct care while outsourcing non-care tasks to technology or contract workers.

Singapore "Flavoured" Solutions

The issue of manpower challenges is prevalent across various service and care industries in Singapore, including eldercare, childcare, retail, and food and beverage. Similar to the healthcare and eldercare sectors, these sectors grapple with the dual pressure of limited salary budgets and the necessity for long, regular working hours.

These are some of the solutions which these sectors have adopted:

Foreign Talents

Singapore's workforce significantly benefits from the contribution of foreign talent. As reported by Reuters in June 2023, foreigners constituted 1.77 million of the population, about 30%, with most working or studying in the country (Kok, 2023). In the service sector, companies can employ foreign workers up to a maximum of 35%, subject to work permit approvals by the Ministry of Manpower. For sectors like construction, the ratio goes beyond 80%; it is not untrue to say that Singapore is constructed by foreign talent (Tan, 2021). Foreign talent solves the challenges of availability, cost, and willingness to work. However, it often becomes a subject of political tension in Singapore's highly dense population when citizens feel that opportunities and social spaces are being crowded out by foreign talent. In 2022, the Minister for Health announced the need to increase nurses from 58,000 to 82,000 by 2030 due to the ageing population and highlighted the need to craft strategies to attract foreign nurses who are in high demand globally (Teo, 2022).

Senior Workers

A report by *The Straits Times* highlighted an uptick in employment among seniors (aged 65 to 69) and mature workers (aged 55 to 64) in 2022. This increase is attributed partly to government incentives like the Senior Employment Credit Scheme, which provides wage offsets to employers hiring Singaporeans aged 60 and above, earning up to S$4,000 monthly. A notable example is Old Chang Kee, a successful Singaporean bakery chain, which has strategically employed a high percentage of older workers, with at least 80% of its staff aged 45 and above. The company's founder, Mr Han, has adeptly combined the employment of senior workers with the adoption of modern technologies, creating a synergy that enhances operational efficiency and provides quality employment opportunities for seniors (Tan, 2018).

Former Prison Inmates

Singapore's commitment to reintegrating former prison inmates into society is evident through the establishment of Yellow Ribbon Singapore, a statutory board focusing on employment and career coaching for this group. In 2023, various initiatives were launched, including training programmes for the food and beverage sector and partnerships with major supermarket chains like NTUC. These initiatives aim to provide job training and placement opportunities, facilitating the seamless reintegration of former inmates into the workforce (Diviyadhaarshini, 2023). It was also reported that 2,969 inmates were available for placements in 2021 and 2,615 inmates in 2022; placements in both years were over 90%, but turnover within six months is about 30% (Tan, 2023a). If we can mitigate risk factors and improve retention, this could be a workforce pool to be considered.

Summary

Applying these solutions to our design thinking statements, we can summarise the ideations in the following table:

Challenge: How Might We Enhance the Productivity of Present Eldercare Workers?	
Usage of information systems to assist and improve decision-making of care workers.	Canada and Japan
Improve the psychosocial work environment of care workers so as to reduce job absence.	Denmark and Singapore
Challenge: How Might We Identify and Tap into New Viable Recruitment Sources?	
Redesign work to allow non-direct care duties to be outsourced to contract workers or technological solutions.	Japan
Increase provision of vocational training in the area of gerontology.	Hong Kong
Develop robotics to take over some basic care duties.	Japan
Attract foreign talents into eldercare.	Singapore
Train seniors/mature workers to work in eldercare.	Singapore
Consider training and hiring former inmates to work in eldercare.	Singapore

For this research, I will scope the solution concepts to the following:

- Develop technology or robotics to take over some basic care duties.
- Redesign work to allow non-direct care duties to be outsourced to contract workers or technological solutions.
- Explore alternative workforce sources.

PROTOTYPING AND TESTING

Field Surveys

To gain insights into the receptivity of various stakeholders towards the strategies developed in the previous stage of design thinking, we conducted three targeted surveys among distinct groups:

Survey A: Current Senior Clients (60 Years and Above):

Participants: 11 adults aged 60 and above.

Methodology: The survey was administered through on-site interviews, where participants were directly asked a series of questions. Unlike Survey A, this approach facilitated a more interactive and guided response process, with participants' identities being noted.

Objective: This survey focused on gauging the direct feedback of the primary beneficiaries of the AACs – the seniors. It sought to capture their comfort level, preferences, and apprehensions regarding the proposed changes and innovations.

Survey B: Younger Adults (Under 59 Years Old):

Participants: 34 adults under the age of 59.

Methodology: The survey was conducted using Google Forms. Participants were provided a link to the form, allowing them to submit their responses anonymously.

Objective: This survey aimed to understand the perspectives and openness of younger adults towards the proposed strategies to ease manpower challenges in the eldercare sector, particularly in relation to technology adoption.

Survey C: Care Workers from HOJ:

Participants: Nine care workers employed at HOJ.

Methodology: The survey was conducted using Google Forms, and participants were asked to provide their input anonymously via a link provided.

Objective: The goal was to collect care workers' views, which are integral to the AACs' operations. This survey intended to understand their attitudes towards implementing new strategies, perceived impacts on their workflow, and any concerns they might have.

Each survey was designed to capture a unique perspective, ensuring a comprehensive understanding of the various stakeholders' views. The anonymity in Surveys B and C was intended to encourage honest and uninfluenced responses. At the same time, the personal interaction in Survey A was designed to cater to the preferences and comfort of the older participants and also to allow the interviewer to describe solutions with which the seniors may not be familiar.

Facial Recognition Technology (FRT) in Active Ageing Centres (AACs)

A field experiment using facial recognition technology was also conducted to explore the use of information systems to assist or replace some of the non-direct care work.

Context

In AACs, a critical administrative task is the identification and registration of seniors, which is essential for programme evaluation and reporting to the Ministry of Health for grant tracking and performance indicators. While digitalisation has transitioned record-keeping from paper to digital formats in centre management systems (CMSs), there remains room for innovation.

The Case for Facial Recognition

Given the older clientele at AACs, traditional self-service kiosks requiring ID cards or passwords may not be ideal due to the potential for memory loss associated with ageing. Therefore, FRT emerges as a promising solution. It simplifies the identification process, allowing for a more seamless and user-friendly experience for seniors.

Growth and Deployment

FRT is becoming increasingly prevalent in everyday life, notably in smartphones. Its integration into healthcare is increasing, facilitating tasks like touch-free appointment check-ins and accurate patient identification. The advancement of machine learning, where algorithms learn to recognise patterns from large datasets, has enabled FRT to diagnose rare genetic disorders and identify subtle correlations between facial morphology and genetic conditions (Martinez-Martin, 2019). Additionally, FRT can potentially predict health characteristics, assess pain, and decipher emotions through facial expressions (Libby, 2021). A study by Lopes et al. (2018) demonstrated FRT's effectiveness in identifying emotions across different age groups, achieving high accuracy rates in detecting facial expressions in younger and older individuals.

Receptivity and Concerns

Public acceptance of FRT is generally favourable for purposes like identity verification and reducing administrative errors, as indicated by a study involving 4,048 respondents in the United States (Katsanis et al., 2021). However, privacy concerns remain prevalent across all age groups. Another potential challenge is legal liability, particularly as FRT diagnostic software evolves to replace physician judgement in some scenarios possibly. This development raises ethical and legal questions about responsibility in cases of diagnostic errors or misjudgements (Libby, 2021). While FRT presents a valuable opportunity for streamlining administrative processes in AACs, careful consideration of privacy, ethical, and legal implications is necessary to ensure its responsible and effective deployment.

Field Experiment Description

Background on Pensees Systems and PesGuard

Pensees Systems, a Singapore-based tech firm specialising in computer vision, Internet-of-Things (IoT) technologies, and integrated industrial application solutions, has developed a product with FRT called PesGuard (PG). This facial recognition door access control device boasts a rapid detection time of less than 0.3 seconds and claims an impressive accuracy rate of 99% and above. PG is currently predominantly utilised in corporate offices for access control.

Field Experiment at HOJ Mountbatten

To explore the applicability of PG in AACs, we conducted a field experiment at HOJ Mountbatten, installing two PG units over a 12-day period. The devices were strategically placed: one at a standard standing height and another at a lower level to accommodate seniors using wheelchairs, ensuring inclusivity and accessibility (Figure 4.1).

Operational Mechanism and Data Comparison

Upon a senior's arrival at the centre, PG will automatically capture their facial image, cross-referenced against a pre-existing database of senior member photographs provided by HOJ. Once a match is found, the system exports their membership identification number for processing. To evaluate the effectiveness of PG, we compared the data captured by this system against the attendance records maintained by care staff using the current CMS. This CMS, an internally developed phone app created with AppSheet, represents the conventional method for tracking attendance and participation in the centre.

Interviews will also be conducted with relevant HOJ care staff and HOJ's AppSheet engineer (AE) to understand user experiences and attitudes towards the adoption of such technology.

FIGURE 4.1   PesGuard in Active Ageing Centres (AACs).

Objective and Anticipated Outcomes

This experiment's objective is to assess PG's efficiency, accuracy, and user-friendliness in an AAC setting. It aims to determine whether this technology can streamline administrative processes, reduce the workload on care staff, and provide a more seamless entry experience for seniors.

DATA ANALYSIS

Survey A & Survey B: Field Survey amongst Adults and Current Senior Clients

A total of 45 adults participated in the survey. Their age composition is summarised in Table 4.1.

Perceived Need (Push Factors)

Table 4.2 aims to uncover what is the greatest perceived need for seniors in Singapore.

The majority of participants under 60 selected "social isolation and loneliness" as the biggest challenge, while participants over 60 chose "poor physical health".

Factors to Visit an AAC (Pull Factors)

Table 4.3 shows the relative importance of factors which could draw a participant towards attending AAPs in HOJ.

TABLE 4.1

Demographic Profile of Survey Participants

	39 Years and Under	40 to 49 Years	50 to 59 Years	60 and above
Number of participants	9	20	5	11

TABLE 4.2
Survey Results for Perceived Need of Seniors in Singapore

What Do You Think Is the Biggest Challenge Facing the Majority of Seniors in Singapore?	39 and Under (9)		40 to 49 (19)		50 to 59 (5)		60 and Above (11)	
Social isolation and loneliness	3	33%	9	47%	2	40%	3	27%
Not enough savings/income for living expenses	3	33%	7	37%	1	20%	2	18%
Digital illiteracy	2	22%	1	5%	1	20%	1	9%
Poor physical health	0	0%	1	5%	1	20%	5	45%
Poor mental health	1	11%	1	5%	0	0%	0	0%

TABLE 4.3
Survey Result of Factors Affecting Participation in AAPs in an AAC

What Are the Five Most Important Factors that Make You Consider attending HOJ Programmes?	39 and Under (9)		40 to 49 (20)		50 to 59 (5)		60 and Above (11)	
Clean and safe environment	4	44%	14	70%	4	80%	11	100%
Attractive environment	2	22%	5	25%	0	0%	1	9%
Proximity of location	5	56%	12	60%	3	60%	9	82%
AAPs suited to interests	5	56%	11	55%	2	40%	8	73%
Ability to provide daily meals	0	0%	1	5%	0	0%	1	9%
Ability to provide regular health checks	1	11%	3	15%	2	40%	2	18%
Friendly care staff	1	11%	3	15%	2	40%	11	100%
Trained care staff	1	11%	2	10%	0	0%	1	9%
Qualified care staff	2	22%	4	20%	2	40%	0	0%
Staffed with professional nurse	0	0%	2	10%	0	0%	1	9%
Staffed with professional nutritionist	0	0%	1	5%	1	20%	0	0%
Staffed with professional fitness trainer	1	11%	0	0%	0	0%	0	0%
Referral from a friend/family member	1	11%	3	15%	0	0%	3	27%
Already have a friend who is a member	1	11%	2	10%	2	40%	4	36%

There is consistency across all age groups for the priority of these factors:

- Clean and safe environment.
- Proximity of location.
- AAPs suited to interests.

The additional factor that stood out for the above 60-year-old participants (who are actual current users of AAC services) is "friendly staff members".

Receptivity to Solutions for Manpower Challenges

Table 4.4 shows the receptivity to possible solutions to mitigate manpower challenges in AAC operations. Participants may select more than one option.

TABLE 4.4
Survey Results for Acceptance of Possible Solutions for Manpower Challenges in AACs

Possible Solutions for Manpower Challenges in AAC	39 and Under (9)		40 to 49 (20)		50 to 59 (5)		60 and Above (11)	
Hiring from neighbouring ASEAN countries and training them.	3	33%	5	25%	0	0%	5	45%
Hiring and training elders to care for other elders.	4	44%	15	75%	4	80%	8	73%
Selecting suitable former prison inmates and training them for eldercare.	2	22%	4	20%	1	20%	7	64%
Invest in technological solutions to provide direct care for elderly (e.g. Japan has robots for affection, companionship, and health monitoring).	3	33%	10	50%	1	20%	1	9%
Invest in technological solutions to allow AAC care staff to be more productive in care work.	4	44%	10	50%	4	80%	5	45%
Create programmes which are suitable for volunteers to manage.	9	100%	14	70%	3	60%	3	27%
Focus on facilitation and allow elders to create and conduct their own Active Ageing programmes.	4	44%	7	35%	2	40%	1	9%

The solution that was least receptive for participants under 50 was "Hiring and training of former prison inmates".

The most unpopular solution for the 50- to 59-year-old group was "Hiring talents from neighbouring ASEAN countries". For the above 60-year-old group, the concepts of "Technological solutions to provide direct care" and "Allowing elders to conduct their own AAPs" had the lowest receptivity.

Further Explorations of the Usage of Technological Solutions

Table 4.5 explores the receptivity to using technological solutions to supplement or enhance eldercare services and operations.

TABLE 4.5
Skills and Knowledge Profile of Staff of HOJ

Possible Care Services to Be Delivered Using Intelligent Machines	39 and Under (9)		40 to 49 (20)		50 to 59 (5)		60 and Above (11)	
Daily monitoring of elderly living alone	5	56%	14	70%	5	100%	10	91%
Providing information to elderly	2	22%	10	50%	1	20%	8	73%
Entertaining elderly	4	44%	10	50%	3	60%	2	18%
Leading exercise programmes	2	22%	9	45%	2	40%	2	18%
Social conversations with elderly suffering from dementia	2	22%	5	25%	3	60%	2	18%
Social conversations with elderly	1	11%	4	20%	1	20%	1	9%
Providing companionship to elderly	1	11%	1	5%	1	20%	1	9%

TABLE 4.6
Skills and Knowledge Profile of Staff of HOJ

	Formal Training in Eldercare Work		Working Experience in Eldercare Sector	
	Yes	No	2 years or less	3 years or more
Number of care staff	3	6	6	3

The concept that was least popular across all age groups was the use of intelligent machines to provide social conversations or companionship to seniors.

Survey C: Field Surveys amongst HOJ Care Staff

A survey was conducted amongst HOJ care staff. The identity of care staff was anonymous as they participated in the survey, which was conducted using Google Forms.

Composition of Staff Members

A total of nine care staff participated in the survey. The list of survey questions can be found in the Annex. Table 4.6 shows the composition of the care staff in terms of their training background and working experience in eldercare. Due to the low sample size of each subcategory of staff, analysis will not be provided for the subcategories.

"Challenging Responsibilities"

Three targeted questions were posed to discern the specific challenges faced by HOJ staff in fulfilling their duties. These questions aimed to uncover the types of responsibilities that care staff found particularly demanding. By analysing the responses, we sought to identify tasks that staff felt less equipped to handle versus those they found more physically, emotionally, or mentally challenging and tasks for which they required additional support.

- The options least frequently chosen in response to the question about "Responsibilities most capable to perform" were taken as indicators of areas where staff felt less proficient.
- Conversely, the options most frequently selected for "Responsibilities most physically, emotionally or mentally challenging to perform" highlighted the tasks that staff found to be the most demanding.
- Similarly, the options most often chosen in response to "Responsibilities requiring support to perform" pinpointed the areas where staff believed they needed further assistance.

The findings are encapsulated in three tables:

- Table 4.7 details the responsibilities that HOJ care staff felt most confident performing, indicating areas of perceived strength or proficiency.
- Table 4.8 outlines the responsibilities considered most challenging by the staff, highlighting tasks that impose significant physical, emotional, or mental strain.

- Table 4.9 lists the responsibilities for which care staff expressed a need for additional support, signalling areas where interventions could enhance performance and well-being.

Responsibilities Most Capable of Performance

TABLE 4.7
Survey Results of HOJ Staff Member's Personal Capabilities Perception

Responsibilities Most Capable to Perform (Can Choose up to 5)	Total
Providing a lookout for their safety	6
Providing basic practical help	5
Facilitating social engagements	4
Facilitating active ageing	4
Facilitating active ageing programmes participation	4
Providing information and knowledge	4
Responding to needs	4
Providing friendship	4
Facilitating overall well-being	3
Providing attention and affection	3
Cultivating hope	1
Advocating needs	1
Providing professional care	0
Fulfilling desires	0

Responsibilities Most Demanding to Perform

TABLE 4.8
Survey Results of HOJ Staff Members' Evaluation Personal Responsibilities

Top Three Responsibilities Which Are Most Physically, Mentally, or Emotionally Demanding to Perform	Total
Providing attention and affection	5
Providing friendship	4
Providing professional care	3
Fulfilling desires	3
Providing a lookout for their safety	2
Responding to needs	2
Advocating needs	2
Cultivating hope	2
Providing information and knowledge	1
Providing basic practical help	1
Facilitating active ageing	1
Facilitating active ageing participation	1
Facilitating overall well-being	1
Creating memories	1
Facilitating social engagement	0

Responsibilities Which Needed Support to Perform

TABLE 4.9
Survey Results of HOJ Staff Members' Responsibilities Which Required Support

Which of the Following Responsibilities Do You Feel that You Need Support to Perform for Your Senior Clients? (Able to Select Multiple)	Total
Providing professional care	5
Facilitating active ageing programmes participation	4
Providing a lookout for their safety	4
Providing basic practical help	3
Facilitating social engagements	3
Facilitating active ageing	3
Facilitating overall well-being	2
Providing information and knowledge	2
Providing attention and affection	2
Responding to needs	2
Advocating needs	2
Providing friendship	2
Creating memories	2
Cultivating hope	1
Fulfilling desires	1

Consistent across all three tables was "providing professional care".

Supporting Challenging Responsibilities

Two solutions, training and technology, were explored to support staff members' capacity and capability to manage responsibilities. The results are listed in Tables 4.10 and 4.11.

Responsibilities that Can Be Developed with Training

TABLE 4.10
Survey Results of HOJ Staff Members' Perception of Skills Development Potential

Which of the Responsibilities Do You Think that, with Training, You Would Be Able to Perform More Effectively for Your Senior Clients?	Total
Providing professional care	6
Advocating needs	4
Providing information and knowledge	3
Facilitating active ageing programmes participation	2
Responding to needs	2
Creating memories	2
Fulfilling desires	2
Facilitating social engagements	1
Providing attention and affection	1
Cultivating hope	1
Providing a lookout for their safety	1
Providing basic practical help	0
Facilitating active ageing	0
Facilitating overall well-being	0
Providing friendship	0

Responsibilities that can Be Improved with Technology

TABLE 4.11

Survey Results of HOJ Staff's Perception of Technology's Aid to Personal Responsibilities

Which of the Responsibilities Do You Think that with Technology, You Would Be Able to Perform More Effectively for Your Senior Clients?	Total
Providing information and knowledge	6
Facilitating active ageing programmes participation	3
Providing a lookout for their safety	3
Responding to needs	2
Creating Memories	2
Facilitating overall well-being	2
Providing basic practical help	1
Facilitating social engagements	1
Facilitating active ageing	0
Providing professional care	0
Providing attention and affection	0
Cultivating hope	0
Fulfilling desires	0
Advocating needs	0
Providing friendship	0

While staff believed that "providing professional care" could be supported by training programmes, they were less optimistic about the usage of technology to improve their delivery of responsibilities; only 8 out of the 15 listed responsibilities were selected, as compared to 11 out of 15 for "training".

Attitudes and Beliefs towards Technology

Two further questions were asked about the care staff's attitudes and beliefs towards technology. Each staff member was given a rating option of 1–5, 1 being "not a single responsibility" and 5 being "every responsibility". The results are summarised in Table 4.12.

The results showed low general receptivity towards technology supporting their work.

Field Experiment at HOJ Mountbatten

Two PG units over a 12-day period in HOJ Mountbatten.

Table 4.13 above shows the number of uniquely identified and matched seniors using PG. Column A represents the actual number of unique seniors recorded using the existing CMS, and column B

TABLE 4.12

Survey Results of HOJ Staff Member's Perception of Technology Being a Core Element in Care Work

Questions	Average Score
To what extent do you think your responsibilities can be *replaced* by artificial intelligence, robotics, and advanced technology?	1.67
To what extent do you think your responsibilities can be *supported* or enhanced by artificial intelligence, robotics, and advanced technology?	2.11

TABLE 4.13

Results of PG's Identification of HOJ Senior Members

Date	Uniquely Identified Seniors from CMS (A)	Uniquely Identified Seniors from PG (B)	Identification Gap (C)	Identification Gap (%) [C/A]
12/8/2023	54	38	16	29.63%
12/9/2023	89	58	31	34.83%
12/11/2023	75	53	22	29.33%
12/12/2023	38	34	4	10.53%
12/13/2023	49	39	10	20.41%
12/14/2023	50	44	6	12.00%
12/15/2023	56	53	3	5.36%
12/18/2023	62	57	5	8.06%
12/19/2023	26	25	1	3.85%
12/20/2023	45	37	8	17.78%
12/21/2023	38	31	7	18.42%
12/22/2023	48	45	3	6.25%

represents the seniors identified via PG. C is the difference between A and B, termed the" identification gap". The identification gap ranges from 3.85% to as high as 34.83%.

Table 4.14 breaks down the reasons for the identification gaps:

As we can see from Table 4.14, the majority of the identification gap arises due to 'Unable to Match D3'. Upon consultation with the project team, we learnt that a large portion of the photos of seniors in the database were photos of their younger selves. Some seniors preferred to give us their photographs to scan instead of taking a live photo upon registration. At that point of data collection, we did not consider the limitations it would place on a facial recognition system. As the test continued, care staff updated the database photos using the photos taken by PG; as such, we can see that D3 had gradually decreased over the course of the test, as shown in Figure 4.2.

TABLE 4.14

Reasons of PG's Non-Identification of HOJ Senior Members

Date	Identification Gap (C)	Identification Gap (%) [C/A]	Missed Scan (D1)	Refused to Scan (D2)	Unable to Match (D3)	Wrong Match (D4)
12/8/2023	16	29.63%	2	0	14	0
12/9/2023	31	34.83%	0	0	31	0
12/11/2023	22	29.33%	0	1	20	0
12/12/2023	4	10.53%	2	0	1	1
12/13/2023	10	20.41%	1	1	8	0
12/14/2023	6	12.00%	3	0	3	0
12/15/2023	3	5.36%	1	1	1	0
12/18/2023	5	8.06%	3	0	2	0
12/19/2023	1	3.85%	0	0	1	0
12/20/2023	8	17.78%	3	1	4	0
12/21/2023	7	18.42%	4	0	2	1
12/22/2023	3	6.25%	2	0	1	0

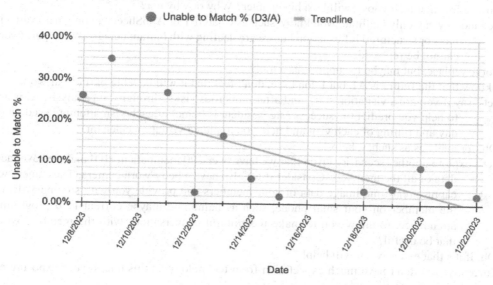

FIGURE 4.2 Graph FEHOJ-1.

From Table 4.14, Missed Scans (D1) and Wrong Match (D4) can be considered as indicators of reliability for the system. Table 4.15 shows that reliability on average is above 95%.

Interviews with HOJ Staff and HOJ AppSheet Engineer

Separate interviews were conducted with the HOJ staff responsible for deploying PG at the centre for the field expert and also the HOJ AE and project manager responsible for cleaning the data and providing statistics from the experiment.

TABLE 4.15
Reliability of PG Scans

Date	Uniquely Identified Seniors from CMS (A)	Uniquely Identified Seniors from PG (B)	Identification Gap (%) [C/A]	Unable to Match % (D3/A)	Reliability of Scans (%) [1-((D1+D4)/A)]
12/8/2023	54	38	29.63%	25.93%	96.30%
12/9/2023	89	58	34.83%	34.83%	100.00%
12/11/2023	75	53	29.33%	26.67%	97.33%
12/12/2023	38	34	10.53%	2.63%	92.11%
12/13/2023	49	39	20.41%	16.33%	97.96%
12/14/2023	50	44	12.00%	6.00%	94.00%
12/15/2023	56	53	5.36%	1.79%	98.21%
12/18/2023	62	57	8.06%	3.23%	95.16%
12/19/2023	26	25	3.85%	3.85%	100.00%
12/20/2023	45	37	17.78%	8.89%	93.33%
12/21/2023	38	31	18.42%	5.26%	86.84%
12/22/2023	48	45	6.25%	2.08%	95.83%

Interview Excerpts with HOJ Staff (S), Interviewer (I)

I: Do you feel that technology will be a big enabler? Why or why not?

S: Technology can only facilitate, you know, such as ChatGPT or AppSheet can only help you with facts or structured information but we are dealing with human beings, so that part (work) cannot be replaced.

I: Not to replace, but maybe to support, as a tool?

S: I know that the future is AI, but I don't see, don't feel that it will be a big part of our work.

I: Let's say if you are a visitation staff, and after numerous visits, through data analysis, AI is able to help you predict the success rate (resident is home) of each visit attempt based on the day and timing of each visit and help you to plan a visitation calendar.

S: Oh, for analysis as such, it helps!

I: Let's explore another scenario, given that we have over 200 members in HOJ and we have facial data of all the members. Consider if we also have a body-worn camera. The camera will capture real-time image data of your members and provide you with summary data of the member on your smartphone, such as name and maybe current HOJ programme attendance, to allow you to make meaningful conversations with the member. Would that be useful?

S: Oh, if it's that extensive, it will help!

I: Currently, you don't have much expectation from technology. Is this due to your exposure and awareness?

S: I would think so.

Interview Excerpts with AppSheet Engineer (AE), Project Manager (PM), Interviewer (I)

PM: Generally, the system had been easy to use, and the ground staff had told us that after some time, the senior members would automatically place their faces and allow the system to capture their facial data. The ground staff told us that only one or two members would ask a bit more about the system's purpose and the duration of time the data would be kept. When we shared with them about the purpose and also that the data would be removed after 30 days, they were okay with using the system. For the cons of the system, there is limited mobility because the system needs to be plugged into a power source to function. Hence it currently cannot be used in spaces without power points. There are also some instances where the system cannot capture the photos accurately due to lighting conditions.

AE: I think this can be solved by adding lighting stands.

I: At an operational level, how do you think this system can be helpful to the HOJ care staff? Supposedly, the system is able to identify new seniors walking into the centre and send an alert prompt to the care staff; this will relieve the centre of a "door host" role without compromising hosting duties.

AE: I think there are still multiple hoops to jump across to actualise this type of capability. Currently, the system does multiple scans and I have to do some backend work to match data with people.

PM: I think the current system can only provide us with a secondary source of attendance data and members' photos, but for real time feedback, I don't think the current system is ready.

AE: I also feel that recognising members should be the primary responsibility of care staff; if the staff really care, they will be able to do it.

LIMITATIONS

The research encountered several limitations that could impact the generalisability and scope of its findings. First, the relatively small sample size of staff members interviewed poses a challenge to the representativeness of the results, potentially limiting their applicability to the broader eldercare workforce in Singapore. Additionally, the senior participants surveyed were exclusively from HOJ centres, omitting perspectives from seniors not affiliated with HOJ. This exclusion restricts the study's ability to capture a more comprehensive view of eldercare needs and preferences across the wider senior population, potentially overlooking diverse experiences and opinions that could enrich the research findings and implications.

FINDINGS

The field surveys and experiments sets out to prototype test the concepts and receptivity of:

- Developing technology or robotics to take over some basic care duties.
- Exploring alternative manpower sources.
- Redesigning work to allow non-direct care duties to be outsourced to contract workers or technological solutions.

RECEPTIVITY AND PERCEPTIONS OF USAGE OF TECHNOLOGY IN BASIC CARE DUTIES

The utilisation of technology in basic care duties reveals diverse perceptions among different stake-holders, as illustrated by recent survey findings. Survey C (Table 4.11) indicates that HOJ care staff possess limited confidence in technology's ability to aid their caregiving tasks, primarily expecting technology to serve as a source of information and knowledge. Contrastingly, the broader adult population presents *varied attitudes*: half of the respondents aged 40 to 49 view technological investment in direct care as a viable solution, whereas only 9% of current seniors and 20% of those aged 50 to 59 share this optimism. This discrepancy might stem from the 40 to 49 age group's first-hand experience with significant technological milestones – from the advent of personal computers and the internet to the development of cloud computing and artificial intelligence (AI) – shaping a belief in the potential of technology to revolutionise care.

This speculation aligns with the imaginative allure of AI like Tony Stark's J.A.R.V.I.S. from the Marvel series, symbolising the ultimate fusion of technology and personal assistance, which, in the physical form of Vision, was even able to fall in love. However, the care staff's scepticism reflects broader concerns similar to those expressed by experts in Japan, as cited by James Wright (2023a), regarding the limitations of robots in tasks requiring physical contact or nuanced human interaction.

In interviews with HOJ clients about employing technology for entertainment or care, a significant majority (91%) expressed reluctance, emphasising a preference for human over machine-mediated care and social interaction. This resistance underscores a deeper sentiment that transcends feasibility, highlighting a profound desire for the human touch in caregiving contexts.

Openness to Alternative Manpower Sources

The divergent attitudes towards employing former prison inmates in eldercare roles highlight a notable generational difference. This proposition received minimal support among individuals younger than 60 years old, with acceptance rates dipping to 22% and below. Conversely, a significant portion of seniors, 64%, expressed openness to this idea, even favouring it over the alternative of hiring foreign labour from ASEAN countries. During interviews, some seniors advocated giving former inmates a second chance, reflecting a more forgiving stance.

This contrast is underscored by the cautious approach of younger generations, possibly reflecting concerns about safety and trust – factors crucial in caregiving scenarios. The scepticism from younger age groups, who may be the children of the seniors in question, presents a formidable obstacle to the widespread acceptance of this solution.

James Timpson, CEO of the British retailer Timpson, offers an insightful case study. With approximately 10% of his workforce comprising former prison inmates, Timpson acknowledges the risks and rewards of such hiring practices. He noted in an interview with the *Journal of Management*

Inquiry: "But there is one proviso, when it goes wrong, it goes really wrong. But when it goes right, I am recruiting much better people than I would normally be able to recruit" (Pandeli, 2019). This perspective highlights the potential benefits and challenges of integrating former inmates into the workforce.

The possibility of a single negative incident, such as a care staff member with a past misusing their position for financial gain from a senior, could cast a shadow over the entire strategy and the organisation involved. While tapping into this manpower source offers readily available workers and the chance to provide individuals with new economic and social beginnings, the potential risks and public perceptions must be carefully weighed. The key lies in balancing the tangible benefits against the possible drawbacks and societal concerns, ensuring that integrating former inmates into eldercare is safe and beneficial for all parties involved.

Another notable discovery from the research was the varied receptivity towards empowering senior members to design and lead their own AAPs. Among participants under 60 years old, acceptance of this concept hovered between 35% and 44%, indicating a moderate level of openness. However, the idea was less popular among the seniors interviewed, ranking equally low as the concept of affection robots. One senior expressed scepticism, stating that "It is not possible, if it's left to us, there will be differences in opinions and quarrels, we would rather the centre staff lead the programmes". This sentiment aligns with observations from a senior residential community group we previously engaged with. Seniors not participating in activities voiced concerns that programme benefits often favoured the organisers' close friends, indicating potential bias or favouritism.

This feedback underscores a challenge in implementing peer-led initiatives within AACs. While allowing seniors to take charge of AAPs aims to foster engagement and autonomy, concerns about internal dynamics, potential conflicts, and perceptions of exclusivity could hinder participation and affect the overall success of such programmes. Addressing these concerns and ensuring equitable access and representation in programme planning and execution are crucial to enhancing outreach and engagement efforts effectively.

The most favoured solution among seniors involved recruiting and training older individuals for caregiving roles. This approach received broad support across different age groups, with approval ratings varying from 80% to 44%. Notably, one of our care staff members, over 60, has consistently been lauded by our senior members for her exceptional care, patience, and reliability. Yet, she faced difficulties with technological integration into her workflow. An instance highlighting this challenge occurred when her supervisor observed her transcribing phone numbers into a notebook, a practice rooted in her unfamiliarity with storing contacts on her smartphone. She manually referenced numbers from her chat messages to her notebook to recognise the message senders.

This scenario underscores the importance of recognising the diverse learning paces and adaptability to technological changes within a multi-generational care team. As we embrace the valuable contributions of elder caregivers, it's equally crucial to ensure tailored support and training in technology use, acknowledging that proficiency varies widely among individuals. This approach not only leverages the strengths of elder staff members but also addresses their challenges in a supportive and inclusive manner, enhancing the overall efficacy and cohesion of the care team.

Supporting Non-care Duties with Technological Solutions

The PG field experiment demonstrated encouraging results, with a reliability level surpassing 95%. However, this achievement does not meet the more stringent reliability standards expected of consumer electronics, such as smartphones, indicating a gap in performance expectations. Furthermore, the care staff's appraisal of technological assistance was notably low, with an average rating of 2.11 out of 5. This suggests a disconnect between the potential benefits of the technology and its perceived utility by the users. Further insights from interviews with the PG experiment team revealed a significant barrier: staff members lacked the necessary knowledge and experience to envision how technology could enhance their caregiving tasks and responsibilities, and their personal negative experiences with technological solutions also generated more caution than excitement when it comes to adopting new technology.

This aligns with observations by James Wright in Japan, where he noted a general reluctance to use technology among hospital and nursing home staff who favoured traditional caregiving methods. This highlights the need for substantial investment in demonstrating the added value and safety benefits of new technologies (Wright, 2023b).

The Covid-19 pandemic and the ensuing social distancing measures necessitated the adoption of telecommuting technologies, such as Zoom, to facilitate remote delivery of HOJ programmes. This period marked significant stress for staff as they navigated unfamiliar recording hardware and video conferencing tools without sufficient support or training. This situation underscored the added burden of mastering new technologies alongside their primary caregiving responsibilities.

Further complicating technology integration into caregiving, Wright observed that introducing care robots in nursing homes offered only temporary benefits. The robots required considerable maintenance, including moving, cleaning, booting, operation, and explanations to residents, leading to increased caregiver workloads. This insight is part of a broader consensus suggesting that, despite their potential, robots often complicate rather than ease the caregiving process (Wright, 2023b).

This presents a critical resource dilemma: do we have the necessary resources to refine and perfect the use of technology in caregiving to the extent required? The development of transformative technologies, like the iPhone, involves extensive iterations and testing before market release. The question arises whether the care industry can afford a similar rigorous product development cycle, both in terms of time and resources. Can HOJ allocate sufficient resources to experiment with technology and ensure its effective and beneficial integration into caregiving practices?

DISCUSSION AND REFLECTION

INCREASING THE CAPACITY TO CARE

The research aimed to address the challenges posed by Singapore's rapidly ageing population by exploring three strategies: leveraging technology to increase available person-hours for care, considering alternative staffing pools for hiring, and examining the nuanced relationship between "care" and time. The study operates under the premise that more time equates to more care, assuming a linear relationship between "care"-related tasks like assisting seniors and "man hours". However, this assumption might be challenged when "care" is understood as showing concern and attention, suggesting that mere availability might not suffice. This is akin to common relationship complaints about the lack of active listening or presence, indicating that the essence of "care" extends beyond physical tasks. The staff survey reveals that what the HOJ staff found most challenging and draining was not task-oriented but "professional care" – the empathetic, attentive aspect of their role. Another survey response highlighted a significant discrepancy between the preferred staff-to-senior ratio between various staff members, raising questions about the inherent capacity for care and whether it varies individually, similar to individual communication preferences. This inquiry challenges the straightforward conversion of freed-up time from non-care to care work, suggesting a more complex interplay between time, care capacity, and the nature of care.

CHANGE MANAGEMENT AND TECHNOLOGY ADOPTION

The PG field experiment and subsequent staff surveys at HOJ illuminated the exploration of technology adoption and change management within elder care. Despite the promising results of the PG experiment, the primary role of HOJ care staff centred around ensuring the operational status of devices and aiding seniors through the facial recognition process, with the HOJ AE managing the backend data processing. Post-experiment interviews, however, needed a more comprehensive grasp among the staff on the potential applications and advantages of facial recognition technology, signalling a significant communication gap emanating from the executive director, the pivotal figure in driving technological change.

The AE's concerns were twofold: questioning the technology's stability and reliability to harness its benefits effectively – fears possibly rooted in past encounters with flawed technology solutions – and the cultural expectations placed on care staff to remember seniors' names personally, hinting at deeper cultural nuances that accompany technological shifts.

These findings echo a McKinsey report which suggests that 70% of technology transformations stumble over change management hurdles, including insufficient team buy-in or skill gaps (Robinson, 2019). According to Yoav Kutner of the Forbes Technology Council, surmounting these obstacles necessitates defined objectives, anticipatory measures for resistance, and proper resource allocation (Kutner, 2022).

These challenges are magnified for non-profits like HOJ, operating under tighter resource constraints. Yet, the imperative for such organisations to adopt technology strategically – to enhance operational efficiency and amplify their impact despite limited resources – highlights the essential role of adept change management and technology integration practices in these environments.

FACILITATING COMMUNITIES OF CARE

The analysis of the three proposed strategies revealed significant potential for enhancement and refinement. A novel perspective involves reimagining the role of HOJ care staff from primary caregivers to social architects who cultivate care-centric communities. This concept draws inspiration from the principles of Web 2.0, which emphasises platforms that facilitate user-generated content. Similarly, the initial design of Singapore's public housing with long communal corridors aimed to encourage social interaction among residents, underscoring the value of designed environments in fostering community bonds.

HOJ care staff could amplify the centre's care capacity by adopting this community-building approach. This shift aligns more closely with staff preferences for facilitating active ageing programmes rather than directly delivering professional care. The crucial development required is for staff members to recognise that AAPs promote individual wellness and serve as vital conduits for social engagement, fostering a sense of team identity and nurturing a cohesive care community. Through training and development, staff can learn to harness these programmes as tools for community building, enriching the care environment's social fabric and enhancing seniors' overall well-being.

ETHICAL CONSIDERATIONS

The 1988 Childcare Centres Act and its accompanying subsidies were designed to empower women by facilitating their entry or re-entry into the workforce, deeming childcare centres an accessible option for families. This policy was significantly reinforced in 2008 with a twofold increase in universal childcare funding to mitigate the escalating costs of childcare operations, positioning employment as economically more advantageous for mothers than remaining homebound (Choo, 2010).

While these policies may be viewed as essential levers for economic advancement, they concurrently sculpt a specific social narrative, implicitly valuing workforce participation over traditional caregiving roles. This shift in social norms and values is profound, suggesting a re-evaluation of societal priorities and the role of women in both the workforce and family life. Fast-forward to 2023, and we see a substantial national investment in eldercare.

Contrary to the expectation that meritocracy might correlate with reduced loneliness, findings from Duke–NUS Medical School indicate an opposite trend within the Singaporean context. The study underscores a heightened risk of loneliness among the elderly, particularly those facing limited resources, in a society where worth and recognition are closely tied to individual productivity (Maulod et al., 2023).

Such observations invite a deeper inquiry into the long-term effects of policies aimed at care productivity. There is a potential ethical dilemma in leveraging technology to extend care coverage while minimising resource allocation. Could these strategies inadvertently devalue the elderly in our society? Moreover, the financial implications of enhanced state expenditure on eldercare – through increased taxation – raise concerns about its potential effect on the disposable income of Singaporeans. This scenario poses the risk of indirectly reducing the incentive or ability of family members to engage in direct caregiving, challenging the balance between state-supported care and familial responsibilities.

These considerations highlight the complex interplay between policy, societal values, and ethical considerations, urging a balanced approach that respects both economic imperatives and the intrinsic value of caregiving within the family and community.

CONCLUSION

Exploring strategies to address the manpower challenges in Singapore's eldercare sector and, in this research context, within HOJ through technology adoption, alternative manpower sources, and work redesign has unveiled a rich tapestry of insights and opportunities for enhancement.

The divergent perceptions of technology's role in caregiving underscore the complexity of integrating innovations that resonate with all stakeholders. While technology holds promise for increasing care efficiency, it confronts cultural norms and expectations around personal touch in caregiving. The exploration of alternative workforce sources, such as employing former prison inmates, reveals generational divides in acceptance, highlighting societal concerns over safety and trust. The proposition to empower seniors in leading AAPs encounters scepticism, suggesting a preference for professionally led care initiatives despite potential community-building and engagement benefits.

The findings warrant and inspire an alternative approach: fostering a community of care where HOJ staff evolve from direct caregivers to social architects. This shift aligns with broader societal trends towards valuing participatory and community-centric approaches, similar to the evolution seen in Web 2.0. Such a transformation necessitates a redefinition of 'care' beyond task execution to include the creation of meaningful, engaging communities that enhance the well-being of seniors.

However, this journey is mired in ethical considerations. Policies encouraging workforce participation among mothers and technology-driven care productivity raise questions about the societal valuation of caregiving roles and the potential devaluation of the elderly. The delicate balance between economic efficiency and the intrinsic value of human connection stands at the forefront of this discourse.

In conclusion, addressing the capacity challenge in eldercare requires a multifaceted approach considering technological innovation, workforce diversification, and reimagining caregiving roles. Success hinges on navigating the ethical landscape with sensitivity to societal values and the diverse needs of caregivers and care recipients.

The core of design thinking lies in empathic considerations. We must not just design care solutions for our elderly clients; we must ask ourselves how we want to age and be cared for.

REFERENCES

Chia, H. X. *et al.* (2022). Blueprint for Developing Assisted Living: The Future of Ageing in Place in Singapore June 2022. rep. Singapore, Singapore: National University of Singapore.

Choo, K. K. (2010). The shaping of childcare and preschool education in Singapore: From separatism to collaboration. *International Journal of Child Care and Education Policy*, 4(1), 23–34. https://doi. org/10.1007/2288-6729-4-1-23

Chua, G.P. (2020). Challenges confronting the practice of nursing in Singapore. *Asia-Pacific Journal of Oncology Nursing*, 7(3), 259–265. https://doi.org/10.4103/apjon.apjon_13_20

Clausen, T. Nielsen, K., Carneiro, I. G., & Borg, V. (2011). Job demands, job resources and long-term sickness absence in the Danish eldercare services: A prospective analysis of register-based outcomes. *Journal of Advanced Nursing*, 68(1), 127–136. https://doi.org/10.1111/j.1365-2648.2011.05724.x

Diviyadhaarshini, B. (2023). Prison inmates to receive training for food and beverage sector, and help with securing jobs. *The Straits Times*. https://www.straitstimes.com/singapore/prison-inmates-to-receive-training-to-secure-them-jobs-in-food-and-beverage-sector

Eun-young, K. (2023). How does Japan cope with manpower shortage at nursing homes? *KBR*. https://www.koreabiomed.com/news/articleView.html?idxno=20650

Fraser, K., Baird, L. G., Laing, D., Lai, J. & Punjani, N. S. (2019). Factors that influence home care case managers' work and workload. *Professional Case Management*, 24(4), 201–211. https://doi.org/10.1097/ncm.0000000000000320

Katsanis, S. H., Claes, P., Doerr, M., Cook-Deegan, R., Tenenbaum, J. D., Evans, B. J., Lee, M. K., Anderton, J., Weinberg, S. M., & Wagner, J. K. (2021). A survey of U.S. public perspectives on facial recognition technology and facial imaging data practices in health and research contexts. *PLoS ONE*, 16(10), e0257923. https://doi.org/10.1371/journal.pone.0257923

Khalik, S. (2023). Singapore set to transform healthcare system to cater to rapidly ageing population. *The Straits Times Article*. Retrieved from: https://www.straitstimes.com/singapore/singapore-set-to-transform-healthcare-system-to-cater-to-rapidly-ageing-population

Koh, W. T. (2023). S$800 million earmarked for improvements to active ageing centres as part of new Age Well SG initiative. *CNA*. https://www.channelnewsasia.com/singapore/age-well-sg-healthier-seniors-assisted-living-facilities-active-ageing-3922391

Kok, X. (2023). Singapore's population grows 5% as foreign workers return post-pandemic ..., Singapore's population grows 5% as foreign workers return post-pandemic. *Reuters*. https://www.reuters.com/world/asia-pacific/singapores-population-grows-5-foreign-workers-return-post-pandemic-2023-09-29/

Kutner, Y. (2022). Council post: Five steps to successful technology change management. *Forbes*. https://www.forbes.com/sites/forbestechcouncil/2021/05/03/five-steps-to-successful-technology-change-management/?sh=69bc09766f0a

Kwok, S. S., Wong, K. W., & Yang, S. (2014). Challenges facing the elderly care industry in Hong Kong: The shortage of Frontline Workers. *SpringerPlus*, 3(S1). https://doi.org/10.1186/2193-1801-3-s1-p1

Libby, C., & Ehrenfeld, J. (2021). Facial Recognition Technology in 2021: Masks, bias, and the future of Healthcare. *Journal of Medical Systems*, 45(4). https://doi.org/10.1007/s10916-021-01723-w

Lindquist, M. (2023). The real cost of turnover in healthcare. *Oracle Singapore*. https://www.oracle.com/sg/human-capital-management/cost-employee-turnover-healthcare/

Lopes, N. *et al.* (2018). Facial emotion recognition in the elderly using a SVM classifier. *2018 2nd International Conference on Technology and Innovation in Sports, Health and Wellbeing (TISHW)* [Preprint]. https://doi.org/10.1109/tishw.2018.8559494

Martinez-Martin, N. (2019). What are important ethical implications of using facial recognition technology in health care?. *AMA Journal of Ethics*, 21(2), E180–E187. https://doi.org/10.1001/amajethics.2019.180

Matsuyama, K., & Singh, S. (2023). Expanding labor shortage pushes Japan to find new ways to fill jobs. *The Japan Times*. https://www.japantimes.co.jp/business/2023/12/04/economy/japan-labor-shortage-elderly-workers/

Maulod, A., Ravindran, M., & Chan, A. (2023). "You become less of a person": Being unneeded and the political economy of loneliness in later life. *Innovation in Aging*, 7(Suppl 1), 187. https://doi.org/10.1093/geroni/igad104.0617

Min, C. H. (2023). 'I felt really exhausted': How a shuttle service has kept these Malaysian nurses at their jobs in Singapore. *CNA*. https://www.channelnewsasia.com/singapore/shuttle-bus-jb-singapore-malaysia-nurses-hospitals-3718026

Ng, A. (2023). Singapore's total fertility rate drops to historic low of 1.05. *CNA*. https://www.channelnewsasia.com/singapore/singapore-total-fertility-rate-population-births-ageing-parents-children-3301846

Ng, S., & Yang, C. (2024). Active ageing centre operators plan to Double Pool of workers, offer more services for seniors. *CNA*. https://www.channelnewsasia.com/singapore/active-ageing-centre-operators-plan-double-pool-workers-offer-more-services-seniors-4020596

Pandeli, J., & O'Regan, N. (2019). Risky business? The value of employing offenders and ex-offenders: An interview with James Timpson, Chief Executive of Timpson. *Journal of Management Inquiry*, 29(2), 240–247. https://doi.org/10.1177/1056492619836167

Prakash, R. J. (2023). Reaching out to the elderly living alone: Tightening the nodes. In S Vasoo, B. Singh, & S. Chokkanathan (Eds.), Singapore Ageing (pp. 133–148). World Scientific Publishing. https://doi.org/10.1142/9789811265198_0008

Robinson, H. (2019). Why do most transformations fail? A conversation with Harry Robinson. *McKinsey & Company*. https://www.mckinsey.com/capabilities/transformation/our-insights/why-do-most-transformations-fail-a-conversation-with-harry-robinson#/

Singh, B. (2023). Singapore and the politics of ageing: An overview. In S Vasoo, B. Singh, & S. Chokkanathan (Eds.), Singapore Ageing (pp. 41–59). World Scientific Publishing. https://doi.org/10.1142/9789811265198_0003

Tan, A. (2018). Old Chang Kee: From Kopitiam stall to popular chain with over 70 stores in s'pore and beyond. *Vulcan Post*. https://vulcanpost.com/641135/old-chang-kee-founder-history/

Tan, C. (2023). Career coaches help ex-offenders stay in their jobs for the Long Haul. *The Straits Times*. https://www.straitstimes.com/singapore/courts-crime/career-coaches-help-ex-offenders-stay-in-their-jobs-for-the-long-haul

Tan, J. (2021). BT Explains: Singapore's love-hate relationship with foreign workers. *The Business Times*. https://www.businesstimes.com.sg/international/bt-explains-singapores-love-hate-relationship-foreign-workers

Tan, J. (2023). Life expectancy of Singapore population rose in last decade but fell during Covid-19. *The Straits Times*. https://www.straitstimes.com/singapore/life-expectancy-of-singapore-population-rose-in-last-decade-but-fell-during-covid-19

Teo, J. (2022). 24k more nurses, healthcare staff needed by 2030 as S'pore ages. *The Straits Times*. https://www.straitstimes.com/singapore/politics/with-aged-care-critical-spore-will-need-more-foreign-nurses-ong-ye-kung

Wright, J. (2023a). *Inside Japan's long experiment in automating elder care. MIT Technology Review*. https://www.technologyreview.com/2023/01/09/1065135/japan-automating-eldercare-robots/

Wright, J. (2023b). *Robots won't save Japan an ethnography of eldercare automation*. ILR Press, an imprint of Cornell University Press.

5 Project Mnemosyne
Bridging Communication Gaps through AI and Oral History in Interdisciplinary Teams

Kenny Png

INTERDISCIPLINARY SOLUTIONS FOR A VUCA WORLD

The current global landscape is characterised by volatility, uncertainty, complexity, and ambiguity (VUCA), presenting many challenges across various spheres. One striking example of volatility is evident in the financial markets, where rapid fluctuations and unforeseen events can trigger cascading effects worldwide (Varma & Salfiti, 2023), as witnessed during the 2008 financial crisis (Wolf, 2014) and more recently during the Covid-19 pandemic-induced economic disruptions (Laborda & Olmo, 2021). Uncertainty pervades geopolitical tensions, such as ongoing conflicts, trade disputes, and diplomatic crises, where shifting alliances and unpredictable outcomes contribute to instability and unpredictability (Katsos & Miklian, 2021). The complexity of modern technological systems, from cybersecurity threats to the intricate interconnectivity of global supply chains, underscores the multifaceted nature of contemporary challenges, where a single disruption can have far-reaching consequences (Kavanagh, 2019). Additionally, ambiguity permeates societal issues, such as the ethical dilemmas posed by emerging technologies like artificial intelligence and genetic engineering, where the boundaries of regulation and ethical considerations remain nebulous (Floridi & Cowls, 2019; Zhang, 2019). As advocated by the writer Nassim Taleb, the concept of "antifragility", where embracing uncertainty and chaos can lead to greater resilience in such a VUCA world, through flexibility, experimentation, and decentralised decision-making (Taleb, 2012).

In this chapter, I posit that an effective way to realise the concept of "antifragility" may be in implementing interdisciplinary teams. But what differentiates an interdisciplinary team from a multi-disciplinary one?

INTERDISCIPLINARY VERSUS MULTIDISCIPLINARY

Used extensively in healthcare, the clearest definition for the author is "Interdisciplinary versus Multidisciplinary Care Teams: Do We Understand the Difference?" by Rebecca Jessup published in the Australian Health Review in 2007. In this article, Jessup examines the definitions and characteristics of interdisciplinary and multidisciplinary teams within healthcare. She highlights the importance of understanding the distinctions between these two types of teams, as they have different approaches to collaboration and problem-solving.

Jessup defines multidisciplinary teams as individuals from different disciplines who work independently within their own domains and come together to share information and make decisions. In contrast, interdisciplinary teams are characterised by a deeper level of integration, with team members actively collaborating and engaging in shared decision-making processes to address complex problems (Jessup, R. 2007).

This approach goes beyond simply bringing together individuals with diverse backgrounds, expertise, and perspectives so that these teams possess the agility needed to navigate turbulent

DOI: 10.1201/9781003469551-5

environments (Nyugen, 2023). Beyond just the healthcare sector, interdisciplinary research and learning are already trending in higher education (Jacob, 2015), with a view that the approach can be effective in helping to solve complex problems by synergising different experiences to enhance innovation (Goh, 2023).

CHALLENGES OF INTERDISCIPLINARY WORK

Despite its potential, what are the challenges of such an integrated approach? Especially when giving decision-making stakes to team members from varying disciplines and backgrounds.

In research and training, there are practical issues of developing a common language, knowledge, and even a shared understanding of the task at hand (Domino et al., 2007).

Julie Thompson Klein, a scholar of interdisciplinary studies, goes further and categorises the key challenges facing interdisciplinary work:

1. Epistemological Differences: Interdisciplinary work often involves bringing together individuals from different disciplines with varying epistemologies, methodologies, and ways of knowing. These differences can lead to misunderstandings, conflicts, and challenges in integrating diverse perspectives into cohesive frameworks.
2. Communication Barriers: Effective interdisciplinary collaboration requires clear and effective communication among team members from different disciplinary backgrounds. However, disciplinary jargon, communication styles, and assumptions can create barriers to understanding, hindering effective collaboration and knowledge exchange.
3. Power Dynamics: Interdisciplinary collaborations may involve unequal power dynamics among team members, particularly when certain disciplines hold more influence or prestige within the group. Negotiating power dynamics and ensuring equitable participation and decision-making can be challenging but crucial for fostering inclusive and productive collaborations.
4. Institutional Constraints: Institutional structures and policies within academic or organisational settings may not always support or incentivise interdisciplinary work. From tenure and promotion criteria to funding mechanisms and administrative processes, institutional constraints can pose significant barriers to interdisciplinary collaboration.
5. Boundary Work: Interdisciplinary work often requires crossing disciplinary boundaries and navigating the tensions between specialisation and integration. This process of boundary work involves negotiating differences in disciplinary norms, values, and identities while striving to create new knowledge and insights that transcend disciplinary silos (Klein, 2015) (Klein et al., 2001).

Analysing this list, there is a strong emphasis on the need for effective communication on many different levels of interdisciplinary work. This goes beyond cultural or linguistic backgrounds to disciplinary training and cognitive processes within a team member's expertise.

It has been argued that for an interdisciplinary team to operate effectively and grow, there must also be a sense of psychological safety, space for reflection, and the ability to overcome defensive interpersonal dynamics (Edmondson, 2012).

In her research, Amy Edmondson identifies several key elements of psychological safety, including mutual respect, trust, openness, vulnerability, and non-defensive communication. These elements create an atmosphere where team members feel valued, respected, and supported authentically expressing themselves and taking interpersonal risks (Edmondson, 2018).

I acknowledge that the challenges facing interdisciplinary work, as outlined by thinkers such as Julie Thompson Klein, are legion. However, the author interprets the concept of psychological safety and the need to overcome defensive interpersonal dynamics as crucial to narrowing down what can be solved.

NARROWING DOWN ON THE NEED TO BRIDGE UNDERSTANDING

This interpretation by the author hinges on the idea that if individuals have more ways to express themselves or be understood in a psychologically safe environment, then perhaps some obstacles, such as epistemological differences and communication barriers, can be more effectively overcome.

Moving forward, the author posits the following questions:

1. How can we help members of an interdisciplinary team understand each other better so that they can formulate a common language?
2. What do these members need to know about each other and themselves to facilitate a common understanding of a shared goal?
3. Can technology, especially machine-learning systems, help them create a platform where their perspectives can align and build a psychologically safe space where they can communicate better?

Research on the dynamics of multidisciplinary/interdisciplinary teams, such as Google's Project Aristotle, has revealed how some effective teams' norms contrast greatly with those of other effective teams (Duhigg, 2016). This suggests that much nuance is needed to bridge understanding between team members so that teams become effective and that not one size fits all.

Relying on machine-learning systems and AI technology to augment this process may also be difficult if there is not enough nuance to better recognise patterns across the kinds of teams involved.

In attempting to find an angle through which to devise a possible communication solution for interdisciplinary work, the author first looked into the complexities of using technology towards this aim.

THE PARADOX OF TECHNOLOGY

One of the earliest definitions of this phrase comes from cognitive scientist, Donald Norman, in 1985. Norman describes the paradox of technology as "the more advanced technology becomes, the more complex and difficult it is to use" (Norman, 1985). He argues that while technology has the potential to simplify our lives, it often ends up adding more complexity and confusion, as users are required to navigate complex interfaces and systems in order to accomplish even simple tasks.

What can the paradox of technology mean for this effort to use technology to augment interpersonal communications within interdisciplinary work teams, especially when the author posits that the nuances required to make technological solutions helpful go beyond professional and technical knowledge? There is a possibility that such a tech-based method may dive into an individual's personal sphere. And there are many problems here.

Sherry Turkle argues that "our devices are not just technological objects, but also psychological ones that mediate our relationships with ourselves and others", meaning that technology is Janus-faced. While enabling communication, it may also lead to feelings of disconnection and isolation as people turn away from face-to-face interactions (Turkle, 2011). If we consider this an important lens to navigate the real-world impact of the tech paradox, we may need to go one step further and investigate the effects on our personal lives.

Writer and sociologist Nancy Jo Sales takes this question further by researching the complications of integrating technology into our love lives. Using accounts from interviews and her own experiences, Sales explores the darker side of online dating and the impact that dating apps are having on our relationships and culture, arguing that dating apps are contributing to the objectification of women and a culture of casual sex, and that they are eroding the boundaries between public and private life (Sales, 2021).

It becomes more complex if we factor in the views of writers like Nicholas Carr, who argues that the constant use of technology is rewiring our brains, making it harder for us to focus and reflect

(Carr, 2010), and Jean Twenge, who believes that today's young children may grow up disconnected and will struggle to fit into society (Twenge, 2017).

Underlying all these concerns is the inescapable fact that employing technology to solve the communication challenges facing interdisciplinary work will require vast amounts of data, another area that requires careful deliberation.

ETHICS AND DATA

In 2021, the US House of Representatives passed the Algorithmic Accountability Act, which would require companies to assess the impact of their algorithms on privacy, fairness, and bias. Meanwhile, the European Union's General Data Protection Regulation (GDPR), enacted in 2018, strengthens privacy protections for individuals and imposes strict rules on how companies can collect, use, and share personal data. Similarly, the Australian government passed the Privacy Amendment (Notifiable Data Breaches) Act, which requires companies to notify individuals if their personal data has been compromised.

Commentators attribute these developments to the public awareness raised by advocates such as Shoshana Zuboff.

Zuboff's work takes the dialogue further in a way that is both broader and personal. She coined the term "surveillance capitalism" and was influential in arguing that it is a new economic system in which technology companies profit by collecting and analysing personal data on individuals and using it to predict and control their behaviour. She also argued that this has a detrimental impact on human agency and freedom (Shoshana, 2019).

It is no surprise that Zuboff counts as one of the strongest advocates for stronger privacy protections, greater user control, and more public oversight of tech companies, as well as the creation of new models of governance that are more responsive to the challenges of the digital age.

The work *Anatomy of an AI System* (2018) by Kate Crawford and Valadan Joler explores the intersection of human labour, data, and resources through the Amazon Echo device. It also reflects the tip of this iceberg. They argue that the Echo user is a chimaera, the "user is simultaneously a consumer, a resource, a worker and a product", and that the deployment of AI technology is not merely technical; it also has social, economic, and environmental consequences (Crawford, 2018). Yet, it becomes an issue with regulation or even ethical discussion when the lines are blurred between users and creators.

As Julie Thompson Klein has defined them, the above perspectives are crucial when defining the scope and use of technological solutions to overcome epistemological differences and communication barriers.

The ethical considerations of ensuring the lack of bias, discrimination, and traceability in data collection for this purpose must also be weighed heavily. This is especially true when data imperfection allows algorithms to inherit society's prejudices and widespread bias (Barocas & Selbst, 2016).

From this starting point, I decided to examine a traditional form of data collection that could help circumvent some of these dilemmas and provide a counterweight.

ORAL HISTORY METHODOLOGY AS A COUNTERWEIGHT

Oral history methodology can be defined as a research approach that involves the systematic collection, preservation, and analysis of spoken narratives to document personal experiences, memories, and perspectives related to historical events, social phenomena, and cultural practices (Abrams, 2016).

The key importance of this project is that oral history methodology emphasises the importance of preserving diverse voices and viewpoints, particularly those of marginalised or underrepresented groups. It also provides a means of accessing historical knowledge that may not be available through traditional archival sources alone (Baker, 1998).

As outlined by the National Heritage Board of Singapore's oral history guidebook, the basic process of an oral history interview involves an open-ended form of questioning centred on a subject matter or historical event that avoids leading questions and value judgements.

Projects such as "The Voices of Civil Rights" in the US (Library of Congress, n.d.) documented the personal stories and experiences of individuals involved in the civil rights movement to create a comprehensive record of the diverse voices and perspectives within the movement. Through these interviews, the project uncovered untold stories of grassroots activists, community organisers, and everyday people, offering a more nuanced and inclusive understanding of the civil rights movement beyond the well-known figures and events.

Oral history may also help us understand emerging businesses by providing insights into why certain investors invest in certain industries and not others and into bureaucratic practices across territories (Jones & Comunale, 2019).

More importantly, oral history methodology is increasingly embracing the digital tools of the 21st century, extending to mediums such as audio, video, and even digital archiving (Boyd & Larson, 2014). A tacit example of this is the 1947 Partition Archive Project (Bhalla, 2011–2024), which aims to preserve the memories and stories of individuals who lived through the partition of British India in 1947, which led to the creation of India and Pakistan and the displacement of millions of people. The collection combines video diaries, blogs, traditional oral history interviews, and photos.

In many ways, the digital form of this method encompasses text, video, images, and audio – areas that can yield high caches of data and information for building datasheets and datasets for machine-learning systems.

The oral history methodology has its drawbacks, however.

Decades ago, there was already discussion of the unreliability of human memory and subjectivity. There is also the role of the interviewer in shaping narratives and the possibility of bias when interpreting oral testimonies (Portelli, 1991).

The interview process can also be time-consuming, and requires careful cataloguing of data captured to be codified effectively for further interpretation or usage (Ritchie, 1995).

The OHMS (Oral History Metadata Synchronizer), developed by Douglas Boyd at the University of Kentucky Libraries, however, shows the potential for synergy between oral history and technology to circumvent some of these drawbacks at least. The OHMS is an open-source, web-based application that allows better search capability for key content within video interviews and transcripts, allowing users to find information more easily (Synchronizer, 2024).

Such a digital solution inspires the author to consider:

1. If oral history methodology can be adapted into a data collection tool to bridge understanding between members of an interdisciplinary team, what kind of data can we collect that can be codified into effective datasets for machine learning systems?
2. Can the interview/data collection approach be tailored specifically to the goals of a specific organisation, minimising the time taken during the collection process?
3. Can there be a system of interpreting the data collected that allows for a more accurate yet nuanced way of gathering stories and perspectives to help overcome epistemological differences and communication barriers?

To answer these questions, the author decided first to break down the kinds of data and nuances present in an oral history video interview and research deeper into the capabilities of current machine-learning systems on the market.

WHAT DATA IS PRESENT IN AN ORAL HISTORY VIDEO INTERVIEW?

A single oral history video interview recording encapsulates a wealth of data, ranging from verbal narratives to non-verbal cues and environmental context. Personal stories, historical accounts, cultural insights, and perspectives on significant events lie within the spoken words. Concurrently, non-verbal cues such as facial expressions, gestures, and body language provide additional layers of meaning, conveying emotions, attitudes, and cultural nuances. Moreover, the surrounding environment captured in the video offers contextual clues, reflecting the interviewee's surroundings, lifestyle, and social dynamics (Boyd, 2024; Frisch, 1990; MacKay, 2007). This abundance of data presents opportunities for deep analysis and interpretation. However, a clear breakdown of the different kinds of data available will help match it with existing machine-learning technologies already on the market.

Visual Data: Visual data captured in oral history video interviews includes facial expressions, body language, gestures, and the physical environment. These visual cues convey emotions, attitudes, and cultural contexts, enriching our understanding of the interviewee's story. For example:

- Facial expressions: A smile, a furrowed brow, or tears can convey a range of emotions, from joy and sadness to frustration or determination.
- Body language: Posture, gestures, and movements can reveal confidence, nervousness, or discomfort, providing additional layers of meaning to the spoken words.
- Physical environment: The setting of the interview, whether it is a home, a workplace, or a public space, can offer insights into the interviewee's daily life, social status, and cultural background.

The interpretation of such visual data will depend greatly on cultural variability and context (Hall, 1973). There may also be ethical concerns about reading facial expressions, race, and body language, as relying solely on any interpretation without corroborating evidence can raise questions about the fairness and reliability of the assessment process, potentially leading to unjust outcomes and stereotypes (Nakamura, 2008).

As such, the author recognises that visual data may be the most difficult information to process.

Audio Data: The audio component of oral history video interviews captures spoken words, vocal tone, accent, and background sounds. These auditory elements provide essential context and detail to the interviewee's narrative. Examples include:

- Spoken words: The interviewee's stories, memories, and reflections are conveyed through their spoken words, offering first-hand accounts of historical events, personal experiences, and cultural practices.
- Vocal tone and inflexion: The tone of voice, pitch, rhythm, and emphasis can convey emotions, attitudes, and nuances of meaning, enriching the narrative with subtleties that may not be captured in the text alone.
- Accent and language variation: The interviewees' accents, dialects, and language choices reflect their cultural background, regional identity, and linguistic heritage, contributing to the authenticity and diversity of the oral history record.
- Background sounds: Ambient sounds, such as traffic, birdsong, or the hum of conversation, provide a sense of place and atmosphere, grounding the interview in its physical context.

Transcript Data: Transcripts of oral history video interviews provide a written record of the spoken words, organised into a readable format. Transcripts capture the interviewee's narrative's content, structure, and nuances, facilitating analysis and interpretation. Examples of data present in transcripts include:

- Verbatim dialogue: The transcript preserves the interviewee's words exactly as spoken, capturing the rhythm, pacing, and phrasing of their speech.

- Annotations and notes: Transcripts may include annotations, notes, or timestamps to indicate non-verbal cues, pauses, or significant moments in the interview.
- Metadata: Transcripts often include metadata such as the interviewee's name, date of the interview, the interviewer's name, and other contextual information to aid in cataloguing, indexing, and retrieving the interview.

Based on this list of categories, the next step would be to examine what is already available on the market.

WHAT ARE EXISTING TECHNOLOGIES TO LEVERAGE ON?

Before developing a new machine-learning system from scratch, surveying the landscape of existing systems and technologies is prudent. By exploring the diverse range of machine-learning systems already available, valuable insights can be gleaned from the current state of the technology to identify possibilities.

This approach is very much in line with the idea of Jugaad, or frugal innovation, where the first step is to work with existing resources and technologies to create simple, affordable solutions (Radjou et al., 2012).

Examining existing systems helps to gauge the capabilities and limitations of machine-learning approaches for the next step of this research project.

The aim is eventually to build a minimum viable product as quickly as possible to validate or invalidate key assumptions and gather feedback from potential users for future iterations (Ries, 2011).

A survey of the current market produced the following list of relevant technologies:

1. Facial Expression Recognition Systems:
   - OpenFace: An open-source facial behaviour analysis toolkit that provides tools for facial landmark detection, head pose estimation, and facial expression recognition (Amos et al., 2024).
   - DeepFace: Developed by Facebook, DeepFace is a deep learning-based facial recognition system capable of accurately identifying human faces in digital images (Taigman et al., 2014).
   - Microsoft Azure Face API: A cloud-based facial recognition service offered by Microsoft Azure, which provides capabilities for face detection, verification, and identification, as well as emotion recognition.
2. Body Language Analysis Systems:
   - OpenPose: An open-source system for real-time multi-person key point detection and pose estimation, capable of analysing body language and movement in video data.openpose (Simon et al., 2019).
   - BodyPix: Developed by Google, BodyPix is a machine-learning model for real-time person segmentation and body part detection in images and video streams. It enables the analysis of body language and gestures.
   - IBM Maximo Visual Inspection: A cognitive visual inspection solution offered by IBM, which uses computer vision and machine-learning techniques to analyse images and video data for defects, anomalies, and patterns, including human body movements and gestures (IBM, 2024).
3. Multimodal Systems:
   - Affectiva Emotion AI: Affectiva offers a range of emotion recognition and analysis tools based on machine learning and computer vision techniques. These tools can analyse facial expressions, vocal intonation, and other non-verbal cues to infer emotional states (Affectiva, 2024).

- DeepMind's multi-object representation learning from object interactions (MORL) model: DeepMind has developed a machine-learning model capable of learning representations of objects and their interactions from video data, including human gestures and body language.
4. Topic Modelling:
   - Latent Dirichlet Allocation (LDA): LDA is a generative statistical model that allows sets of observations to be explained by unobserved groups called topics. It can identify topics or themes within a collection of text documents, grouping related ideas together (Maklin, 2022).
5. Speech Recognition and Transcription:
   - IBM Watson Speech to Text: IBM Watson offers a speech-to-text service that converts audio files into text transcripts, analysing spoken language and extracting meaningful insights.
   - Descript: Descript is a multilingual translation and transcription app that allows users to edit videos via text. The service can also search for keywords across transcripts (Descript, 2024).

The author posits that if the technology already exists in the market, then it is possible to imagine a solution that can be built by combining different systems.

It is important to note that during this market survey, the author discovered cross-capability features augmenting older software platforms. For instance, Adobe Premiere, a video editing software, now employs AI features that allow translation, transcription, and video editing via topic word grouping, all within its main software (Adobe, 2024).

The last discovery was crucial in helping the author formulate a hypothesis.

PUTTING TOGETHER A HYPOTHESIS

The market survey shows machine-learning systems can extract visual, audio, and text data from video oral history interviews. Various products on the market can interpret specific kinds of data according to their designed algorithms.

The challenge may lie in understanding how these different technologies can be synthesised into an integrated approach for a specific use case. The case of Adobe Premiere's AI features on its traditional video editing application suggests that cross-medium/platform usability is already on the horizon.

As suggested by Steven Johnson's adaptation of Stuart Kauffman's theory of "The Adjacent Possible" (Johnson, 2011; Kauffman, 1993), where innovations can emerge from the combination and exploration of possibilities available to an individual at any given moment; the proliferation of technologies and methodologies for data collection, analysis, and interpretation makes it conceivable to envision an amalgamation capable of harnessing different kinds of data harvested from an adapted form of oral history methodology.

The next step is to compare these learnings against the questions raised by this research project so far:

1. Is it possible to improve communication between members of an interdisciplinary team?
2. Given the broad spectrum of obstacles facing such communication, should the focus be narrowed down to epistemological differences and communication barriers (Klein, 2015) (Klein et al., 2001), to enable an achievable research outcome?
3. Can oral history methodology help overcome these obstacles and help build psychological safety (Edmondson, 2012) among team members?
4. Can this methodology be augmented by machine-learning systems for greater impact and scalability?

With this understanding of capability and need, the following hypothesis was formulated:

It is possible to build a digital device that helps members of an interdisciplinary team better communicate with each other by building a system powered by data gleaned through oral history video interviews.

The caveats are:

a. The hypothesis lacks the "how" this digital device can do this, which will become a new research goal of this project.
b. I acknowledge that the broad hypothesis must be proven, disproved, adjusted, or iterated one level at a time.
c. There is no measure of how much data or nuance is needed to train a system like this.
d. The parameters of this hypothesis have to be further narrowed to allow for actionable research and experimentation.

EXPERIMENT 1: PROCESSING EXISTING INTERVIEWS

The first experiment aims to prove if it is possible for an application to organise and put together a coherent reflection of one subject matter from characters with different professional backgrounds. These characters may come from different disciplines but should be connected to one hobby or passion, mirroring the members of an interdisciplinary team.

The key is to see if the simple task produces a compilation of perspectives, reveals commonality in thought, or exposes opportunities for nuances not seen previously.

The parameters of this experiment are as such:

1. There will only be one main subject matter identified.
2. Three video interviews, each over 60 minutes long, will form the raw input data.
3. The three interviews should come from characters with different backgrounds but a common interest.
4. The programmes in use will have AI capabilities such as Adobe Premiere and Descript.
5. The output should be a condensed video that can be watched within the space of five minutes.
6. The process should be recorded, including the time it takes for the application to ingest the "data" and process the required output.

Source Material

Drawing from one of the author's personal documentary film projects, three video interviews with three different backgrounds were selected.

The subject matter of their interviews related to popular music in Singapore from the 1960s to the 1980s.

Of the three interviewers, one was a musician born in the early 1970s, one was an Australian sound archivist who researched the subject matter, and the last was a banker/heritage food author who knew the original producers involved in the era's pop hits. They all have their own experiences and understanding of the subject matter.

The key search word was "Malay Pop Music", a side topic within the larger context of what the characters were interviewed about.

The second is a screenshot of the resulting video.

Do see Annex1a_Experiment1Output_Video.mp4, and Annex 1b_Experiment1OutputTranscript. pdf for the actual video and transcript in the Annexes Folder of this project's submission master folder.

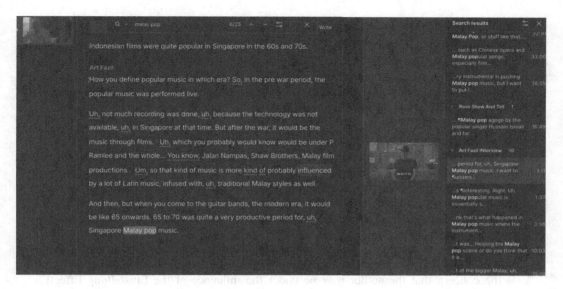

FIGURE 5.1 Screen capture of the process.

A second step was added to further refine the output – eliminating repetitive content or content that lacks context to provide a more coherent message.

This step used a combination of human judgement and machine ordering to mimic what it may be like for a trained algorithm to perform a similar task. See Figure 5.1 for a screen capture of the process.

a. Keywords were identified by machine and counted in frequency of appearance manually.
b. Every third repetition was deleted, together with the context they appeared in using human judgement.

Do see Annex2a_Experiment1Output_Video.mp4, and Annex 2b_Experiment1OutputTranscript. pdf in the Annexes Folder of this project's submission master folder.

The end video was shown in a screening to a small select group of four individuals with no knowledge of the subject matter. They included an artist (painting), a florist, an academic, and a professional working in a tech company. They were not informed about the backgrounds of the interviewees, although the musician was recognised as he was formerly a prominent pop star in Singapore back in the 1990s.

The evaluation aimed to determine whether they gleaned any new perspectives and whether these perspectives revealed anything about the three different characters.

A survey system was eschewed in favour of a verbal roundtable. The general findings were:

a. They found the videos slightly confusing as there was no "start and end" to the stories, and the videos were a multi-angle ramble about a subject.
b. They found the perspectives corroborating because there was an alignment on the broader subject matter between the interviewees.
c. They found that each interviewee added a different angle to the subject matter.
d. They recognised that each interviewee came from a different working background from their attire and manner of speech.

The last step of this experiment involved feeding the transcripts of the finished video to CHAT GPT and requesting that it produce a summary paragraph that encompasses the key points of all the interviewees' lines.

This was the result.

ANALYSIS OF EXPERIMENT 1

The experiment revealed the ease of extracting and collating similar perspectives from the available data/video recordings. The process helped create a single work (video/audio/text) that combined different perspectives on the same subject matter. The experiment also showed how even a non-main subject matter within the interview could contain enough nuance to be compiled into a message.

What is concerning about the experiment is:

i. Many steps require human curation and intervention.
ii. The current process relies on proprietary technology that can be expensive to purchase or engineer from scratch (e.g. language, text grouping).
iii. Compiling this raw data for such a small output was time-consuming, especially when the questions were open-ended, in the style of oral history interviews. The volume of data required to manage a working prototype may be unachievable within the timeframe of this research project.
iv. In many ways, the process is similar to the human editorial process present in documentary film-making, on which the author has based his career in the last two decades. There is the concern that the author may be under the influence of the Einstellung Effect, a cognitive bias for familiar methods, where a person's existing mental habits, strategies, or approaches to problem-solving can prevent them from finding a better solution to a problem (Luchins, 1942).

The key learnings in the author's opinion that can help further this research project are:

1. The research project must be more concise for it to be effective. The hypothesis is too broad, and there are too many layers to study and prove.
2. The oral history method must be balanced with a more streamlined, goal-oriented approach. The open-ended nature of questioning and the flow of content gathered must follow a more structured approach so that it will be easier to codify and organise.
3. More thought must be put into whether the research project is outward messaging or interactive communication.
4. There may be a need to build and label categories, even before the data collection process/ oral history interviews

To solve some of the dilemmas arising from the experiment, the research project needs to be anchored in a real-world use case for a real-world client. There must also be a conscious effort to mitigate any possible influence of the Einstellung Effect, where the author's professional experience as a documentary film-maker may cloud any attempts to look at other perspectives present regarding the research project's methodology.

John Dewey argues for a philosophy of pragmatism, suggesting that knowledge and ideas should be actively applied and tested in real-world situations (Dewey, 1966). Similarly, educator and philosopher Donald Schön argues that professionals, including researchers, should engage in a process of "reflection-in-action" where they actively seek out and learn from real-world experiences and challenges (Schon, 1983).

At the same time, the author posits that there must also be a stronger agency of storytelling involved rather than simply collecting vast amounts of data through oral history interviews. Peter Senge argues, for instance, that storytelling is an important tool for guiding organisational change and that narratives can help align individual actions with larger goals and foster a sense of purpose and commitment among team members (Senge, 2006). Similarly, communications should pivot around a central narrative arc, with clear stakes, conflict, and resolution to create emotional connections and inspire engagement (Duarte, 2010).

With these learnings in mind, a call for collaboration was sent out through the author's personal and professional networks, and a match was eventually found in a German start-up based in Munich, Germany.

ABOUT THE CLIENT: VIRIDIS

VIRIDIS is a German start-up aiming to build a green tech eco-system/cluster that connects actors from business, industry, scientific research, politics, and civil society. They attempt to do this using a multi-layer work approach that leverages development projects and investors, centred around a physical hub known as the Kesselhaus, an 85,000 square metre building near the former sites of Dachau and Auschwitz concentration camps. Work has already started on the building, and a few partners within the cluster are taking up residency in the property with vertical farms and green research labs.

The organisation's main goal can be summarised by the heading on its website: "The focus of action and investment is the development of bioeconomic potentials for the efficient and sustainable utilisation of natural resources" (VIRIDIS, 2024).

Many elements of the company resonate with the goals of this research project:

a. The cluster has over 30 investors and collaborators/partners at the moment. These members include medical companies, green researchers, a game company, and even a company working on a biodegradable alternative to aluminium for the food and beverage industry. While it may seem like they are running parallel operations, they have begun integrating their work to create a specific organisational output. This fits the description of an interdisciplinary organisation that this research project revolves around.

b. The company plans to use Web 3 and blockchain technology to build its ecosystem and internal economy, reflecting its openness to studies involving digital assets.

c. The company has two arms that need to operate in tandem, a for-profit side and a not-for-profit side, each led by a separate leader with expertise in different fields.

d. The company has a multi-layered narrative arc, presenting a good research challenge for this project.

THE CLIENT BRIEF

In conversations with the CEO and co-founder, Josef Kohl, some of the challenges facing the company were outlined as:

i. The number of partners is increasing, but there has been difficulty getting them to understand the central idea behind the VIRIDIS cluster.

ii. There has been no alignment of core organisational goals with current partners and investors due to point (i) above.

iii. With seed investors in place, the company is now gearing up to attract Series A investors, who will require the ability to represent and market the constituents of its cluster.

iv. There are many ground initiatives being carried out across the different arms of the company and its partners, and the core team is losing track of what is happening due to the speed of its expansion.

v. There is an ambition to turn the cluster into a decentralised autonomous organisation (DAO), but beyond the tech build-up, there is not enough transparency or understanding between partners about what they do, as is ideal within such a blockchain community.

Given these challenges, the client put forward the following brief for this research project:

Can the author put together a solution that can assist the C-Suite team in solving some of the communication challenges? The solution and research process should also be low-cost at the moment so that they can be actionable for both research and implementation.

PROPOSAL, FEEDBACK, ITERATION

Formulating a Proposal

The author considers the brief a complex challenge, given its open-ended nature, lack of quantifiable performance indicators, and the possibility that the project's final output may be subject to subjective appraisal.

To address this hurdle, the principle of "agents", where basic building blocks or processes interact to become an intelligent system (Minsky, 1988), was adopted to help deconstruct the complexity of our challenge and identify the essential components central to its resolution.

The following were identified as crucial to building a workflow:

i. A process of gathering the different understandings of what the VIRIDIS cluster means for partners, investors, and collaborators.
ii. A process where these gathered perspectives can be quantified, either by identifying keywords, or more resonant phrases/descriptions.
iii. A process where their gathered perspectives can be compared and contrasted against a central "gospel" of what the organisation stands for from the perspective of its core founders.
iv. A process of cataloguing the different ground initiatives under the cluster's umbrella.
v. A process where to synthesise the different understandings of what VIRIDIS presents into a coherent message.
vi. A process that ensures nuances are present despite the project having core objectives.
vii. A process to synthesise all these processes into a system that can be implemented.

From this breakdown, a proposal was formulated.

The following is a redacted compilation of key points of the concept paper presented to VIRIDIS:

"PROPOSAL"

A propriety digital library, not unlike the Greek goddess, Mnemosyne, that is the sum of the perspectives and views held by partners within the VIRIDIS cluster.

A prototype data collection system encompassing text, audio, and video, modelled on oral history methodology, will affect this.

This first phase of experimentation will focus on collecting interviews from VIRIDIS' current partners and team across diverse disciplines. Their stories and consented sharing will provide the foundation and required participation to trial this system.

The goal is to have a workable analogue prototype to aid a long-term databank-building process.

THE MECHANICS: ADAPTING ORAL HISTORY METHODOLOGY

The key to its project is its focus on the data categorisation process before the data collection begins. This includes formulating keywords, imaging devices, and aural/visual capture settings that can allow for quicker data harvesting and subsequent building of usable data sheets.

The data management approach will be explained further in the next part of this chapter. However, it will contain keywords/images/aural indicators in every individual's recording to facilitate easier codification into datasheet building.

In reverse, this data management system will *inform* the data collection/recording process. This process currently consists of video recording, rich picturing on camera using a blank slate/whiteboard, imaging device, audio recording, and subsequent text transcription. (See the data collection part of this proposal further on.)

Focusing on the VIRIDIS community and wider ecosystem, the recording process will narrow down to these key subjects in no order of importance. It must be noted that the various facets of each subject matter have a complex relationship and are not linear.

a. PROBLEM-SOLVING LOGIC
   Individual's heritage and background – how it informs their unique problem-solving and task management style.
   Identifying cultural legacies that may shape this worldview and life approach.
   Wider world view and how it interweaves with personal experiences and life journey.
   Identifying unique perspectives that may fall out of business textbooks or scientific journals.

b. PERSONAL MOTIVATION
   Individual's history and background – how it creates a personal drive and motivation.
   Identify clear goals and personal perspectives.
   Present a picture of an individual's self-identification to causes or personality types.

c. MAIN WORK TO BE DONE
   Explore the individual's main pursuit/project focus during recording.
   Identify the individual's practical approach to reaching goals.
   Understand the connection of an individual's main work project to the wider world at large and its placement within the VIRIDIS cluster.

d. CONNECTION TO THE VIRIDIS IDEALS
   Establish personal connections between individuals' belief systems and the shared goals of the VIRIDIS initiative.

e. BLUE SKY IMAGINATIONS
   Hypothetical imaginations – what would an individual pursue if given unlimited resources?
   Mine possible blue-sky ideas that may be tangible possibilities with emerging technology or future social trends.

f. TANGIBLE FUTURE PREDICTIONS
   Logging the individual's most pragmatic approach to problems that can be solved today or in the near future.
   Quantify what such a pursuit may need (e.g. x resources, y personnel) – that may become important when the data is used to build Tier 1 use cases (see further on in this proposal).

End of redacted compilation: key points of the concept paper to VIRIDIS:

The experiment phase relevant to this research project called for video interviews with VIRIDIS partners in Munich, Germany. This video interview will be scheduled for 60–90 minutes per interviewee.

The following permutations form the basic required sample size for the video interviews, with the company deciding which permutation was more possible to arrange:

Permutation A

8 x decision-makers (four from the core structure of VIRIDIS, four from partnering organisations).
   8 x scientists/tech operatives (four from core organisation, four from partners).
   Permutation A attempts to gain and compare data between big-picture thinkers and specific task-oriented thinkers.
   They are even numbers, so four will be interviewed at a time, and the data will be analysed to adjust the system before the next four, and so on.

Permutation B

Organisation development focused.

6 x physical hub-related individuals who should include a planner; an engineer, two decision-makers.

4 x investors (if possible) – we would probably uniform this group by focusing mainly on decision-makers.

6 x projects related individuals – two decision-makers, four scientist/tech operatives.

Client Feedback and Iteration 1

The client was receptive to the idea of having a human resource library and was keen to gather the myriad personal perspectives of the partners and operatives under the VIRIDIS umbrella.

However, they asked for the following:

a. What are the varieties of data that they can hope to obtain from this exercise?
b. More details about how this data can be organised and used for machine-learning systems in future.
c. A better sense of how this information can be employed now in practical uses as there is considerable effort required to bring all the partners to one place across the space of a week, not to mention the time commitment required on the part of the interviewees.

As a non-programmer professional not trained in data science, I had to rely on a basic logic of input and output to create a system of categories. The whole idea is to "satisfice" or work towards achieving an acceptable rather than an optimal outcome (Simon, 1996).

The following excerpt was part of the paper sent to VIRIDIS in this round of iteration:

STAGE 1: BASIC CATEGORIES

The *most basic form* of data grouping can be seen in the following flow chart.

This flow chart marks out the details that will be gleaned from each individual to be recorded and lists key indicators of categorisation. Note that each "box" is not mutually exclusive from any other boxes within the diagram (see Figure 5.2). Notice how open-ended each subject category can potentially be. That factors into the next stage of data handling. This flow chart can also be employed as the first data grouping, based on box categories across various individuals – e.g. investors from Asia with a passion for renewable energy and a philosophy for food security.

Important Notes

The categories dictate the questions and formatted answers that must be gathered from the individual recorded. For instance, questions must be phrased back to the interviewer or interviewing machine so as to produce specific text/verbal indicators such as the city, e.g. Berlin; the role, e.g. research fellow; and the field, e.g. retail banking.

At the base level, this allows for a text-based harvesting of content that can be grouped into a query that requires the following data. As an example, we will use a hypothetical match-making application (see Applications: Tier 1 in later part of this chapter).

These are possible user queries that will activate the key indicators:

I would like to find researchers, working on terraforming technology within the vicinity of Berlin.

Or

I would like to source for possible investors who are very passionate about food security, and believe that we can do away with animal husbandry, and willing to put up $200k euros for a prototype crop farm in the city of Munich.

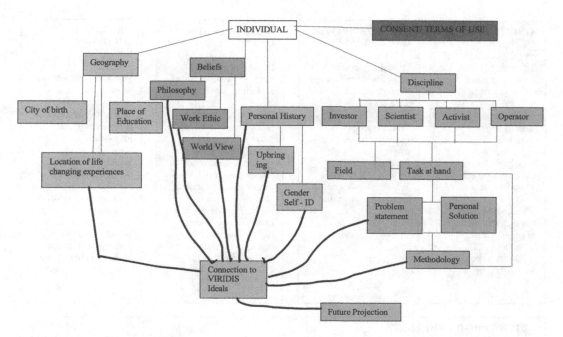

FIGURE 5.2 Data grouping flowchart

The most important of these categories, however, is the "consent" box. An on-camera explanation of possible use cases and for interviewees' direct consent can later be translated into the following crude example, say to be used in the programming of an interactive onboarding application for new employees (see Applications: Tier 1 in later part of this chapter).

query: consent to use content for operator onboarding: yes/no – yes
action: next layer
query: consent to reveal identity: yes/no – yes
action: next layer
query: consent to share personal methodology: yes/no – no
action: group data to general advice application, no usage for operator level 2 training.

The data categorisation described above is a nascent iteration of how this data can be grouped and harvested.

STAGE 2: BEYOND TEXT AND THE VERBAL

The next stage of data grouping takes the basic forms above and manages them into the following imagined sets of data for more advanced applications (see Applications: Tier 2 in the later part of this chapter). This is a starting point and is not exhaustive (Figure 5.3).

With regard to possible "real-world" uses, the following were devised as Tier 1 products from this research/experiment phase.

Tier 1:

a. Short-form video series that can be used for marketing or workshop purposes (see document "CHANGING THE GAME@VIRIDIS").
   This video series can capture a diverse group of investors, collaborators, researchers, and activists with their myriad personalities, geographies, and ideas/proposed solutions. The

AUDIO DATA:	VISUAL/BEHAVIOU RAL DATA:	RICH PICTURES DATA:	PERSONAL SOLUTIONS DATA:
Which voice recordings fall into the range of smoothing green noise spectrums?	Does the interviewee dress like a creative, businessperson, or field worker?	In the rich picturing exercise captured on camera:	Which interviewee has an investment friendly project based on correlating interests from investor interviewees/ data?
Which accents relate most to target markets?	Which interviewer maintains at least 60% of eye contact with interview interface?	Which interviewee used more circles than lines?	Which interviewee has the most pragmatic solution that is possible with current levels of technology?
Which voice timbre reinforces stereotypes of decisiveness etc?	Which interviewees sit wear tech accessories?	Which interviewee preferred using more colours and which used monochromatic tones?	Which solution has immediate resonance with prevailing social trends and zeitgeist?
Which interviewee has the most number of unsure pauses (measured in 2 seconds and beyond per pause)		Which interviewee's rich picturing exercise has the best clarity and potential for training applications?	

FUTURE PROJECTION DATA:

Which interviewee has future projections of processes of technology that has resonance/ relevance in current ongoing research in the wider world?

Which interviewee has future projections that sound the most plausible with current levels of science and the predicted curve of development?

FIGURE 5.3   Data grouping beyond texts and the verbal.

aim of this series is to provide a look beneath the hood of the very people that VIRIDIS has been able to put together into its community. The key is to unify their diverse perspectives with the ideals behind the VIRIDIS initiative.

b. People Directory

As an organisation that spans different disciplines and cities, it may not be physically possible for everyone to be in tune with each other. A people's directory of their thoughts, perspectives, and disciplines could effectively foster communication. Such a directory can be catalogued into scientific divisions, on-ground operators, and even investors and decision-makers. An account manager, for instance, could immediately understand a scientific researcher's ethos by accessing their transcripts or video files. By searching through this database, a new collaborator can peruse the people's directory to get their footing within the larger community.

c. Match-maker

This is targeted towards potential investors or for project managers seeking investments. Interviewees will reveal investment requirements, timelines, or interest levels in specific areas in the interviews. This match-making application can link projects to investors within the eco-system or serve as a search directory for individuals hoping to invest in the projects they believe in. The investments can be small or large, organisational, or personal.

End of part 2 of redacted compilation: key points of the concept paper to VIRIDIS

CLIENT FEEDBACK AND ITERATION 2

In this round of feedback, the client accepted the basic map of possible kinds of data to be obtained according to the original breakdown of content to be targeted.

The client understood that any use cases now are hypothetical and need to be reassessed after onsite interviews have been collected and analysed. The client was reassured, however, with the promise of using the video interviews to create a short video series for internal communication, as that would.

However, the client raised the following concerns from stakeholders who required more information about the questions and how they can tacitly direct the content towards the company itself.

To address this feedback, the following question list and breakdown were formulated. The interviews were also to include a rich picturing exercise that could yield data beyond what we may be looking for.

Below is a snapshot of the structured question list (Figure 5.4). The full document can be found as Annex2_ Annex3_ProjectMnemosyne_QuestionList.pdf.

	QUESTIONS	INTENTION	KEY WORDS FOR CATALOGUE	ADDITIONAL NOTES
	ABOUT SELF/ IDENTITY			
1	Can you tell us about yourself, where you grew up and any interesting stories about yourself from your youth/ childhood? Please include the words "City", "Village", "Town" and "country" in your story as well as the decade (80s, 90s, 2000s) that you grew up in.	To capture social/ geographical context and demographic	City, Year, Emotional labels: "Most embarrassing.." "Happiest Moment	
2.	Was there an event or a moment in your early/ younger years that changed the way you viewed the world? Please include the phrase "The event OR moment that changed the way I looked at the world was...." In your answer	To understand personal motivation, history and social/ political context if any. To cross reference against other interviewees to explore at what age can an individual be best influenced towards a cause or a perspective.	Event, Moment,	
3.	When did you begin paying attention to climate or earth related matters and why? Please include "I started paying attention matters regarding climate change OR Sustainability OR... around the time (year or period)" in your answer Ich habe angefangen mich mit nachhaltigkeit zu beschaeftigen.	To obtain insight on what can cause change in the individual. This will also be important to compare against other interviewees to see what connections or bonds can be woven. Questions 2 and 3 will provide context as to the kind of individuals who would be drawn to the VIRIDIS cluster and approach.	Year, period. Specific Events Emotions if any – e.g. "I was very distressed... I was very inspired" OR "I was very amazed"	
	ABOUT VOCATION/ PERSONAL CALLING			
4.	a. Tell us more about what you do. Start by saying "I am an engineer.... Or a scientist doing... OR... I am an account manager..." Ich bin ein	To hear how the individual identifies with their discipline or profession.	Profession/ discipline – e.g. "Engineer, Scientist, investor" etc.	

FIGURE 5.4 Structured question list.

FIELD STUDY: PROJECT MNEMOSYNE (MUNICH, GERMANY)

Codenamed "Mnemosyne", the video interviews took place on site in Munich, Germany, during the last week of January 2024.

The main aim of this field study was to:

i. Test the efficacy of a structured adaptation of oral history methodology for extracting data and perspectives.
ii. Stress test the idea of pre-categorisation in the interview process in its efficacy when it comes to obtaining targeted information/perspectives.
iii. Collect data to see if they can be codified enough to help synthesise an aligned vision for the client company.
iv. Test out mechanics for capture of information beyond just verbal and audio.
v. Set the foundation for a human resource library for the client company.
vi. Explore if the interview process can be automated without a human interviewer.

The sessions garnered 22 active participants from Munich, Cologne, and Dresden, representing six partner companies, seed investors, scientists, and engineers working on the construction of the Kesselhaus physical cluster, as well as the company's C-suite (Figure 5.5).

The interviews were captured on two cameras, to ensure dual usage:

a. To capture enough facial and body gestures.
b. To ensure that they can be used for internal communication videos.

FIGURE 5.5   Video recording interviews.

The only complications on site were:

1. The interview sessions were limited to under 30 minutes per participant given the schedules of everyone involved.
2. For many participants, the interviews had to be conducted in German, as they felt more honest and expressive in their native language.
3. The video captures did not include the rich picturing exercise, which proved too time-consuming and challenging when employed on the first two participants.

The key observations of the field study exercise were that the partners and investors were open enough to share as much as they could, whether it was their business models or even their own perspectives of what the VIRIDIS cluster constitutes. The diversity in perspective was evident when listening in on the on-site interviews. Using the VIRIDIS cluster idea as a rallying point for the content helped to align the responses even if the participants came from different backgrounds.

The structure of the questioning for each participant was executed 70% according to plan, but deviation was required to coax additional nuance out of particular participants. This would prove to be a problem later during the data sorting phase, post-field study.

The client provided positive feedback that the process helped align and bring together partners who were present and listening to each other. The feedback reinforced the idea that the Socratic method of dialogue and inquiry can foster collaboration and capture and transfer tacit knowledge (Dixon, 2000; Schwarz, 2013).

That was a positive indicator in favour of one of the research project's goals: to understand if we can facilitate better communication between members of an interdisciplinary team by helping them understand each other's backgrounds and training more deeply.

ANALYSIS: POST-FIELD STUDY

Given the limited time frame between the completion of the field study and the submission of this chapter, this project narrowed its focus to answering one requirement of the client brief, as a means of answering the aims of this research endeavour.

1. Can we communicate the communal understanding of what the VIRIDIS cluster is according to our different partners and operatives?

The following method in Table 5.1 to extract the required answers from the raw data/video interviews was devised based on a hypothetical application with multiple capabilities:

TABLE 5.1

Analytical Approach in Distilling Insights from Raw Data/Video Interviews

Step	Task	Application Used in Place	Notes
1.	Translate videos from German to English	DESCRIPT/NOTTA.AI	
2.	Transcribe dual language of German and English for text analysis	DESCRIPT/ ADOBE PREMIERE PRO	
3.	Extract relevant content for subject matter	ADOBE PREMIERE PRO/ DESCRIPT	Human curation required
4.	Order/Group Content	ADOBE PREMIERE PRO/DESCRIPT	Human Curation Required
5.	Analyse group content for following:	wordcounter.net (databasic.io, 2024), VOYANT TOOLS	Human curation required
	1. Most common perspectives (via text analysis of words and phrases)		
	2. Most unique/least common perspectives (via text analysis words and phrases)		
6.	Arrange content into singular coherent message	human intervention	
7.	Output content as video messages	ADOBE PREMIERE PRO, human intervention	

To answer the first requirement, the interviews were designed such that every interviewee had to start this part of the content with the phrase "The VIRIDIS cluster is …".

The process ran into difficulty immediately on execution:

a. Language translation models on the selected applications for German were not accurate or concise enough for a non-German speaker like the author.
b. The main categories of the questioning process help in grouping content across all the participants. However, given the open-ended nature of the interview process, there had to be additional effort to comb through resonating perspectives in other parts of the interview process, away from the question categories.

   The participants' diversity, based on their disciplines, meant that specific viewpoints were unique to their understanding. This was, however, a plus as it provided additional dimensions to the conversation.

   For instance, one of the participants, Nicholas Wild, is a programmer and holds CTO positions in many MNCs; his perspective considered corporate structures and the need for large organisations to be more fluid in addressing challenges. Whereas the cluster's communication manager, Ini Ana Christine, who dropped out of school to pursue small start-ups, focused her perspectives and approach at the individual level of responding to major challenges facing the environment and her personal stake in being part of such a green start-up.

   Another interesting observation was when the English transcripts of all the interviews (whatever could be translated) with regard to how the participants understood the operations of the VIRIDIS cluster, the following word map was obtained:

At first glance, it may seem that the algorithm simply narrowed in on "connecting words". Another way to interpret the reading could be that there is an emphasis on "Think", "also", and "would", which may translate to the fact that the members of the organisation still need to concretise what they are doing, given that the start-up is relatively young and very different from the organisation of a traditional German corporate entity (Figure 5.6).

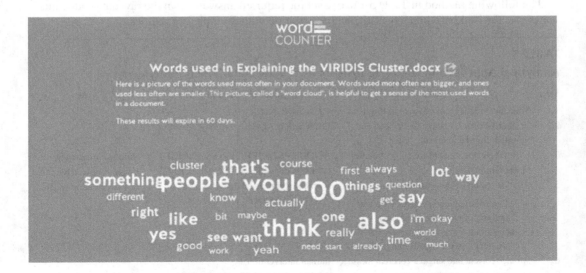

FIGURE 5.6   (a, b) Word map of translated qualitative data from interviews.

TOP WORDS ⊕		BIGRAMS ⊕		TRIGRAMS ⊕	
Word	Frequency	bigram⊖	Frequency	trigram⊖	Frequency
00	514	in the	364	a lot of	168
think	449	i think	311	a little bit	59
also	438	a lot	246	and i think	53
would	419	and i	207	you have to	49
people	385	i would	201	i want to	47
that's	333	have to	189	and then i	45
like	303	you can	175	what do you	43
yes	298	lot of	171	would like to	41
say	286	and then	168	you want to	39
something	278	if you	165	and so on	38
lot	249	of the	162	i would like	36
one	219	want to	158	we have to	28

FIGURE 5.6 (*Continued*)

The basic organisation of the video interview process did, however, enable an across-the-spectrum collection of perspectives about this subject matter, even on a first search of the keyword across all videos and transcripts (Figure 5.7).

This enabled the output of a single video file for the client to understand the thread running through their partners and operatives regarding what the organisation stands for.

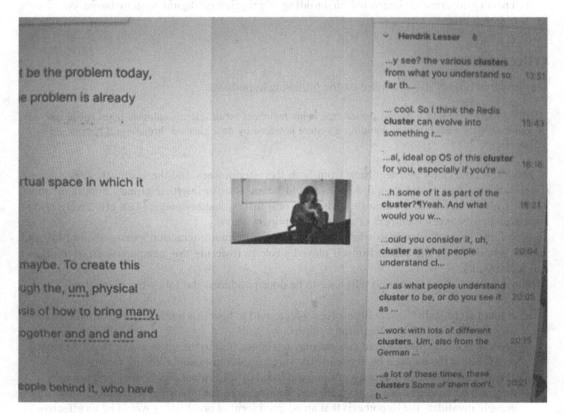

FIGURE 5.7 Screen capture of across-the-spectrum collection of perspectives.

Human judgement was needed to determine the common touch points and any unique ideas to build a *coherent story* within a shorter video file.

The following video was sent as one of the deliverables to the client close to the submission date of this chapter. It can be found as Annex4a_WhatIsViridis.mp4 in the "Annex" folder within this submission master folder for this research project. The transcript is labelled as Annex4b_WhatIsViridis.pdf.

Client Feedback (At Time Of This chapter's Submission):

a. The client felt that the short video was very helpful in synthesising the communal perspective about the VIRIDIS cluster, and it intends to use it to onboard new investors and partners.

b. The client reiterated that the video interview process itself was very beneficial. It brought people from different backgrounds and cities together and strengthened rapport and understanding among the team.

c. Participants have given feedback that they found some of the questions difficult to answer but that they were triggered to think deeper about their roles within the organisation and how they can express themselves.

d. There is interest in exploring further how this process can lead to an automated way of gathering perspectives from across the cluster in the next one or two years.

e. They want to begin building a human directory/resource library with the collected material. Each participant's full interview will be tokenised, and the participants will personally own it as a non-fungible token.

f. The organisation is interested in building a proprietary digital system based on this research project.

CONCLUSION

This research project was premised on the following hypothesis:

> It is possible to build a digital device that helps members of an interdisciplinary team better communicate with each other, by building a system powered by data gleaned through oral history video interviews.

From the interviews and feedback from both the interviewees and the client, there is a strong suggestion that data-based digital devices would be an asset to interdisciplinary communication. There is also a strong suggestion that adapting oral history methodologies can effectively bridge understanding between members of an interdisciplinary team.

It remains inconclusive as to whether the factors of human interaction, personal curation, and narrative building by the author himself played a role in fostering this perspective from the client and participants.

Thus, further experimentation will have to be done to address the following loopholes:

a. A lot of technicalities were unresolved with regard to how this method can obtain data that can codified for machine-learning systems.

b. Pre-categorising the required content may introduce bias to the interview process. At what point should the balance be between structured questioning and allowing for nuance?

c. The video interviews have a common issue faced by traditional oral history practitioners – they can be cumbersome to navigate and require more technological assistance. This can gravely invalidate the hypothesis that an adapted form of oral history would be an effective data-gathering tool for machine-learning systems.

The research project was started as a further iteration of a business/tech idea I had started when the Pandemic hit in early 2020. It was a time of upheaval for my industry (film-making), which proved even more tumultuous with the advent of generative AI.

In many respects, the "old way" of doing things has been turned on its head. Where once creative and editorial professionals like myself were the key resources in any project, the landscape has transformed into one where we find ourselves valued out of a production chain increasingly governed by numbers and data. After all, even Netflix commissions look to algorithms to decide for its content makers what is a marketable story and how to tell it.

This dramatic change prompted me to understand the role of human agency today in relation to a basic requirement of human society: communication.

Many of my colleagues and collaborators face a difficult present and uncertain future. Music composers I know who used to display their Grammy Disc Awards proudly across the wall of their studio are now forced to live with their parents because an algorithm can now generate the music for shampoo ads, their basic bread and butter.

The motivation for this project deepened with a conversation I had with an American acquaintance running his own tech start-up. He said AI was never intended to put artists and creatives out of work; it was simply safer to test the technology on the realm of art than to put it through the paces in a more regulated sector. I found this insight both illuminating as well as irresponsible. I wondered what it would be like if artists and creatives took the lead and looked at technology instead of technopreneurs taking that first leap for us at our expense.

The case for exploring how oral history methodology can be adapted to improve dialogue in an interdisciplinary team (as a film team always is) is also an attempt to see if we can collect data and use it more responsibly.

The process of this project brought me through many discussions and writings about the ethical usage of data, especially data that pertains to creators and builders. I have an impending sense that in two or three short years, regulation will bear down on the indiscriminate usage of books, writings, films, music, and even personal problem-solving matrices. And when that happens, how do we power the algorithms that we would have been so used to employing?

More importantly, is there a way for technology to be balanced once more with human agency and perspective? Throughout this project, I have found that there is still no algorithm for human ingenuity or problem solving, at least not in this era of technology.

My conversations with the partners and collaborators of our client, who make up entrepreneurs, scientists, engineers, and doctors, reaffirmed my faith in the "human algorithm". That is, the best products and solutions always start because some human somewhere is convinced that he/she/they can solve a problem they feel passionately about. By doing so, they create new systems and methodologies. An algorithm cannot do that if all it works on is the data of the past.

To create something iconoclastic requires dialogue and trust between collaborators. For that to happen, we have to use both technology and human instinct to build ever-wider canals for thoughts to flow like rivers.

REFERENCES

Abrams, L. (2016). *Oral history theory* (2nd ed.). Routledge.

Adobe. (2024). *Adobe premiere pRO*. adobe.com. https://www.adobe.com/sg/products/premiere.html

Affectiva. (2024). *Humanizing technology with emotion AI*. https://www.affectiva.com

Amos, B., Ludwiczuk, B., & Satyanarayanan, M. (2024). *Open face*. https://cmusatyalab.github.io/openface/

Baker, A. (1998). *Voices of resistance: Oral histories of Moroccan women*. State University of New York.

Barocas, S., & Selbst, A. D. (2016). Big data's disparate impact. *California law Review, 104*(3), 671–732.

Bhalla, G. S. (2011–2024). *1947 partition archive*. https://in.1947partitionarchive.org

Boyd, D. (2024). Digital Omnium: Oral history, interviewing, archives and recording techniques. *Digital Omnium*. https://digitalomnium.com

Boyd, D. A., & Larson, M. A. (2014). *Oral history and digital humanities*. Palgrave Macmillan.

Carr, N. (2010). *The shallows: What the internet is doing to our brains.* W. W. Norton & Company.

Crawford, K. J. (2018). *Anatomy of an AI system.* https://anatomyof.ai

databasic.io. (2024). *WordCounter.* https://www.databasic.io/en_GB/wordcounter/

Descript. (2024). *https://www.descript.com.*

Dewey, J. (1966). *Democracy and education: An introduction to the philosophy of education.* The Free Press.

Dixon, N. M. (2000). *Common knowledge: How companies thrive by sharing what they know.* Harvard Business School Press.

Domino, S. E., Smith, Y. R., & Johnson, T. R. B. (2007, March 16). Opportunities and challenges of interdisciplinary research career development: Implementation of a women's health research training program. *Journal Women's Health, 16 2,* 256–261.

Duarte, N. (2010). *Resonate: Present visual stories that transform audiences.* Wiley.

Duhigg, C. (2016, February 25). What google learned from its quest to build the perfect team. *The New York Times Magazine.*

Edmondson, A. (2012). *Teaming: How organizations learn, innovate, and compete in the knowledge economy.* Jossey-Bass.

Edmondson, A. (2018). *The fearless organization.* John Wiley & Sons.

Floridi, L., & Cowls, J. (2019, July 02). A unified framework of five principles for AI in society. *Harvard Data Science Review.* 10.1162/99608f92.8cd550d1.

Jessup, R. (2007). Interdisciplinary versus multidisciplinary care teams: do we understand the difference? *Australian Health Review,* 31(3), 330–331.

Frisch, M. (1990). *A shared authority: Essays on the craft and meaning of oral and public history.* State University of New York Press.

Goh, A. (2023, October 14). Interdisciplinary studies and the future of work. *Medium.*

Hall, E. T. (1973). *The silent language.* Double Day & Company.

IBM. (2024). *IBM Maximo visual inspection.* IMB.com. https://www.ibm.com/products/maximo/visual-inspection

Jacob, J. (2015). Interdisciplinary trends in higher education. *Palgrave Communications.*

Johnson, S. (2011). *Where good ideas come from: The natural history of innovation.* Riverhead Books.

Jones, G., & Comunale, R. (2019). Oral history and the business history of emerging market. *Enterprise & Society,* 20, 19–32.

Katsos, J., & Miklian, J. (2021, November 17). A new crisis playbook for an uncertain world. We're entering a period of unprecedented instability. Is your business prepared? *Harvard Business Review.* https://hbr.org/2021/11/a-new-crisis-playbook-for-an-uncertain-world

Kauffman, S. (1993). *The origins of order: Self-organization and selection in evolution.* Oxford University Press.

Kavanagh, C. (2019, August 28). *New tech, new threats, and new governance challenges: An opportunity to craft smarter responses?* https://carnegieendowment.org/2019/08/28/new-tech-new-threats-and-new-governance-challenges-opportunity-to-craft-smarter-responses-pub-79736

Klein, J. T. (2015). *Interdisciplining digital humanities boundary work in an emerging field.* University of Michigan Press.

Klein, J. T., Häberli, R., Scholz, R., & Grossenbacher-Mansuy, W. (2001). *Transdisciplinarity: Joint problem solving among science, technology, and society: An effective way for managing complexity.* Birkhäuser.

Laborda, R., & Olmo, J. (2021). Volatility spillover between economic sectors in financial crisis predic- tion: Evidence spanning the great financial crisis and Covid-19 pandemic. *Research in International Business and Finance, 57,* 101402. https://doi.org/10.1016/j.ribaf.2021.101402.

Library Of Congress. (n.d.). Voices of civil rights project collection. https://www.loc.gov/item/2012655454/

Luchins, A. S. (1942). Mechanization in problem solving: The effect of Einstellung. *Psychological Monographs,* 54(6), i -95. https://doi.org/10.1037/h0093502

MacKay, N. (2007). *Curating oral histories: From interview to archive.* Left Coast Press.

Maklin, C. (2022, August 2). *Latent Dirichlet allocation.* medium.com. https://medium.com/@corymaklin/latent-dirichlet-allocation-dfcea0b1fddc

Minsky, M. (1988). *Society of mind.* Simon & Schuster.

Nakamura, L. (2008). *Digitizing race: Visual cultures of the internet.* University of Minnesota Press.

Norman, D. (1985). *Living with complexity.* MIT Press.

Nyugen, T. S. (2023, July 26). *The power of cross-disciplinary problem-solving and collaboration. FORBES.*

Portelli, A. (1991). *The death of Luigi Trastulli, and other stories: Form and meaning in oral history.* State University Of New York.

Radjou, N., Prabhu, J., & Ahuja, S. (2012). *Jugaad innovation: Think frugal, be flexible, generate breakthrough growth.* Jossey-Bass.

Ries, E. (2011). *The lean startup: How today's entrepreneurs use continuous innovation to create radically successful businesses.* Crown Business.

Ritchie, D. A. (1995). *Doing oral history.* Twayne Publishers.

Sales, N. (2021). *Nothing personal: My secret life in the inferno of dating apps.* Legacy Literature.

Schon, D. A. (1983). *The reflective practitioner: How professionals think in action.* Basic Books.

Schwarz, R. M. (2013). *Smart leaders, smarter teams: How you and your team get unstuck to get results.* Kindle Edition. Jossey-Bass.

Senge, P. (2006). *The fifth discipline: The art and practice of the learning organization.* Random House Books.

Shoshana, Z. (2019). *The age of surveillance capitalism: The fight for a human future at the new frontier of power.* PublicAffairs.

Simon, H. A. (1996). *The sciences of the artificial* (3rd ed.). MIT Press.

Simon, T., Wei, S.-E., & Sheikh, Y. (2019, May 30). *OpenPose: Realtime multi-person 2D Pose estimation using part affinity fields.* arvix.org. https://arxiv.org/pdf/1812.08008.pdf

Synchronizer, O. O. (2024). *OHMS: Oral history metadata synchronizer.* https://www.oralhistoryonline.org

Taigman, Y., Yang, M., Ranzato, M. A., & Wolf, L. (2014, June 24). *DeepFace: Closing the gap to human-level performance in face verification.* https://research.facebook.com/publications/deepface-closing-the-gap-to-human-level-performance-in-face-verification/

Taleb, N. (2012). *Antifragile: Things that gain from disorder.* Random House.

Turkle, S. (2011). *Alone together: Why we expect more from technology and less from each other.* Basic Books.

Twenge, J. (2017). *iGen: Why Today's super-connected kids are growing up less rebellious, more tolerant, less happy—and completely unprepared for adulthood.* Simon & Schuster.

Varma, A., & Salfiti, Z. (2023, March 17). Fear of financial crisis unleashed chaos across markets this week. These 7 charts show how the shockwaves engulfed stocks, bonds, and commodities. *Business Insider.*

VIRIDIS. (2024). *Viridis cluster 4.0.* https://viridis.info/en

Wolf, M. (2014). *The shifts and the shocks: What we've learned—and have still to learn—from the financial crisis.* Penguin Press.

Zhang, P. (2019, January 21). China confirms birth of gene-edited babies, blames scientist He Jiankiu for breaking rules. *South China Morning Post.*

6 Legal Practices Redefined
Transforming Law Firm Operations and Management with AI

Siok Khoon Lim

INTRODUCTION AND BACKGROUND

LEGAL TECH'S "CHATGPT" MOMENT

In the past few years, the rise of AI and its potential to disrupt the legal industry have been forecast multiple times, yet the industry needs to be faster to adopt technology. While legal futurists have long anticipated technology's transformation of the legal industry, the impact to date can best be described as evolutionary rather than revolutionary (Perlman, 2023, p. 1).

However, when OpenAI released a chatbot called ChatGPT in November 2022, many in the legal industry started to notice, seeing it as the "beginning of a paradigm shift" (Stokel-Walker, 2023). Within a few months of ChatGPT's release, legal tech companies announced new ways of using generative AI tools (Morris, 2023). As large language models (LLMs) started to become more advanced, showing increasing proficiency in law-related tasks such as first-year law school exams, the uniform bar exam, and even statutory reasoning (Dahl et al., 2024 p. 1), it is undeniable that the legal industry could finally be on the cusp of a technological metamorphosis.

ABOUT THE CLIENT

My client is a leading business law firm in Oslo, Norway. It is a large organisation with almost 200 lawyers and 40 support staff.

I worked closely with the digital programme manager from the client's digital department for this research study. The digital programme manager is responsible for (1) finding and adopting new solutions and driving digitalisation in the organisation and (2) implementing new core systems in the other departments. From the outset, the client was excellent to work with, was open to collaborating and coming up with new ideas, and provided constructive feedback at each stage of the process.

The client is tech-forward and cognisant of the recent rapid rise in legal tech. In June 2023, it engaged a consulting firm to conduct a feasibility study to explore the scope and potential of using generative AI in its business processes.[1,2] The goal of the feasibility study was to identify strategic opportunities for the client's use of generative AI, evaluate relevant tools for implementation in the near future, and uncover potential tools that might be used later.

As part of the feasibility study, the consultant carried out the following:

- Conducted in-depth interviews with 11 of the client's employees. These interviews identified areas of opportunity for the use of generative AI.
- Identified and analysed more than 60 AI specialist legal generative AI tools and solutions,[3] and recommended those that could add the most value to the client.

DOI: 10.1201/9781003469551-6

Based on the insight gained from the interviews, the consultant identified the following tasks with the highest priority for using generative AI:

Category	Task
Contract management	Draft legal documents
	Analyse legal documents to identify key information
	Compare transactions to identify similarities, differences, and potential contradictions
Classification, search, summary	Classify documents based on relevant categories and content
	Search internal data sources to screen relevant information
	Summarise documents to screen relevant information

Based on the above-identified high-priority tasks, the consultant recommended that the client carry out the following:

- Build its legal chatbot using OpenAI's ChatGPT.
- Launch clear usage guidelines for its employees' use of ChatGPT.
- Test up to three of the recommended specialist legal generative AI tools.

For this chapter, I will be discussing my involvement in the implementation of two AI tools: (1) the chatbot; and (2) a legal drafting tool.

PROBLEM STATEMENT: HOW CAN AI TOOLS BE EFFECTIVELY IMPLEMENTED IN A LAW FIRM?

IMPLEMENTATION IS THE BOTTLENECK

The most challenging task organisations can arguably undertake is planned organisational change. Some authors suggest that two-thirds of planned initiatives fail, with failures attributed to the organisation's failure to incorporate the human side of organisational change (Bakari et al., 2017, p. 155).

In the context of AI, while the technology is poised to have a transformational impact, the bottleneck is not only in the maturity and capability of the technology itself but also in its management and implementation (Brynjolfsson & McAfee, 2017). In a survey conducted by Deloitte in 2017, 47% of respondents said that the top challenge with working with AI is the difficulty in integrating AI with existing processes and systems (Deloitte, 2017, p. 12). As Plastino and Purdy point out, AI's potential competitive advantage is "equalled by the challenges of integrating AI technologies effectively" into an organisation's business model (2018, p. 16).

While change management is always an uphill task, implementing AI tools in an organisation seems more challenging than other types of change. In this chapter, I will explore the reasons for this.

Therefore, the challenge the client faced was implementation. The client was aware of this challenge. In the digital programme manager's words, you only have one shot with technology adoption. If adoption rates are not high at the point of introduction, it will be challenging to get users to use the tools at a later stage. Raising the project's stakes is the fact that time is a valuable and sparse commodity for legal professionals because lawyers are always busy and because of the law firm's billable hour revenue model. This means that failed technology adoption will be costly for the client.

Change ultimately results from people adopting new skills and demonstrating new capabilities (Hiatt & Creasey, 2012, p. 3). One cannot assume that change will happen automatically on an individual level. One cannot expect everyone in an organisation to use a tool immediately and effectively simply because it was introduced to them. The client was careful to avoid making

FIGURE 6.1   Lewin's field theory (Cameron & Green, 2012, p. 121).

these assumptions and wanted to avoid investing money and time in AI tools, only to see a low adoption rate.

Therefore, it is important to apply a set of processes and tools to manage the people side of change from a current state to a new future state to achieve the desired change results and expected return on investment (Hiatt & Creasey, 2012, p. 9).

LEWIN'S THREE-STEP MODEL: UNFREEZE, MOVE, REFREEZE

Using Kurt Lewin's three-step model of organisational change as a framework, I will examine the people side of the change management process in terms of implementing AI tools in the client.

According to Lewin's Field Theory (which underpins the three-step model), to understand any situation, it is necessary to view the present situation, i.e., the status quo, as being maintained by certain conditions or forces (Burnes, 2004, p. 981) (Figure 6.1). If one can identify, plot, and establish the potency of these forces, then it would be possible not only to understand why individuals, groups, and organisations act as they do but also what forces would need to be diminished or strengthened in order to bring about change (Burnes, 20045, p. 981). The underlying principle of field theory is that the driving forces must outweigh resisting forces in any situation for change to happen (Cameron & Green, 2012, p. 121).

Following his field theory analysis, Lewin proposed the following three steps to bring about change at a group, organisational, and societal level (Figure 6.2):[4]

Step 1   *Unfreeze* the current state of affairs.
Step 2   *Move* to a new state through the participation and involvement of all stakeholders.
Step 3   *Refreeze* and stabilise the new state of affairs by setting policy, rewarding success, and establishing new standards.

Cameron and Green (2012, p. 122).

I will elaborate more on the three steps in the context of this project in the section headed "Mapping Change Management Aspects, Using Lewin's Model as a Framework" below.

For two reasons, I chose Lewin's three-step model as a framework (instead of other organisational change models). First, Lewin's model is simple and easy to understand. While the model has been criticised for over-simplifying the change process (Cummings et al., 2015, p. 34), I agree with

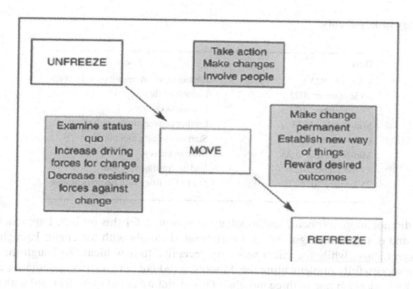

FIGURE 6.2   Lewin's three-step model

Source: Cameron & Green (2012, p. 122).

the view that any approach to creating and managing change necessarily begins with a process of "unfreezing" or challenging the conditions of the existing situation (Burnes, 2004, p. 986).

Second, and more importantly, Lewin's model concerns changing and directing human behaviour (Burnes, 2020, p. 35). Given that I joined the client's project mid-way at a relatively late stage (as elaborated in The Experiments below), I was not involved in any preliminary stages such as "vision setting" (a step in Kotter's eight-step model) (Cameron & Green, 2012, p. 126) or "exploration" (Bullock and Batten's phases of planned change) (Cameron & Green, 2012, p. 124). As such, I decided to focus on what I could realistically do, i.e., design experiments to influence behaviour on an individual and group level – the focus of Lewin's model.

THE EXPERIMENTS

EXPERIMENTAL DESIGN AND METHODOLOGY

At this juncture, I moved from Singapore to Oslo, Norway, in September 2023. The relocation was why I chose to work with an organisation in Oslo for this chapter.

Before I left Singapore, on 7 August 2023, I started preliminary conversations with the client. At that stage, the client was in relatively advanced stages of the project – a final prototype of the chatbot was already available to test in a development environment. The client planned to roll out the chatbot firmwide in September 2023. At that time, I also had no visibility on the groundwork the client had done. At that stage, I assisted the client with the communications on the chatbot (which will be explained in further detail below).

The client launched the chatbot firmwide on 20 September 2023, with the communication material I helped prepare.

I finally met the client physically when I arrived in Oslo in October 2023. After further conversations, I learnt about the feasibility study and its plans to implement other specialist legal generative AI tools. As the client had yet to start the implementation process with these tools, the chatbot could be the first phase of my experiment. Using what we learnt from the chatbot implementation, the second phase of my experiment would be the implementation of the drafting tool. The client agreed to this approach.

Chronology of key events

Date	Event
7 August 2023	Preliminary conversations with client
20 September 2023	Launch of chatbot
27 September 2023	I arrive in Oslo
Mid-October 2023	Further conversations with client
6 November 2023	Start of drafting tool pilot
Mid-November 2023	Chatbot survey
27 November 2023	End of drafting tool pilot
25 January 2024	Start of drafting tool implementation

While I did not adopt a specific methodology or approach for this project, I approached it with an iterative and design-thinking mindset. I collaborated closely with the client: I sought feedback at every project stage, while the client was very receptive to new ideas. Although the client was interested in successfully implementing the AI tools, it did not have a strict quantitative assessment metric (e.g. 95% adoption rate in three months). This meant we could move forward with the experiments with an open mind and learn and tweak accordingly throughout the process.

EXPERIMENT PHASE 1: THE CHATBOT

About the Chatbot

The client partnered with a software developer to develop the chatbot. The chatbot is the client's ChatGPT interface – instead of going to chat.open.ai, the client has its application, directly connected via API to Microsoft Azure OpenAI Service. This was a conscious decision on the client's end to ensure it could (1) meet its privacy and confidentiality obligations and (2) customise the user journey and experience.

The solution is set up within the client's cloud environment to ensure privacy and security. All data and documents entered into the chatbot stay within the client's environment and premises and are unavailable to OpenAI and Microsoft.

When the chatbot launched on 20 September 2023, it used GPT-3.5, and had the following functionalities (Figure 6.3):

- As with ChatGPT, the ability to interact with users conversationally.
- Document uploader – upload a document (<2,000 words) and get the chatbot to perform tasks with the text in the document, for example "convert this memo to create a draft of a newsletter article".
- Language toggle – English and Norwegian.
- "Personality" toggle: creative, neutral and precise – ability to customise the tone and voice of chatbot output.

On 10 November 2023, the client introduced an updated version of the chatbot, enhanced with GPT-4 technology, and the following new functionalities:

- GPT-4 capabilities, such as improved context window, more accurate and human interactions, and faster responses.
- Document translator – upload a document for the chatbot to translate (Figure 6.4).
- Improved document uploading – no upper limit on the number of words contained in an uploaded document.

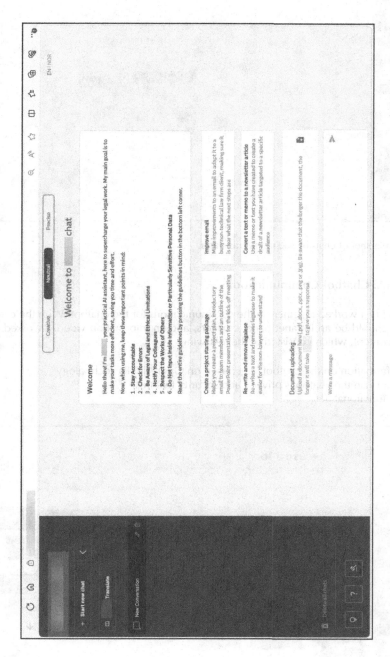

FIGURE 6.3 Screenshot of chatbot landing page.

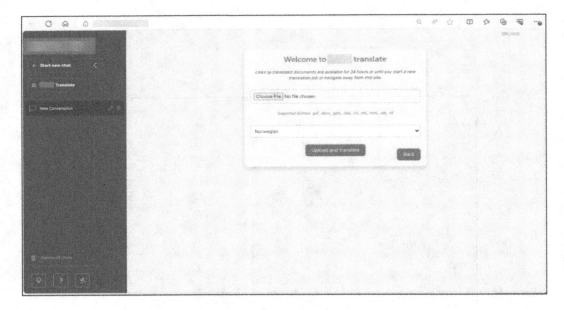

FIGURE 6.4   Screenshot of chatbot document translator.

Experiment 1A: Chatbot Communications

The client and I first worked together on the communications for the chatbot. With the client, we decided there should be an intranet page, emailers, and an information video. I worked on the intranet page content, which consisted of the following:

- Basic information on the chatbot, e.g. based on OpenAI GPT technology.
- Information on the chatbot's privacy and security features.
- Chatbot functionalities.

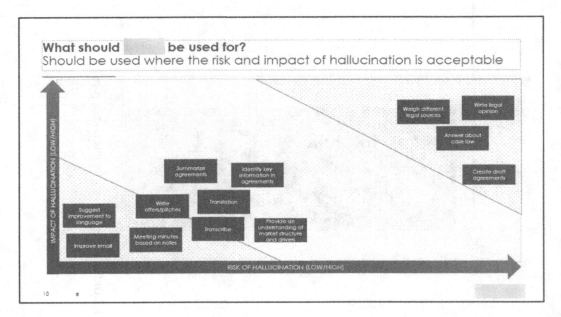

FIGURES 6.5   Extracts of chatbot communication material.

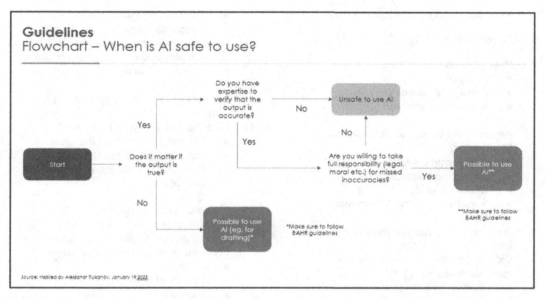

Guidelines
Flowchart – When is AI safe to use?

Source: Inspired by Aleksandar Tiulkanov, January 19 2023.

FIGURES 6.6   Extracts of chatbot communication material.

- Examples of what users may use the chatbot for, e.g. "Explain the concept of internal rate of return to someone who has no financial background".
- Basic information on prompting.
- Chatbot use guidelines.

To help users get started with the chatbot, there are prompting tips on the chatbot landing page. I also assisted the client with preparing these prompting tips.

Experiment 1B: Chatbot Survey

In order to gauge and measure chatbot adoption, I suggested to the client that the employees complete a qualitative feedback survey on their experience using the chatbot. I prepared the survey questions (with input from the client), and we released the survey[5] to the employees approximately two months after the launch of the chatbot, in mid-November 2023.

The survey covered the following areas:

- If they had used the chatbot and, if so, how often.
- Tasks users used the chatbot for and which tasks it helped the most with.
- If users found the communication and training helpful material, and if there were any areas in which they would like more training.
- If users had any prior concerns about using AI solution/chatbot, and if they still had those concerns now.
- Overall experience with using the chatbot: speed, quality of output, and usefulness.
- If they had not used the chatbot, why not, and what would encourage them to use the chatbot.

Through the chatbot survey, I wanted to learn what people were using the chatbot for, how comfortable users were with AI, what would get people more comfortable using AI and other tech tools, and overall sentiment towards AI. With this intention in mind, I kept the questions in the chatbot survey rather general. This was an important step for the project because I was working with a completely new organisation in a new country and working culture, and because it would give me a better idea of how to run the next phases of the experiments.

1. Have you used the Chatbot? (Y/N)

If you have used the Chatbot

2. How often do you use the Chatbot?
    - Everyday
    - A few times a week
    - A few times a month
    - Other
3. What are the tasks that you use the Chatbot for?
    - Translating documents
    - Explaining concepts
    - Checking work
    - Summarising long documents
    - Other
4. What tasks did you find the Chatbot helped the most with?
5. Did you use the "creative", "neutral" and "precise" features? (Y/N)
6. Did you find the prompts helpful? (Y/N)
7. What would encourage you to use the Chatbot more?
    - New features
    - Faster Chatbot
    - More training on Chatbot's features
    - Seeing that other colleagues are also using the Chatbot
    - Other
8. Are you aware of the following communication and training material: (Y/N)
    - Chatbot use guidelines
    - Information video
    - Intranet page
    - Emails from the Digital department
9. Which did you find most useful in providing information about the Chatbot?
10. Are there any areas in which you would like more training?
    - More information on the Chatbot
    - More information on prompting and crafting good questions
    - More information on how to make the most out of the Chatbot
    - More information on AI in general
    - Other
11. Before using the Chatbot, were you concerned about using an AI solution/chatbot such as the Chatbot? (Y/N)
12. What were your concerns?
    - I did not think that the Chatbot would help me with my work

- The Chatbot might provide inaccurate answers
- The Chatbot might be too slow
- Security and confidentiality concerns
- I did not know how to use AI
- Other

13. Do you still have these concerns now? (Y/N)
14. What helped with your concerns?
    - Learning more about the Chatbot
    - Having tried out the Chatbot
    - Speaking to colleagues about the Chatbot
    - Other
15. Based on your experience with the Chatbot, please rate the following: (with 1 being the most negative experience and 5 being the most positive experience)
    - Speed
    - Quality of output
    - Usefulness
    - Overall experience

16. Do you have any comments on the Chatbot or AI tools or any feature requests to the Digital team?

If you have not used the Chatbot

17. Why have you not used the Chatbot?
    - I did not know about the Chatbot
    - I did not know where to find the Chatbot
    - I did not know how to use the Chatbot
    - The Chatbot will not help me with my work
    - Other
18. Why do you feel the Chatbot will not help me with your work?
19. What would encourage you to use the Chatbot?
    - More information on Chatbot features
    - More training on the Chatbot
    - More training on IT tools
    - If the Chatbot had more features
    - If I see more of my colleagues using the Chatbot
    - Other
20. Do you have any comments on the Chatbot or AI tools or any feature requests to the Digital team?

FIGURE 6.7   Chatbot survey questions.

Summary of Results of Chatbot Survey

Overview

- 34 responses out of 268 invitations.
- 85% of respondents had used the chatbot.
- Most (58.6%) used the chatbot a few times a week, with 20.7% using it daily.
- Respondents found the chatbot most useful for translating documents, checking work output and improving/refining texts and emails.
- 24% of respondents said that seeing other colleagues using the chatbot would encourage them to use the chatbot more.
- Generally, respondents had an overall positive experience with using the chatbot.

Communication and training

- 60% of respondents requested more training, specifically more information on how to make the most out of the chatbot, prompting and crafting good questions and more information on AI in general.
- Almost 100% of respondents were aware of all the communication and training material, with the information video and use guidelines being the most useful sources of information.

Concerns

- Most (65.5%) had no concerns about using a chatbot Those who had concerns were mainly concerned about security, confidentiality, and the accuracy of the chatbot's answers. Learning more about the chatbot and trying it out helped with these concerns.
- Reasons given by respondents who had not used the chatbot "would not help me with my work", "would take longer to check quality of output", and "did not know how to use". However, 100% of respondents who had not used the chatbot said that more training on the chatbot would encourage them to use it.

Since we did not make it compulsory for everyone in the organisation to respond to the survey, the results of the chatbot survey are only a voluntary response sample. This means that, while the responses were largely positive, it could be that only people who used the chatbot and/or felt positively towards the chatbot took the time to respond to the survey. That said, 15% of the respondents had yet to use the chatbot and provided their reasons for not doing so, which provided valuable insight.

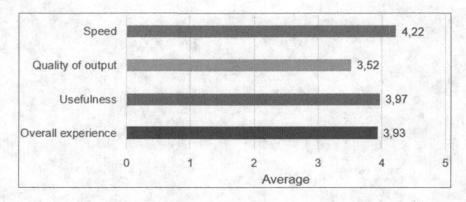

FIGURE 6.8   Overall user experience with the chatbot.

Chatbot User Metrics

During my preliminary conversations with the client, I asked the client if it was possible to track the following user metrics:

- Number of log-ons.
- Number of unique user interactions.
- Type of requests made.
- Type of documents uploaded.
- Most used features.

The client spoke to the developer, who promised to provide these metrics on a dashboard. However, sometime in November 2023, when we followed up with the developer, the developer changed tack and said it could only provide the requested metrics at a very high additional cost. The developer could provide the total number of requests over a certain time, as shown in Figures 6.9 and 6.10 below.

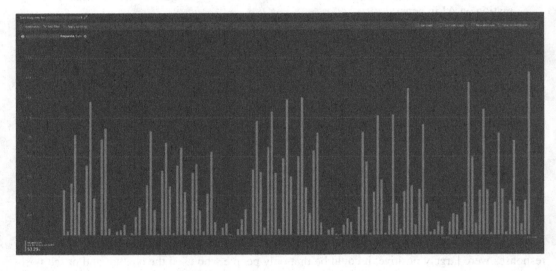

FIGURE 6.9   Total requests from October 2023 to November 2023.

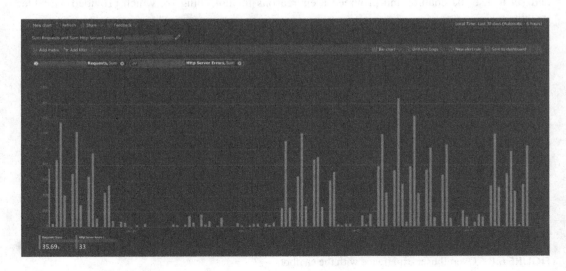

FIGURE 6.10   Total requests from December 2023 to January 2024.

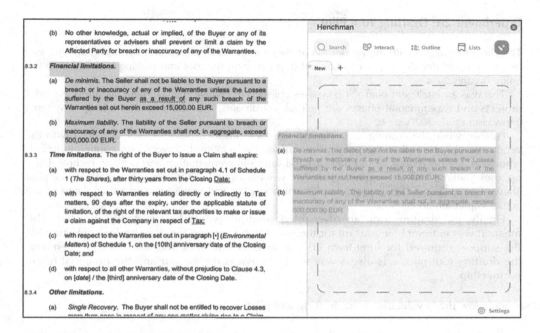

FIGURE 6.11 Example of drafting tool search functionalities: "Drag and drop" search.

EXPERIMENT PHASE 2: THE DRAFTING TOOL

About the Drafting Tool

The vendor of the drafting tool is a Belgian company.

The drafting tool is a contract drafting enablement tool that improves the contract drafting process.[6,7] It accesses and indexes agreements from the client's contract repository to create a database of clauses from these agreements. Users can easily search for previously written clauses in the client's database using the drafting tool's Microsoft Word plug-in, thus streamlining the contract drafting process. the drafting tool also has drafting capabilities, such as summarising and translating functions.

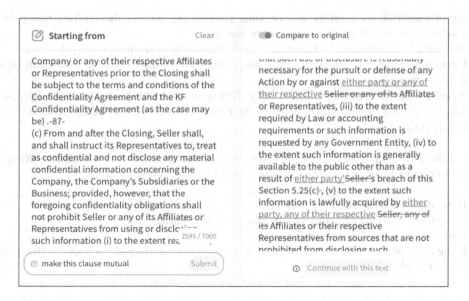

FIGURE 6.12 Example of drafting tool drafting functionalities: "Help me write".

Experiment 2A: Drafting Tool Pilot

The client negotiated with the vendor to conduct a three-week pilot phase to test the drafting tool before it would decide whether to commit to purchasing the tool and making it available to the entire organisation.

This was an intelligent plan – it is well-established in change management literature that pilot projects and experimental phases are helpful instruments, especially in introducing AI systems (Stowasser et al., 2020, p. 4).

Further, while the drafting tool is an off-the-shelf product, the client's interactions with the vendor would not end at the point of purchase. The relationship would be continuously collaborative, as the vendor would have a hand in the implementation and provide technical set-up, user training, and ongoing analytics. As I will elaborate further below, the drafting tool pilot allowed us to assess the vendor's implementation plan and make improvements ahead of the actual implementation.

Finally, the drafting tool, an AI-enabled solution, would be constantly updated. This meant it was extremely important for the client to be assured that the vendor could provide the support required for long-term critical success (Mittal & Solomon, 2023). Therefore, the drafting tool pilot was also a way for us to assess the vendor and the potential future partnership.

I set out below the key details of the drafting tool pilot:

- Ten participants (one partner and nine associates) from two different practice groups were selected by the practice group leaders.
- The vendor conducted two separate training sessions, two weeks apart. The first session covered the basic search functionalities of the drafting tool. The second session covered the drafting tool's more advanced features.
- Pilot users had a total of three weeks to test the drafting tool.

Periodically, through the drafting tool pilot, the vendor sends the pilot users nudge emails, reminding them of key features and including tips and tricks to optimise the use of the drafting tool.

Apart from nudging, we took the following steps to make the entire process as frictionless as possible for the pilot users, i.e., to "reduce sludge" (Dooley, 2021):

- An official kick-off meeting was organised before the start of the drafting tool pilot to communicate timelines and expectations clearly.
- Started a Teams channel to facilitate communication between us and the pilot users.
- Created a matter number where pilot users could put in their time for time spent on the drafting tool pilot.
- To ensure that the pilot users fully tested all functionalities of the drafting tool, I prepared a task checklist to guide them through the process. An extract of the task checklist is in Figure 6.13.

At the end of the three weeks, the pilot users were asked to respond to a feedback survey[8] that I prepared (with input from the client). The survey covered the following areas:

- How often the users used the drafting tool.
- If the users used the drafting tool on a live ongoing matter.
- Ease of use of the drafting tool features.
- If the drafting tool helped the users (a) work more efficiently and (b) improve the quality of their work.

1. Have you completed the task checklists? (Y/N)
2. How often did you use the tool?
   - Once a week
   - A few times a week
   - Once a day
   - A few times a day
   - Other
3. What type of data did you use?
   - In connection with a real ongoing matter
   - On old and/or demonstration documents
   - Other
4. How easy to use were the different features?
   - Search: Keyword search, Drag & Drop, filters
   - Compare: Smart Compare, redline compare
   - AI: Help Me Write, Ask questions about documents, convert clauses to tables, Smart Replace, Translate
   - Curation: Label clauses, save clauses in lists
5. Did the different features help you work more efficiently?
6. Did the different features help you improve the quality of your work?
7. If you did not use some of the tool's features, please state what they are and why you did not
8. Please rate your overall experience with using the tool.
9. Do you find that the updated interface is an improvement to the tool?
10. How could the tool be improved?
    - More features
    - Easier to find features
    - Quicker responses
    - Better search results
    - Larger repository of indexed documents
    - More training on how to use
    - Others
11. Would you continue to use the tool?
12. Purely from an end-user perspective, in your view, should the Client continue with this tool?
13. Do you have any other comments on the tool or the pilot program?

FIGURE 6.13 Drafting tool pilot survey questions.

- Overall experience with using the drafting tool.
- How the drafting tool could be improved.
- If the users would continue to use the drafting tool after the drafting tool pilot.
- Purely from an end user perspective, if the client should continue with the drafting tool.

Results of Drafting Tool Pilot Survey
Overview of results

- All pilot users, except the partner, tested the drafting tool and responded to the survey.
- 50% tested the drafting tool in connection with a real ongoing matter.
- Average rating of 3.5 on overall experience using the drafting tool (with five being the most positive experience).
- 100% said that they would continue to use the drafting tool.
- Most (90%) believed the client should continue with the drafting tool.

What pilot users liked/did not like

- The drafting tool's search functionalities were the most well-received feature – 100% of pilot users felt that the function helped them work more efficiently and improved work quality.

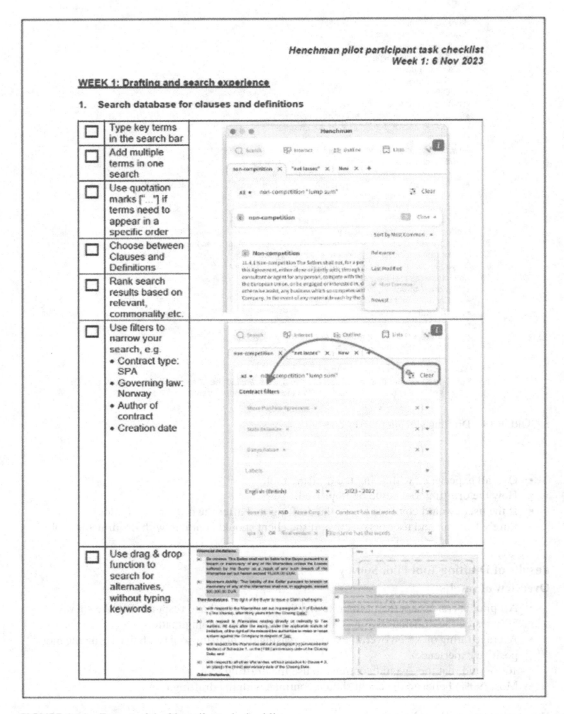

FIGURE 6.14 Extract of drafting pilot task checklist.

- Many were lukewarm about the drafting functionalities – most felt they had no impact on work efficiency and quality, while one respondent felt they decreased work efficiency/quality.
- 22% of respondents requested more training on the drafting tool, with a specific request for training on "specific examples on how to use the tool in different settings (step-by-step guide for different situations)".

Experiment 2B: Drafting Tool Implementation

The client decided to purchase the drafting tool and made it available to 80 other users in the organisation. However, the decision was not made based on the positive results of the drafting tool pilot survey alone. Before the client decided to purchase the drafting tool, it asked the vendor to prepare a draft implementation plan for its consideration for the roll-out of the drafting tool. The client did not just want to purchase a tool that might increase productivity; it also wanted to see that the vendor was committed to a long-term collaborative partnership.

We continued with what we saw worked during the drafting tool pilot:

- The training was delivered across two sessions, one for the basic search functionalities and another for the more advanced features. Research has shown that an overload of information causes cognitive overload on users, preventing them from coping with such a large amount of information (Cowan, 2001, p. 87). Therefore, we liked this phased approach, as it did not inundate the users with too much information at once, bearing in mind that this was a completely new tool that most would be unfamiliar with.
- Regarding communications, we saw that the regular follow-up email nudges effectively reminded the users to use the drafting tool. However, in the drafting tool implementation, we decided that all communications should be internal, i.e., from the client's digital department (as opposed to being sent by the vendor).

Working closely with the vendor, and through iterating the draft implementation plan that the vendor provided based on the lessons learnt from the drafting tool pilot, we tweaked the aspects that we thought could be improved on:

- In the drafting tool pilot, the vendor conducted the training to run through the functionalities of the drafting tool, for example, "the next thing that the tool can do is this". However, based on the feedback from the pilot users, we redesigned the training material to show the user journey of the drafting tool, with a step-by-step guide using a document that the client had worked on in a recent matter. An extract of the training material is in Figure 6.15.

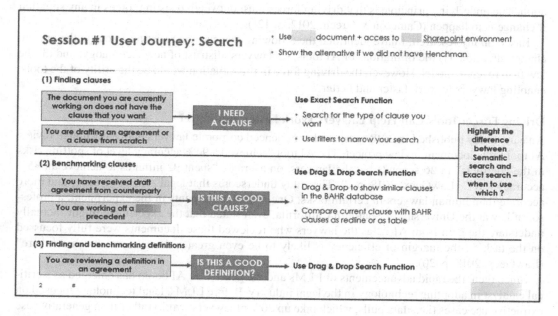

FIGURE 6.15 Extract of drafting tool implementation training material.

- The pilot users also provided feedback that the drafting tool's search function was the best feature. Based on the feedback, we asked the vendor to focus on and emphasise the search function during the first training session. We were trying to encourage the "stickiness" of the tool and to get the users to start with the best feature in the hope that they will interact with the drafting tool regularly. Also, instead of getting users to try the entire tool from the beginning, we wanted to break down the task into smaller, more manageable components, which Heath and Heath call shrinking the change (2010, p. 134). By shrinking the change, the investment required from the users is limited, and the change is small enough such that people "can't help but score a victory" (Heath & Heath, 2010, p. 134).
- While all training for the drafting tool pilot was done virtually, the vendor conducted subsequent training in person at the client's office.
- We appointed three project assistants (members of the back-office staff) as "super users" of the drafting tool. The super users received targeted training ahead of the other users and were equipped to answer any questions that the users might have.

At the point of writing this chapter, we have just concluded the first training session for the first 30 users. The initial responses during the training were very positive. The redesigned "user journey" training approach effectively showcased the tool's potential and demonstrated its use cases. Based on the latest analytics provided by the vendor, following the first training session there is a 64% activation rate (i.e., users logging in and activating their account), with a 43.1% search conversion rate.

MAPPING CHANGE MANAGEMENT ASPECTS, USING LEWIN'S MODEL AS FRAMEWORK

In this section, I will describe the steps taken to implement the chatbot and the drafting tool within the framework of Lewin's three-step model.

APPLYING LEWIN'S FIELD THEORY ANALYSIS TO THE CASE AT HAND

As stated above, Lewin's field theory examines the driving and resisting forces in any change situation. The underlying principle is that driving forces must outweigh resisting forces in any situation if change is to happen (Cameron & Green, 2012, p. 121).

Based on my research, I have identified the following areas as the resisting forces in the present change situation: (1) the limitations of AI tools; (2) lawyers' distrust of new technology; and (3) the law firm revenue model. However, the driving force in this situation would be the utility of the tools, enabling lawyers to work faster and better.

Driving Force: Tools Can Help Lawyers Work Faster and Better

An early study published in 2018 compared experienced corporate lawyers with a document review AI tool for a document review project. The AI tool "achieved a 94% accuracy level of spotting risks in the contracts" in 26 seconds but the lawyers, on average, "spent 92 minutes to achieve an 85% accuracy level" (LawGeex, 2018, p. 2). The study underscores that legal AI can be faster and more accurate than human lawyers in certain tasks. One of the participants in the experiment, a professor of law at the University of Southern California, even stated that the experiment might "actually understate the gain from AI … as the lawyers who reviewed these documents were fully focussed on the task … the margin of efficiency is likely to be even greater than the results shown here" (LawGeex, 2018, p. 20).

Since then, the rapid advancements of LLMs and the generative AI boom has emerged as a critical catalyst in adopting technology in the legal industry. Before LLMs, legal technology focused on extractive use cases (i.e., data pulls, which take up 20% of lawyers' work) rather than generative use cases – drafting and summarisation, which take up 80% of lawyers' work (Han, 2023). AI-powered software offers a tenfold improvement in the day-to-day experience of a lawyer, as compared to a twofold improvement before, saving lawyers ten time the time and costs, thus improving efficiency drastically (Han, 2023). There are multiple potential use cases for generative AI in the legal context, from textual content generation – creating content based on input such as drafting legal contracts and classification – assigning a category to a given input such as tagging data to create a smart searchable knowledge management database and extraction – deriving specific information from a given input, for example in legal due diligence.

The qualitative feedback from the chatbot survey and the draft tool pilot shows that many legal professionals think AI tools can help them work better and faster. For example, 100% of the pilot users felt that the drafting tool's search function helped them work more efficiently and improved work quality.

Resisting Force: Limitations of AI Tools

While the new technology has the potential to transform the nature of legal work, there are notable limitations to what these tools can do. One obvious problem would be hallucinations, or the tendency of LLMs (tone of the underlying technologies that ChatGPT is based on) to produce content that deviates from actual legal facts or well-established legal principles and precedents (Dahl et al., 2024a). A recent Stanford study found that hallucination rates range from 69% to 88% in response to specific legal queries for LLMs such as ChatGPT 3.5 and Llama 2 (Dahl et al., 2024b, p. 2). In an infamous example, in May 2023, a Manhattan lawyer submitted a legal brief generated largely by ChatGPT, littered with "bogus judicial decisions … bogus quotes and bogus internal decisions" (Weiser, 2023).

The qualitative feedback from my experiments also reveals the limitations of the technology. For example, a common complaint in the chatbot survey was that the chatbot would skip translating parts of the document. Even though the pilot users found the search function very useful with the drafting tool, there was still feedback that the tool showed too many irrelevant search results, sometimes making it "hard to see the forest for the trees".

Hallucinations and made-up facts are problematic in any industry but are even more amplified in the legal context – what some commentators refer to as an "asymmetric downside for errors in legal work" (Han, 2023). Predictability and consistency are commonly identified as core components of the rule of law, with aims to reduce arbitrariness through concepts like the doctrine of precedent and principles of natural justice (Tang & Foley, 2014, p. 1198). In the realm of legal work, one of the critical aspects of the profession is accuracy, with legal work demanding precision, meticulous research, and thorough analysis of statutes and regulations. An overriding adherence to source text (such as case law) means that unfaithful or imprecise interpretations of the law can lead to nonsensical, or even harmful and inaccurate legal advice or decisions (Dahl et al., 2024, p. 1). For example, in a common law system where the principle of *stare decisis* or precedence requires attachment to the "chain" of historical case law, any misstatement of the binding content of that law would make an LLM lose any professional or analytical utility (Dahl et al., 2024, p. 4). Due to these features of legal practice, the legal industry is exceptionally intolerant of any technological errors or shortfalls. Law firms like the client, which pride themselves on their ability to provide the highest level of professional services, cannot afford to have hallucinations contaminate legal advice.

Resisting Force: Distrust of New Technology and Resistance to Change

Technological change is more than just incorporating a new tool or piece of software. It also entails changing the behaviour of employees who can be content with a given way of doing things and resistant to changing what they are used to (Delaney & D'Agostino, 2015, p. 9).

Professional service industries, such as the legal industry, are exceptionally resistant to change (Kronblad et al., 2023, p. 218).[9] On an individual level, lawyers are distrusting and prudent by nature, generally sceptical and resistant to adopting new technology (Han, 2023). A prudent perspective, which requires caution, scepticism, and "reality-appreciation", is, in fact, an asset in the legal profession, as it enables a good lawyer to see pitfalls that might conceivably happen in, for example, a transaction (Seligman et al., 2005, p. 5). This extremely critical lens is applied when using new tools – lawyers will always be on the lookout for errors or shortfalls in the technology. Once an error is spotted, it could turn the lawyers off using the tool altogether. The probabilistic nature of generative AI and the lack of explanation and uncertainty around its output (especially where deterministic answers are needed) compounds the distrust and resistance to the technology (Bendor-Samuel, 2023), especially when working with legal professionals who are used to very "strong black and white areas" (Mittal & Solomon, 2023).

The findings of my qualitative research mirror the general sentiment of distrust and scepticism. For example, in the chatbot survey, a respondent who had not used the chatbot said that the reason for not doing so was that the Chatbot "would not help me with my work", an interesting response because it appears to prejudge the functionality of the chatbot without any prior testing or experience. In the drafting tool pilot, a pilot user remarked that they did not use the AI drafting functions because "it is easier just to draft the clause myself".

Of course, this distrust and risk aversion are closely linked to the limitations of current AI tools. As highlighted in the section above, generative AI has obvious risks. In this context, scepticism is well-founded, and it would be unwise and wildly irresponsible to ignore the technology's limitations.

In practical terms, it is all the more important to manage the change closely and carefully. For example, tool and vendor selection processes should be stringent, given that the user requirements for the tools are so high. More time and effort must be invested in the implementation process. Clear communication and extensive education are paramount, especially when dealing with risk-averse lawyers, who need to be equipped with knowledge before using the tools. This is backed up by the findings of the chatbot survey: of those who had not tried out the chatbot, 100% said that more training on the chatbot would encourage them to use it. Even though we had published comprehensive training material in various forms, 60% of respondents (who had used the chatbot) requested even more training.

Nonetheless, a recent Thomson Reuters survey found that the legal industry's attitude towards new technology is shifting, with more than 50% of the law firm leaders and legal professionals surveyed believing that generative AI should be applied to legal work (2023, p.7). Anecdotally, when I spoke to one of the pilot users, he said: "[the client] needs to start using AI; otherwise, we will get left behind".

Resisting Force: Law Firm Billable Hour Revenue Model

The traditional law firm revenue model (and the client's) is based on billable hours. This means that lawyers generate income for the firm by billing clients a fixed hourly rate for their work. In many law firms, individual lawyer performance is measured by the number of hours billed, with incentives contingent on meeting billable hour targets.

This is a structural barrier to change – lawyers are incentivised to exploit the current revenue model and bill many hours instead of working faster or smarter (Kronblad et al., 2023, p. 228). According to a 2021 survey of 200 lawyers in ten leading law firms in the United Kingdom, 40% of trainees and associates said that a concern over loss of billable hours was holding them back from adopting new technology tools (Legatics, 2021, p. 9). The study found that given the lack of trust, the motivation to use technology purely to experience the benefits is weak, so internal motivations like career progression become more important, especially to junior lawyers (Legatics, 2021, p. 9).

However, change may be impending. As the technology gets more advanced and specialised, the inherent incentive to use the tools grows stronger (i.e., the driving force of utility increases). After all, no lawyer enjoys mundane work such as document review or discovery and would rather be liberated from these routine tasks to focus more on complex, high-value work, which translates to significant growth opportunities for the law firm (Deloitte, 2023, p. 4). Also, historically, the primary driving factor in getting lawyers to change their behaviour is for their end clients to put pressure on their practice (Han, 2023). As clients expect lawyers to work more efficiently due to the rise of LLMs, this will put similar pressure on lawyers to act more efficiently to retain their clients (Han, 2023).

Step 1: Unfreeze

As mentioned above, Lewin believed that the stability of human behaviour was based on a quasi-stationary equilibrium supported by a complex field of driving and restricting forces. According to Lewin, the equilibrium must be destabilised (unfrozen) before old behaviour can be discarded (unlearnt) and new behaviour successfully adopted (Burnes, 2004, p. 985). This is the first step in the three-step model. In this section, I will explain how these activities in my experiments reduced the resisting forces and strengthened the directing forces in the present situation.

Of the limitations of AI tools, we could only do a little to improve the capabilities of the current technology. However, the way to mitigate this was to communicate the limitations to the users and reduce the impact of the tools' limitations on user adoption. We did not want to end up with a situation where the users expected a super-tool, only to end up dissatisfied with the technology and less likely to integrate the tool into their day-to-day work. Therefore, the communication principle that we adopted was "under-promise, over-deliver". For example, ensuring that users were aware of the potential of the chatbot hallucinations and guiding them on specific scenarios when it would not be appropriate to rely on the chatbot (see Figures 6.3 and 6.4 – Extracts of Chatbot Communication Material). With the drafting tool, we ensured that the vendor clearly emphasised during the training sessions the shortfalls of the drafting tool. For example, only documents within a certain timeframe were searchable, i.e., a limited data set. Ultimately, since hallucinations of some kind are generally inevitable when using generative AI (Dahl et al., 2024, p. 14) and any AI legal tool will have its shortcomings, the best approach to take is one of transparency and honesty (Booz Allen Hamilton, 2021). The communication and training also served to remind the users to remain vigilant in their use of the AI tools and verify the accuracy and quality of the outputs, mitigating the inherent risks when using AI.

Clear communication and well-designed training also emphasised the utility of the tools (driving force) and reduced the distrust surrounding the technology. If the users do not have a clear vision of how the tools can help them, this would begrudge the effort required to adapt. For example, the step-by-step "user journey" training delivered during the drafting tool implementation showcased the full potential of the drafting tool. Using an actual contract on a recent matter that the client worked on allowed the users to see how the drafting tool worked in real practical terms, i.e., "what's in it for me". Since all behaviour is self-centred, people will only accept changes that they understand and believe to be beneficial to themselves (Delaney & D'Agostino, 2015, p. 12). Therefore, answering and demonstrating the "what's in it for me" is critical to any change management process (Delaney & D'Agostino, 2015, p. 15).

Step 2: Move

The second stage of Lewin's model involves moving the organisation to a new equilibrium level or desired behaviour. Due to the relatively short span of this project and the client's internal timelines, I do not believe that I reached this stage of Lewin's model in this project.

However, I will comment on the law firm's billable hour model. Change management literature clearly states that to move to a new equilibrium effectively, a reward system must be in place to incentivise people to change their behaviour (Richter & Sinha, 2020). As discussed above, due to its revenue model, there is no such reward system in the law firm context.

Short of overhauling the traditional law firm model, some ways to incentivise lawyers to explore and adopt technology might be to provide dedicated "innovation hours", which would count towards billable hour targets and allow lawyers to view the time spent learning new skills as beneficial. This is an extension of what we did in the drafting tool pilot, where we provided a matter number where the pilot users could allot their time spent on the drafting tool pilot.

Separately, in light of the risks of using generative AI, any push to use AI tools has to be managed carefully. Lawyers should be educated on when it is appropriate or not to use AI in their work. Deloitte has designed the "digital artifact generation/validation" method to help determine whether AI should be used in legal work (2023 p. 11). At the core of this method are two critical elements: the human effort required to complete the task without AI and the necessary effort to validate or fact-check the output. Further, part of practising law is about understanding the client's particular circumstances and the case context, so raw output from the tools will rarely be optimal (Stokel-Walker, 2023). This means that AI tools are not a substitute for actual human lawyers, and any application must always feature "human-in-the-loop" safeguards, whether a professor, supervisor, or partner (Katz et al., 2024, p. 11).

STEP 3: REFREEZE

The refreezing process describes the changes necessary "to bring about the permanence of the new situation" (Lewin, 1951, p. 48). It seeks to stabilise and institutionalise new behaviours at a "quasi-stationary equilibrium" to ensure they are relatively safe from regression (Burnes, 2020, p. 50).

As mentioned above, my experiments for this chapter did not advance beyond the "unfreeze" and "move" stages of Lewin's three-step model. As such, I am unable to make any extensive comments here.

However, an interesting observation about the chatbot can be made here. Apart from a communication emailer in November 2023 announcing the upgrade, the client still needs to remind the employees to use the chatbot consistently. Compared to the drafting tool, much less effort had been made to encourage the use of the chatbot. Nonetheless, there has been a consistent, sustained use of the chatbot since its launch in September 2023, as seen in the chatbot user metrics (see Figures 6.1 and 6.2). Apart from a dip in December 2023, which can be attributed to the holidays, the number of users has remained more or less constant since the launch.

Since user behaviour seems to have stabilised, the chatbot implementation might be at the "refreeze" stage of Lewin's model. Interestingly, the chatbot implementation reached this stage quickly, with relatively minimal intervention from the client. A possible explanation might be that the chatbot is a technology familiar to most users. Natural language is, for many, the default mode of interaction online and the preferred interface for interacting with digital services (Følstad & Brandtzag, 2017, p. 38). Most people know how to use Google and do so daily. Effectively, this means that the behaviour change required on the part of the users is minimal (as compared to learning a new way of searching for information, like in the case of the drafting tool). Change is thus facilitated more quickly because the resisting forces are minimised.

Therefore, in the client's case, implementing the chatbot is low-hanging fruit in familiarising employees with new tools and AI. This might be a potential area for future research – for organisations keen on exploring the potential of generative AI, an internal chatbot (for example, in the HR function) might be a good starting point.

KEY TAKEAWAYS

LEARNINGS

The following practical learnings can be distilled:

- Always conduct a pilot or experimental phase, in which experience can be gathered before a comprehensive introduction to the rest of the organisation and possible need for adaption (in terms of implementation, training, or communication) can be determined.

- Shrink the change: Break up the desired behaviour into smaller components, such as smaller tasks (in the case of the drafting tool), or introduce some familiarity into the new situation, like in the case of the chatbot, where users were already comfortable with natural language.
- Transparent clear communication is key – "tell it like it is" or even "under-promise, over-deliver". Because AI-enabled solutions are not deterministic, showcasing examples of their weaknesses and where they may not meet requirements helps to build trust and mitigate the impact of the limitations of the tools on user adoption.
- Apply a human-centred design mindset. When we redesigned the drafting tool training material using the user journey approach (i.e., focusing on the "jobs to be done" and not just how "cool" the tech was, which was the vendor's approach), we witnessed how much more receptive the users were to the drafting tool. Remember that you are dealing with humans and human behaviour and that you need to channel empathy and understanding to meet the needs of the end user.

LIMITATIONS OF EXPERIMENTS

The parameters of this chapter were defined by the parameters set by the client in the project. While working on a real project was valuable to seeing change management in a real-life scenario, it meant that realistic and pragmatic restrictions had to be set for its success. There were multiple instances where the focus of the client's project did not completely align with the information I would need for this chapter. For example:

- Budgetary constraints prevented me from testing as comprehensively as I would have wished to obtain as much data as possible. It would have been useful to have the chatbot user metrics I had first requested (e.g., number of unique user interactions, types of requests made). This would have given me a more accurate picture of how many people were using the chatbot and a better understanding of what people were using it for.
- More detailed user metrics might also have allowed me to spot some patterns, such as a spike in usage following a communication emailer. However, it is hard to identify any correlation and causation with the data I had on hand.
- Testing relied on employee participation, and there were varying degrees of engagement. For example, it was not feasible to make it mandatory for everyone to complete the chatbot survey, and since it was optional, there was a potential for voluntary response bias.
- The short timeframe of this project and client's internal timelines meant that we had not fully completed implementation of the tools at the time of writing this chapter. Frequent analysis of user adoption metrics and continuous solicitation of feedback from the users must be sought to ascertain if the implementation was successful or not. After all, change is a capability that happens continuously in an agile fashion and evolves.

FUTURE WORK

Building on what we have done so far, some potential areas for future work are:

- Explore the social aspect of driving change, for example, appointing early adopters as "change champions" and using peer-group-informed messaging (Mohahan et al., 2016, pp. 182–183). We did not manage to do this in the experiments.
- Build AI-related competencies across the organisation to build trust in the technology and user confidence. For example, provide general training on tech literacy, more information on generative AI (news, latest updates), and training on prompting.

CONCLUSION

ETHICAL ISSUES

As with any discussion around AI and technology, one must carefully consider the ethical considerations surrounding the procurement and use of legal technology. The law is a profession, so lawyers must comply with rules of professional conduct and ethical duties. In Norway, the Norwegian Bar Association's Code of Conduct applies. These rules are intended to protect the public and maintain the integrity of the legal profession. With the proliferation of legal technology, law firms and lawyers must adhere to these professional duties while deploying new technologies.

For example, under Rule 2.3 of the Code of Conduct, the lawyer's duty to treat information confidentially is "a basic and cardinal right and obligation for the lawyer". These requirements played a big part in the client's decision to develop its chatbot interface set-up with the client's cloud environment. Other measures that the client took and is taking include ensuring that all data submitted into the AI tools will not be used as training data (by the vendor or Microsoft, for example) and issuing guidelines (and constant reminders of these guidelines) prohibiting the input of sensitive client personal data. From my observation, the client takes these confidentiality and privacy obligations very seriously. In separate dealings with a third-party vendor (not *the* vendor), the client considered the fact that the vendor might use the client's data inputs as part of the vendor's training model as a dealbreaker and did not continue further commercial negotiations until the vendor provided more clarity on the matter.

Because the technology is so new and AI-driven tools are advancing so fast, the laws have yet to catch up with the technology. There are currently many open questions to which no one has the answers. For example, is it lawful to use a piece of software built on a mass data scraping foundation (Stokel-Walker, 2023)? Another example: Rule 3.2 of the Code of Conduct also requires lawyers to avoid conflicts of interest and exercise professional independence, i.e., act and exercise professional judgement free from external pressures. Suppose lawyers overly rely on technology or AI programmes, for example, to search for relevant documents in an investigation. In that case, it is arguable that the lawyers are not exercising independent professional judgement. This is complicated by the fact that lawyers are unlikely to be able to evaluate the functioning of the technology independently due to their lack of technical knowledge, or a lack of transparency in an AI system's functioning, such that no one (not even the developers) may fully understand how the outputs were generated (Legg & Bell, 2022). Yet another interesting question – if AI can eventually perform a task better than humans, is it negligent not to use the technology? In the legal context, lawyers are expected to promote the interests of the client to the best of their ability (see for example, Rule 1.2 of the Code of Conduct), and if a lawyer makes a blunder that could have been avoided by incorporating technology into the legal work (and a reasonable lawyer would have used the technology), would it be malpractice to use the tech?

With the proposed Artificial Intelligence Act, the European Union is working to catch up with the technology. The proposed Act would introduce a framework to regulate the providers of AI systems and entities using AI and aims to ensure that fundamental rights, democracy, the rule of law, and environmental sustainability are protected from high-risk AI (European Parliament, 2023). Until then, it is up to organisations and individuals to strike a balance and rely on their internal ethical compasses.

WHAT THE LAW FIRM OF THE FUTURE WOULD LOOK LIKE

With the legal tech boom, law firms face additional pressures from alternative legal service providers (non-law firm providers that perform legal work at a lower cost) (Gartner, n.d.) and clients demanding better, faster, and less expensive solutions (Henderson, 2014).

Hopefully, the staggering potential time savings derived from LLMs and AI-driven legal tech will finally prompt much-needed business model changes in the legal industry (Han, 2023). Lawyers must understand that their value proposition can extend beyond knowledge and expertise in the law and that clients require more holistic solutions. Further, because work can be done in less time but with more value, the billable hour revenue model will no longer be viable, and law firms should price their services based on the value delivered instead of the time taken to carry out the task. A revenue model that bases its fees on outcomes, such as fixed fees for an engagement or the achievement of designated milestones, would be more appropriate. Changing the revenue model from billable hours to fixed pricing will have significant implications for the law firm's incentive structure – for example, it will motivate lawyers to be more efficient, either by adopting new technology or otherwise (Boston Consulting Group, 2016, p. 9).

Finally, the law firm will need to alter its organisational model. The traditional partnership model is one of a pyramid: few partners at the top and many junior lawyers and associates at the bottom. However, this model has misled customers' value propositions (Cohen, 2018). The model worked when law firms sold the single product of legal expertise (which firms had a monopoly on) and churned out legal work on high billable hourly rates (Cohen, 2018). A more suitable organisational structure would be shaped more like a rocket than a pyramid, with fewer junior lawyers and associates for each partner. Legal tech solutions to handle standard low-skill tasks can reduce the ratio of junior lawyers to partners by up to 25% (Boston Consulting Group, 2016, p. 11). This structure will also be supplemented by other employees who are not lawyers, for example, project managers, data analysts, and legal technicians.

NOTES

1 I was not involved at any stage of the Feasibility Study.
2 Due to confidentiality obligations, I am unable to reproduce the Feasibility Study in its entirety in this chapter.
3 These are generative AI tools that specifically cater to lawyers, for example, by being trained on legal content, and specifically designed to support the provision of legal services. Examples of specialist legal generative AI tools include tools used to generate and negotiate legal documents (for example, Luminance: https://www.luminance.com/) and analysis and drafting of contracts (for example, Spellbook: https://www.spellbook.legal/).
4 I say "group, organisational, and societal level" because although the three-step model is Lewin's key contribution to organisation change, he was not thinking only about organisational issues when he developed the three-step model (Burnes, 2004, p. 985). For most of his professional life, Lewin was best known as a renowned child psychologist and sought to apply his research to resolving social conflict such as, for example, religious and racial intolerance (Burnes, 2020, p. 33)
5 The survey was conducted on Questback, a survey solution and feedback platform.
6 I was not involved in the decision-making surrounding the Vendor and/or the Drafting Tool.
7 More information on the Drafting Tool and a full list of its functionalities may be found on the Vendor's website: https://henchman.io/)
8 This survey was also conducted on Questback.
9 When legal futurist Richard Susskind predicted in 1996 that email would someday replace the telephone as the dominant communication for lawyers, senior officials in the Law Society of England and Wales said that he should not be allowed to speak in public, as *"prudent and ethical lawyers would never succumb to such an insecure method of client communications"* (Henderson, 2014, p. 1123). Yet today, lawyers, as with other professionals, are chained to their computer devices and smartphones.

REFERENCES

Bakari, H., Hunjra, A. I. H., & Niazi, G. S. K. (2017). How does authentic leadership influence planned organizational change? The role of employees' perceptions: Integration of theory of planned behavior and Lewin's three step model. *Journal of Change Management*, 17(2), 155–187.

Bendor-Samuel, P. (2023). Key issues affecting the effectiveness of generative AI. *Forbes*. Retrieved January 25, 2024, from https://www.forbes.com/sites/peterbendorsamuel/2023/12/05/key-issues-affecting-the-effectiveness-of-generative-ai/

Booz Allen Hamilton. (2021). *Navigating AI change management like a boss*. Retrieved January 25, 2024, from https://www.boozallen.com/insights/ai/change-management-for-artificial-intelligence-adoption.html

Boston Consulting Group and Bucerius Law School. (2016). How legal technology will change the business of law. Retrieved January 25, 2024, from http://media-publications.bcg.com/How-legal-tech-will-change-business-of-low.pdf

Brynjolfsson, E., & McAfee, A. (2017) The business of artificial intelligence. *Harvard Business Review*. Retrieved January 25, 2024, from https://hbr.org/2017/07/the-business-of-artificial-intelligence

Burnes, B. (2004). Kurt Lewin and the planned approach to change: A re-appraisal. *Journal of Management Studies*, 41, 977–1002.

Burnes, B. (2020). The origins of Lewin's three-step model of change. *The Journal of Applied Behavioral Science*, 56(1), 32–59.

Cameron, E., & Green, M. (2012) *Making sense of change management* (3rd ed.). Kogan Page Limited.

Cohen, M. A. (2018, November 8). New business models – Not technology – Will transform the legal industry. *Forbes*. Retrieved January 25, 2024, from https://www.forbes.com/sites/markcohen1/2018/11/08/new-business-models-not-technology-will-transform-the-legal-industry/?sh=32732aab18cc

Cowan, N. (2001). The magical number 4 in short-term memory: A reconsideration of mental storage capacity. *Behavioral and Brain Sciences*, 24(1), 87–114.

Cummings, S., Bridgman, T., & Brown, K. G. (2015). Unfreezing change as three steps: Rethinking Kurt Lewin's legacy for change management. *Human Relations*, 69(1), 33–60.

Dahl, M., Magesh, V., Suzgun, M., & Ho, D. E. (2024a). Hallucinating law: Legal mistakes with large language models are pervasive. Retrieved January 25, 2024, from https://hai.stanford.edu/news/hallucinating-law-legal-mistakes-large-language-models-are-pervasive

Dahl, M., Magesh, V., Suzgun, M., & Ho, D. E. (2024b). Large legal fictions: Profiling legal hallucinations in large language models. *arXiv preprint arXiv:2401.01301*.

Delaney, R., & D'Agostino, R. (2015). The challenges of integrating new technology into an organization. *Mathematics and Computer Science Capstones*, 25, 1–37.

Deloitte. (2017). Bullish on the business value of cognitive. *The 2017 Deloitte State of Cognitive Survey*. Retrieved January 25, 2024, from https://www2.deloitte.com/content/dam/Deloitte/us/Documents/deloitte-analytics/us-da-2017-deloitte-state-of-cognitive-survey.pdf

Deloitte. (2023). Generative AI: A guide for corporate legal departments. Retrieved January 25, 2024, from https://www2.deloitte.com/content/dam/Deloitte/nl/Documents/legal/deloitte-nl-legal-generative-ai-guide-jun23.pdf

Deloitte. (2023). *'Generative AI is all the rage'*. Available at: https://www2.deloitte.com/content/dam/Deloitte/us/Documents/deloitte-analytics/us-ai-institute-gen-ai-for-enterprises.pdf (Accessed 25 January 2024).

Dooley, R. (2021). You can't nudge if you've got sludge. *Forbes*. Retrieved January 25, 2024, from https://www.forbes.com/sites/rogerdooley/2021/09/29/you-cant-nudge-if-youve-got-sludge/

European Parliament. (2023). *Artificial Intelligence Act: Deal on comprehensive rules for trustworthy AI*. Retrieved January 25, 2024, from https://www.europarl.europa.eu/news/en/press-room/20231206IPR15699/artificial-intelligence-act-deal-on-comprehensive-rules-for-trustworthy-ai

Følstad, A., & Brandtzag, P. B. (2017). Chatbots and the new world of HCI. *Interactions*, 24(4), 38–42.

Gartner. (no date). Alternative legal service providers. Retrieved January 25, 2024, from https://www.gartner.com/reviews/market/alternative-legal-service-providers-alsps

Han, L. (2023). Legaltech x AI: The Lightspeed view. Retrieved January 25, 2024, from https://lsvp.com/legaltech-x-ai-the-lightspeed-view/

Heath, C., & Heath, D. (2010). *Switch: How to change things when change is hard* (1st ed.). Broadway Books.

Henderson, W. D. (2014). Letting go of old ideas. *Michigan Law Review*, 114(6), 1111–1131.

Hiatt, J. M., & Creasey, T. J. (2012). *Change management: The people side of change* (2nd ed.). Prosci Inc.

Katz, D. M., Bommarito, M. J., Gao, S., & Arredondo, P. (2024). GPT-4 passes the bar exam. Retrieved January 25, 2024, from https://papers.ssrn.com/sol3/papers.cfm?abstract_id=4389233

Kronblad, C., Pregmark, J. E., & Berggren, R. (2023). Difficulties to digitalize: Ambidexterity challenges in law firms. *Journal of Service Theory and Practice*, 33(2), 217–236.

LawGeex. (2018). Comparing the performance of artificial intelligence to human lawyers in the review of standard business contracts. Retrieved January 25, 2024, from https://images.law.com/contrib/content/uploads/documents/397/5408/lawgeex.pdf

Legatics. (2021). Barriers to legal technology adoption. Retrieved January 25, 2024, from https://www.legatics.com/barriers-to-legal-tech-adoption/

Legg, M., & Bell, F. (2022). Artificial intelligence and solicitors' ethical duties. Retrieved May 17, 2023, from https://lsj.com.au/articles/artificial-intelligence-and-solicitors-ethical-duties/

Lewin, K. (1951). *Field theory in social science* (1st ed.). Harper and Row.

Mittal, A., & Solomon, J. (2023). Change management is hard, but even harder with AI. Retrieved January 25, 2024, from https://www.wolterskluwer.com/en/expert-insights/podcast-legal-leaders-exchange-episode-16

Mohahan, K., Murphy, T., & Johnson, M. (2016). Humanizing change. Developing more effective change management strategies. *Deloitte Review*. Retrieved January 25, 2024, from https://www2.deloitte.com/tr/en/pages/human-capital/articles/developing-more-effective-change-management-strategies.html

Morris, C. (2023). A major international law firm is using an AI chatbot to help lawyers draft contracts: 'it's saving time at all levels. *Fortune*. Retrieved January 25, 2024, from https://fortune.com/2023/02/15/a-i-chatbot-law-firm-contracts-allen-and-overy/

OpenAI. (2022). Introducing ChatGPT. Retrieved January 25, 2024, from https://openai.com/blog/chatgpt https://openai.com/blog/chatgpt

Perlman, A. (2023). The implications of ChatGPT for legal services and society. *Suffolk University Law School Research Paper* No. 22–14.

Plastino, E., & Purdy, M. (2018). Game changing value from artificial intelligence: Eight strategies. *Strategy & Leadership*, *46*(1), 16–22.

Richter, F., & Sinha, G. (2020). Why do your employees resist new tech? *Harvard Business Review*. Retrieved January 25, 2024, from https://hbr.org/2020/08/why-do-your-employees-resist-new-tech

Seligman, M. E., Verkuil, P. R., & Kang, T. H. (2005). Why lawyers are unhappy. *Deakin Law Review*, *10*(1), 49–66.

Stokel-Walker, C. (2023). Generative AI is coming for the lawyers. Retrieved January 25, 2024, from https://www.wired.co.uk/article/generative-ai-is-coming-for-the-lawyers

Stowasser, S., Suchy, O., Huchler, N., Müller, N., Peissner, M., Stich, A., Vögel, H., & Werne, J. (2020). Introduction of AI systems in companies – Design approaches for change management. [White paper]. *Platform Lernende Systeme*. Retrieved January 25, 2024, from https://www.plattform-lernende-systeme.de/publications.html

Tang, S., & Foley, T. (2014). The practice of law and the intolerance of certainty. *University of New South Wales Law Journal*, *37*(3), 1198–1225.

Thomson Reuters Institute. (2023). ChatGPT and generative AI within law firms. Retrieved January 25, 2024, from https://www.thomsonreuters.com/en-us/posts/wp-content/uploads/sites/20/2023/04/2023-Chat-GPT-Generative-AI-in-Law-Firms.pdf

Weiser, B. (2023, May 27). Here's what happens when you lawyer uses ChatGPT. *The New York Times*. Retrieved January 25, 2024, from https://www.nytimes.com/2023/05/27/nyregion/avianca-airline-lawsuit-chatgpt.html

7 Enabling Leaders to Cultivate Awareness and Manage Unconscious Bias Continuously

Cynthia Mak

INTRODUCTION

Bias is inherent in everyone, prompting organisations to invest significant time and resources to improve workplace representation and foster fairness. In the United States, a staggering US$8 billion is spent annually on diversity training (Kirkland & Bohnet, 2017). Yet, some studies suggest that unconscious bias training (UBT) needs to be more effective, and it has the potential to backfire (Gino & Coffman, 2021). This raises critical questions: are organisations implementing UBT effectively? What role does a leader play in influencing its outcomes? How can a leader help?

Given the crucial role of leaders and the limited openness of organisations in sharing their data related to unconscious bias efforts, this chapter focuses on the leader. Leaders are often challenged to navigate complex decisions while maintaining an inclusive and equitable organisational environment. Recognising leaders' pivotal role and the evolving nature of unconscious bias, this research project focuses on "How might we enable leaders to cultivate self-awareness and manage unconscious bias continuously?"

This study explores the strengths and weaknesses of UBT in organisations and leaders' practices in managing their unconscious bias. Drawing insights from the primary and secondary research, the proposed solution involves a design strategy delivered through an AI-powered app. As we move towards a BANI world, which stands for brittle, anxious, non-linear, and incomprehensible (Kraaijenbrink, 2022) world, Adam Grant (2021) advocates for the ability to rethink and unlearn. With this design strategy as the fundamental approach, a checklist of questions was developed in the prototype to get the leaders started on the reflection process. To support a leader *continuously*, an app provides prompts to nudge the leader consistently. As unconscious bias evolves, AI capabilities will be leveraged to scan and personalise prompts and activities for the leader.

The feedback received from the prototype is promising. The leaders think it can help them cultivate awareness and manage unconscious bias, but more work needs to be done. Working with organisations such as Mathison will be advantageous from the viability perspective. This prototype better augments existing UBT practices. The advantage of this prototype is that it has the potential to extend learning beyond unconscious bias towards becoming a more effective leader.

Note that in this chapter implicit bias is used interchangeably with unconscious bias.

METHODOLOGY

This research project was conducted over three to four months on unconscious bias from an organisational and personal perspective. Both perspectives are important, and I hypothesise that they are fundamentally interconnected. I used a combination of primary and secondary research to gather data and insights about unconscious bias. For my secondary research, I used various academic and professional sources.

DOI: 10.1201/9781003469551-7

TABLE 7.1
List of Interviewees

Name	Title	Age	Industry	Profit/ non-Profit	Experience	Team Size
Leader A	Former Director of Supply Chain, APAC	60–70	FMCG	Profit	26	350
Leader B	Vice President, International Business Development	50–60	eLearning (EduTech)	Profit	40	45–50
Leader C	Previously VP, HR Business Partner for APAC	60–70	Publishing	Profit	40+	15
Leader D	Consultant	40–50	Advertising	Profit	20	6
Leader E	Senior counsellor	50–60	Social service agency	Non-profit	30	10
Leader F	Former Vice-President, Sales, APAC	50–60	Technology	Profit	NA	NA
Leader G	DEI for APAC	40–50	Consulting	Profit	25	35
Leader H	Consultant, Global	50–60	Tech Startup	Profit	30	300+

Rather than surveying my primary research, I have opted for in-depth interviews with eight leaders (See Table 7.1). These leaders have global and local team management experience spanning different industries, from profit and non-profit to government, corporate, and start-up organisations. Given the diverse types of organisations, I wanted to understand if there is a difference in how a leader approaches unconscious bias. These qualitative insights were extracted and organised into an affinity diagram for further analysis.

Building on the research, I incorporated a design strategy rethinking by Adam Grant (2021) into my prototype. Given the potentially widespread unconscious bias, the prototype focused on illustrating the management of unconscious bias in hiring decisions as an example. The prototype was tested with a few leaders for feedback. The feedback was analysed and discussed, and recommendations for the next steps were made.

EVERYONE HAS BIASES

The human brain receives an enormous load of information and can process 11 million bits of information every second. However, conscious minds can only handle 40 to 50 bits of information per second. Therefore, our human brain embraces biases and, sometimes, for survival, takes cognitive shortcuts that can lead to unconscious bias. These shortcuts compartmentalise people, things, and events (Agarwal & Kwong, 2020; Fuller et al., 2020).

Our brains evolved to group things mentally. Bias occurs when the information that our brain receives is tagged and grouped with labels like "good" and "bad", and it is then applied to the whole group (McCormick, 2016). The mental grouping allows the brain to make quick decisions on what is safe or not safe and what is appropriate or not. It is a survival mechanism.

'Pronin et al.'s (2002) study found that individuals tend to see much more biases in others than in themselves, a bias blind spot leading to a lack of self-awareness and overestimating one's own objectivity and rationality. Cognitive and motivational bias, perceptions, and accusations of bias in others and a denial of bias in self are inevitable. Therefore, when individuals attribute disagreements and bias to our opponents, finding common ground will be difficult, leading to misunderstanding and mistrust.

WHAT IS UNCONSCIOUS BIAS?

Social psychologists believe that daily, much of what we do is unconscious (Freud, 1965, cited in Wheeler, 2015). In psychology, one of the least controversial propositions is that people are only sometimes aware of what they do (Greenberg et al., 2000, cited in Wheeler, 2015). Despite many empirical theory-testing efforts in the middle third of the twentieth century, Freud's views of unconscious mechanisms never achieved conclusive support amongst scientists (Erdelyi, 1985). Consequently, Freud's psychoanalytic theory of unconscious mental processes was abandoned. This gave way to re-establishing the concepts of conscious control over human behaviour in the last third of the twentieth century.

Greenwald and Krieger (2006) introduced implicit bias as an aspect of the new science of unconscious mental processes. It also has been demonstrated in many reproducible research findings. If the individual is aware of the action being taken, it is a conscious intent and is explicit. Conversely, the science of implicit cognition proposes that individuals may not consistently be conscious and have intentional control over the processes involved in social perception, impression formation, and judgement that drive their behaviours. Implicit bias is attitudes or stereotypes that affect our understanding, actions, and decisions unconsciously, and it is based on implicit attitudes or implicit stereotypes. Implicit bias can be problematic as it can produce a behaviour that diverges from the individual's beliefs or principles.

Unconscious biases are biases when we are unaware of things outside our control. Biases that happen automatically and get triggered by our brain into making quick judgements of people and situations are influenced by our culture, background, context, and personal experiences. They are views and opinions we are unaware of and are automatically activated, operating outside our conscious awareness and affecting our everyday behaviour and decision-making (Atewologun et al., 2018). This means that unconscious bias can profoundly impact individuals and organisations despite the lack of conscious intent.

WHAT IS THE IMPACT OF UNCONSCIOUS BIAS?

Did you know your name may provide you with a significant advantage? In a randomised double-blind study, science faculty participants rated male (John) applicants as significantly more competent and hireable than the female (Jennifer) applicants with the same credentials and qualifications. They would also award "John" a higher starting salary and volunteer to mentor him more than the female candidate, "Jennifer" (Moss-Racusin et al., 2012). You may have unconsciously thought that male candidates are better than female candidates.

Reeves (2014) studied whether attorneys unconsciously believe African Americans produce inferior written work as compared to Caucasians. In this study, all the attorneys were told that the memo provided to them was written by an associate named Thomas Meyer, who graduated from New York University Law School. However, half of the attorneys were told Thomas was Caucasian, and the other half were told Thomas was an African American. The attorneys were tasked to give the memo an overall rating of 1 (poorly written) to 5 (extremely well written), and they were asked to edit the memo for mistakes. The results indicated strong confirmation bias. The African American Thomas Meyer's memo was given an average overall rating of 3.2 out of 5.0 with twice as many spelling and grammatical errors (5.8 out of 7.0) and the Caucasian Thomas Meyer was given an overall rating of 4.1 out of 5.0 and fewer errors (2.9 out of 7.0) for the same memo. Dr Arin also found out that female and ethnically diverse attorneys were as likely as their white male participants to be more vigorous in examining African American Thomas Meyer's memo.

In one of the first studies in Asia, a 2019 study in Hong Kong has provided significant revelations regarding the prevalence, nature, and extent of unconscious bias (Kapai, 2019). The research revealed that unconscious biases related to gender and race are pervasive amongst the diverse social groups, with racial biases exhibiting a more pronounced presence. Notably, the study highlights that generic, one-size-fits-all solutions could effectively address these biases. While interventions can be successful, their effectiveness varies based on specific types of bias, the targets of discrimination, and the characteristics of the social groups involved. Importantly, this study has shown that systemic and indirect discrimination is prevalent at a subtle level and is capable of being more challenging to manage.

A study investigating what is preventing Australian corporations from making gender diversity in senior leadership a key strategic priority has highlighted that women's employment opportunities are affected by social discrimination, sexism, and unconscious bias (Evans & Maley, 2021). It revealed a persistent gender imbalance with men holding dominant positions, while women bear most of the responsibilities for balancing work and family commitments. Additionally, women face significant unconscious bias, particularly in their roles as mothers. There is also a noticeable absence of male leaders in Australia's corporations relinquishing their power, even if it is in the best interest of their business success.

Hence, it can be dire if unconscious bias is prevalent in a workplace. It may potentially lead to unequal opportunities, unfair treatment, and discrimination against employees based on race, gender, and age, impacting morale, productivity, and legal issues. We may also overlook qualified candidates with our preconceived notions, stifling innovation, hindering organisational growth, and potentially leading to a toxic work environment.

UNCONSCIOUS BIAS FOR LEADERS AND IN THE WORKPLACE

There are more than 150 identified unconscious biases, making it tricky to address them. McCormick (2016) shared some of the known unconscious biases that directly impact a workplace:

- Affinity Bias: The tendency to warm to people like us.
- Halo Effect: The tendency to think everything about someone is good because you like that person.
- Perception Bias: The tendency to form stereotypes and assumptions about certain groups that make it impossible to make an objective judgement about members of those groups.
- Confirmation Bias: People seek information confirming pre-existing beliefs or assumptions.
- Groupthink: This bias occurs when people try too hard to fit into a particular group by mimicking others or holding back thoughts and opinions. This causes them to lose part of their identities and causes organisations to lose out on creativity and innovation.

In my interviews with experienced leaders, the majority cited that culture, such as language, belief system, value system, and race, is the most prevalent bias regardless of the industry or organisation. For example, Leader B (2023) shared that, when he first came to Asia, he realised that Americans use contractions significantly more often. It is, however, quite different to British English. Also, he noticed that American sports metaphors do not work in Asian societies. Therefore, if he had continued to use contractions and American sports metaphors, he may have either lost interest in the team or, as team members were unresponsive, assumed that they were a team of introverts or, worse, not very engaging employees.

Nordell (2021) suggests that bias functions like a circuit. It begins when we absorb "cultural knowledge" from everything around us: our families, the media, our classroom, etc. Some of this information will be true, such as statistical differences between the average weight of men and women. Some will not be; for example, boys are better at coding on average than girls. However, this information may unfortunately become deeply embedded as associations and stereotypes over time. Therefore, when we encounter something that triggers these associations, our cultural knowledge affects how we react to these situations. Devine (1989) believes that having these associations does not mean you are wrong. It means you exist in a culture.

Confirmation bias is where our brains are quick to confirm pre-existing assumptions, making us close-minded and dismissive of ideas and suggestions from others (McCormick, 2016). Therefore, in human resources (HR), we often set up processes as guard rails to help work through pre-existing assumptions or stereotypes. Leader C (2023) recounted an incident where a sales leader decided to hire a senior leader within the first three seconds of meeting the candidate. Leader C disagreed with the sales leader's approach and had him follow-through on the organisation's hiring process. Unfortunately, when the individual has decided, any processes developed as guard rails may be superficial (Leader C, 2023).

The interviewees commonly highlighted that a lack of awareness is a contributing factor to the occurrence of unconscious bias in the workplace. As Leader D (2023) pointed out, managing unconscious bias becomes challenging when individuals are unaware of its existence. As noted by Leader G (2023), it is also universally acknowledged that everyone harbours biases, and pretending otherwise is not productive. Additionally, Leader E (2023) shed light on some individuals' mentality, wherein they perceive themselves as entitled. This entitlement mindset can contribute to perpetuating unconscious biases in a professional setting. Pronin et al. (2002) found that people tend to perceive themselves as less biased, and their motivation to see themselves positively may sometimes be thwarted by availability biases.

From an organisational perspective, garnering support from employees and leaders to address bias is only possible by clearly tracking progress (Leader A, 2023; Leader G, 2023). The absence of robust success metrics complicates bias mitigation (Leader C, 2023). How do you hold the leader accountable? For example, if the leader has bias against young mothers, can establishing key performance indicators (KPIs) to hire or promote more mothers be a viable accountability measure (Leader G, 2023)? This question, posed by Leader G (2023), underscores the need for tangible actions to address biases at a leadership level. Unfortunately, Leader C (2023) notes that there seems to be limited evidence linking success in mitigating unconscious bias to company performance improvements that can motivate leaders. In other words, specific data points, milestones, and metrics can effectively hold a leader and an organisation accountable.

When Gianmarco Monsellato was the CEO at Taj, a French firm focused on the intricacies of law, he made changes to ensure the company's practices were fair. He embedded fairness, inclusion, and flexibility in the workplace, provided a foundation for empowerment and advancement of women to senior positions, and a culture that redefined the model for success for all employees regardless of gender. Over his 12-year tenure, the revenue increased by 70%. His methods included mentoring, transparency with clear criteria for promotion, and creating accountability to examine that managers were fair (Nordell, 2021). With time and coordinated efforts between the leader and the organisation, unconscious bias in the workplace can be managed.

From the leadership perspective, leaders should not impose their personal experiences onto others. Given the inherent power imbalance, leaders must exercise prudence when intervening in various situations or discussions. It is also essential for leaders to remain cautious and avoid unconsciously favouring one individual over another due to unconscious biases (Leader B, 2023). Striking this delicate balance is crucial for fostering a fair and inclusive leadership approach.

Moreover, the perpetual challenge for leaders lies in continuously dismantling their biases, especially considering that seasoned leaders may have deeply ingrained perspectives. While such leaders may succeed, there is a risk of developing blind spots in their treatment of individuals and clients. For a leader, the ongoing task is to be cognisant of these potential biases and cultivate an awareness that ensures equitable and unbiased interactions with people and clients (Leader G, 2023).

If the individual is unaware of their biases and not motivated to change, their bias will persist. Therefore, UBT was introduced to increase awareness, reduce unconscious bias, reduce explicit bias, and change behaviour (Atewologun et al., 2018).

How Effective Is UBT?

In America, at least 20% of large US companies provide UBT, expecting that 50% will offer it in the next five years (Lublin, 2014). In China, 64% of organisations provide UBT for hiring managers.

UBT can be done online or in person, and it can take many forms, from taking an implicit association test to attending a presentation to role-playing as a hiring manager to consider how you evaluate candidates from different groups.

In the 2019/2020 Hays Asia Diversity & Inclusion Report (2020), 72% of the employees interviewed consider UBT to impact the selection and hiring of diverse talent positively. The report also emphasised the need for unbiased and diverse leadership. Unfortunately, 59% of respondents across Asia strongly agreed that leaders in the organisation have a bias towards hiring people who look, think, and act like them.

UBT aims to reduce bias in attitudes and behaviours at work, from hiring and promotion decisions to daily interactions. However, research has shown that UBT is ineffective and may even backfire. UBT helps to raise awareness, but more is needed because sending a message that bias is involuntary and widespread may make it impossible to avoid (Gino & Coffman, 2021). The three key findings from a survey that Gino and Coffman ran with 500 working adults from various US organisations are:

- Organisations are worried about a backlash, so UBT is made a voluntary programme. Hence, the training was only embraced by those familiar with it.
- 91% indicated that their firms do not collect information on metrics they care about, such as race, gender, promotions, and employee recognition awards.
- 87% of the respondents indicated that the training did not explain the science behind bias or the cost of discrimination in organisations, and only 10% of the training programmes gave strategies for reducing bias.

Leader G (2023), a diversity, equity, and inclusion (DEI) leader, shared that customising UBT is more effective than relying on generic videos. There should be numerous examples, and the reference points must be relatable and recognisable. For example, using a testimonial from a black woman about going back to work may not resonate if you are training a group of Asian mothers.

Short-term education interventions remain the same as individuals acquire biases over a lifetime of media exposure and real-world experience. UBT can activate stereotypes, making them at the top of the mind after the training, and majority groups may feel left out, reducing their support. People may respond unfavourably to attempts to control them, leading to potential resistance to making UBT mandatory, as it may give them a sense of disempowerment. Training can also make employees complacent about their biases, potentially having a 'moral licensing' effect (Dobbin & Kalev, 2018). The evidence for UBT's ability effectively to change behaviour is limited, and there is potential for back-firing effects when UBT participants are exposed to information that suggests stereotypes and biases are unchangeable (Atewologun et al., 2018). Furthermore, 'Kalev et al.'s (2006) review of more than 700 companies found that the likelihood that black men and women would advance in organisations often decreased after UBT.

In a meta-analysis research by Forscher et al. (2019), evidence from 494 studies (80,356 participants) was synthesised to investigate the effectiveness of different procedures to change unconscious bias and their effects on explicit bias and behaviour. The researchers found that unconscious bias can be changed, but these changes do not necessarily translate into explicit bias or behaviour changes. There needs to be more understanding of what approaches are consistently effective in changing unconscious bias. While the debate on the effectiveness of UBT continues, one thing remains clear: simply raising awareness is not enough. We must delve deeper, explore innovative approaches tailored to specific contexts, and address systemic biases alongside training to unlock their true potential. Doing so can foster lasting changes and create a truly inclusive and equitable workplace.

Can You Detect and Reveal Unconscious Bias?

The implicit association test (IAT) was first published in 1998 by Project Implicit and has since been continuously updated. This test measures the strength of associations between concepts and evaluations or stereotypes to reveal an individual's hidden or subconscious biases. Project Implicit is a non-profit organisation founded by Tony Greenwald of the University of Washington, Mahzarin Banaji of Harvard University, and Brian Nosek of the University of Virginia (Project Implicit, 2013).

Much research has suggested that IAT measures are consistent and stable in measuring implicit attitudes and stereotypes (Greenwald & Krieger, 2006). Siegel's (2006) findings also supported the validity of the IAT as an attitude measure. The results from Siegel's experiments imply that the IAT is sensitive to whether participants endorse the associations they learned.

However, in a critical analysis of the IAT by Blanton et al. (2009), the authors argue that there is weak evidence that the IAT is an informative diagnostic tool on the prevalence of implicit biases. There are several limitations when it comes to predicting individual-level behaviours. They argue that researchers from two prominent studies, McConnell and Leibold (2001) and Ziegert and Hanges (2005), that used IAT must pursue more robust statistical techniques and move beyond simple zero-order correlation tests of implicit attitude-behaviour relations. They should examine the role of implicit attitude after controlling for relevant explicit measures.

Aside from the IAT, the Privilege Walk is a potentially controversial activity that creates awareness of unconscious bias. It is a group activity where everyone stands in a straight line, either blindfolded or with their eyes closed. The facilitator will ask a question afterwards, and everyone moves forward or backward. The activity's objective is to provide an opportunity to understand and create awareness of the intricacies of the privileges of certain social groups. This will allow us to identify what we do not know (Leader D, 2023). Some argued that other than a potentially sensitive activity, the privilege walk itself could not teach how an individual can shift perspectives (Ehrenhalt, 2017). It must be supplemented with the following steps and follow-ups.

The Equality and Human Rights Commission commissioned a report to identify and evaluate available evidence to help determine whether, when, and how UBT works. Their findings indicated a mixed picture. UBT effectively raised awareness using an IAT (followed by a debrief) or other advanced training workshops to reduce unconscious bias. However, it is unlikely to eliminate it (Atewologun et al., 2018). While the IAT offers a glimpse into our unconscious biases, it is the first step to self-awareness. It is important to remember that the IAT should not be the sole indicator of our biases. Critical thinking, engaging with diverse perspectives, reflecting on real-world situations, and potentially AI tools can provide a broader and deeper understanding of one's unconscious bias.

Can Technology Influence Unconscious Bias?

Advancements in technology and AI have played a critical part in addressing unconscious bias and, in part, widening the talent pool. Many organisations have started using technology to reduce unconscious bias by removing potential data points that might be biased. For example, when a resume is submitted for hiring, the technology removes elements that might create bias, such as gender, ethnicity, name, etc. (Sergott, 2021).

Better data, analytics, and AI have the potential to create and serve as powerful tools to help examine human biases. This might involve deploying algorithms in conjunction with human decision-makers, comparing outcomes, and investigating potential reasons for disparities. When organisations realise a flaw in their algorithm, instead of ceasing usage of the algorithm, they should consider how the underlying human behaviours need to change (Silberg & Manyika, 2019). Another finding shows that retroactively adjusting AI algorithms may not solve the problem. The study has demonstrated that bias introduced to a user by an AI model can persist in a person's behaviour, even after they stop using the AI programme (Leffer, 2023).

Wang et al. (2019) found that an AI algorithm, exposed to over 100,00 images sourced from the internet, had developed an association between women and images of kitchens. As the algorithm continued to learn, it became evident that its biased assumptions surpassed those initially present in the data set. Consequently, the outcomes replicated the inherent bias in the images provided and exacerbated and amplified that bias.

Some recent examples of bias in AI include the following, as reported by Niral Sutaria (2022).

- Amazon abandoned an AI-driven hiring tool due to its exhibited bias against women (Dastin, 2018).
- Microsoft issued an apology after its AI-based Twitter account started tweeting racist comments (Lee, 2016).
- IBM had to abandon its facial recognition tool after it was found exhibiting bias towards certain ethnicities (IBM abandons 'biased' facial recognition tech, 2020).

AI algorithms are evolving quickly, and AI systems are not equal in terms of bias risk. For example, an AI system that suggests products for a shopping cart has less risk than an AI system that determines who should move on to the next round of interviews (Sutaria, 2022). It is crucial to incorporate explicit design considerations for sensitive features and other factors when developing an AI model to prevent it from learning to process data in a biased manner from the outset.

Technology has the potential to both amplify and mitigate unconscious bias. While algorithms can perpetuate existing biases if not carefully designed, they can also offer tools for raising awareness, identifying biases, and promoting fairer decision-making. Understanding the complex interplay between technology and bias is crucial for harnessing its potential to create a more equitable and inclusive workplace.

FRAMING THE PROBLEM STATEMENT

From the research and interviews conducted, there are many directions to dive into. From an organisation to a leader's perspective, to identifying and creating measures and outcomes, to studying how to improve or better biases in technology, and to looking at it from very specific situations (e.g., hiring, performance reviews, etc.). In conversations with HR leaders and potential clients:

- Numerous individuals have expressed concern about sharing organisation data points for privacy reasons.
- Many have indicated that addressing unconscious bias is not a current priority for their organisation.
- Many have acknowledged the critical role a leader plays in managing unconscious bias in the workplace.

Therefore, the focus of this chapter is from a leader's perspective. Apart from the difficulties associated with identifying a client, the leader's proactive effort, motivation, and commitment to manage their personal biases, conscious and unconscious, may be fundamental to changing behaviour and managing bias in the workplace. Interestingly, few studies are related to supporting a leader in managing their unconscious bias. Therefore, the problem statement I have identified for this chapter is: how might we enable leaders continuously to cultivate self-awareness and manage unconscious biases?

OPPORTUNITIES: HOW DO LEADERS MANAGE UNCONSCIOUS BIAS IN THE WORKPLACE?

Here are approaches and tools that leaders can and have practised to manage their personal biases. Can some of these practices be augmented?

Being Aware

The most consistent feedback from my interviewees was about being aware and creating awareness. Most interviewees foster awareness by identifying triggers, being mindful, continuously self-assessing, and looking at themselves in the mirror to create deeper awareness. For example, Leader E shared that her trigger is when she starts rejecting cases (her clients): "I can't only take cases that I like; then I will be biased" (Leader E, 2023).

Mindfulness is a state of non-judgemental awareness of the present moment that can facilitate the discontinuation of automatic mental operations. The four related but distinct elements of mindfulness (awareness, attention, focus on the present, and acceptance) contribute to "deautomatisation" through subsequent processes (Kang et al., 2012).

Deautomatisation of mental habits may allow enhancement of adaptive self-control ability and increase well-being, providing us with more influence over our reactions. Alternatives may be possible once our habits are interrupted. For example, in working with a difficult co-worker, although you will continue to feel the tension with mindfulness, you will pause and notice your feelings. Once you notice the feelings, you can decide what to do, rather than reacting automatically (Nordell, 2021).

Many who practise mindfulness use the acronym RAIN (recognise, accept, investigate, and non-identification) as a reminder (Magee, 2021). Rhonda Magee (2021), Professor of Law at the University of San Francisco, shared an example: when we *recognise* privilege is present in the room, and it might be disrupting other voices to be heard, we pause, notice, and *accept* the situation for what it is. Thereafter, we might move on to *investigate* what voices are absent. What are we missing when other voices (racial or otherwise) are left out? Rather than resisting, we investigate with the willingness to be uncomfortable and to learn. This should be done with non-identification and non-attachment.

In another example, where aggression and force are accepted, Eric Russell, a member of the SWAT (special weapons and tactics) team, discovered the effectiveness of mindfulness in helping him regulate his emotions. In his interview with Nordell (2021), Russell initially dismissed the concept of mindfulness as absurd. However, he underwent the mindfulness course and discovered its efficacy. Now, Russell practises mindfulness regularly. He will put a sticker on his rear-view mirror as a reminder to stop and take ten seconds to focus.

Personal Accountability

Kim Scott, author of Just Work, shared that everyone is responsible for fixing workplace bias, prejudice, and bullying. She introduced four different roles and responsibilities: the observers, individuals that are harmed, individuals that cause harm, and the leaders. Although it is fine for those who are harmed to stay silent, Kim suggests the individual be proactive in choosing the response to build solidarity with others by talking to others and even documenting what is happening. Agarwal believes that being quiet and nonracist is not enough anymore, as silence breeds prejudice (Agarwal & Kwong, 2020).

For observers, the role is to be an upstander and intervene. The individual does not have to be the knight in shining armour but an organisation. Right To Be (previously known as Hollaback) has recommended 5Ds for how an individual can intervene: directly, delaying (the situation), delegating (to an individual with more authority), documenting, and, lastly, creating a distraction.

For those that cause harm, the individual should learn how to acknowledge that they have caused harm. Instead of just an apology, the individual should also make amends by understanding what was wrong and how to take steps to make it right. It is important to be part of the solution and not the problem. It is important to note that, often, it is very hard for people to admit or confess. Therefore, Leader G (2023) recommends creating psychological safety will provide a safe space for people to admit or confess without being judged.

As for leaders, Scott (2023) recommended three steps leaders could take in creating bias interruptions, a code of conduct and consequences for bullying. The first step is to create a shared vocabulary, either a word or phrase that the team can use to disrupt bias when they notice it. Secondly, it is to agree on a norm for how to respond when you are the person who caused harm. Thirdly, when you reach the end of a meeting and nobody has said anything, pause for 30 seconds and raise it, what did we miss? Did someone say something biased in the meeting?

Be Curious, Practise Active Listening, and Ask Questions

Being curious, keeping an open mind, and spending more time listening (active listening) are important reminders before going into a meeting or intervening in a discussion or conflict. When Leader E (2023) trains new counsellors, she encourages them to ask more questions and explore more angles. In every case (client) that is managed, you may make certain assumptions that may not be true because there are likely hidden stances. Therefore, for Leader E to remain curious and open, she starts by asking more questions.

In the broader diversity, equity, and inclusive work, it is crucial to embrace intersectionality and understand each individual's unique dimension, rather than making broad assumptions. Therefore, Leader B (2023) uses active listening and courageous conversation tools to uncover and mitigate bias.

In Leader E's (2023) role as a counsellor, it is a requirement that she attend clinical supervision to check in. Additionally, if she senses her bias is creeping in, she will attend personal therapy to check in on her own awareness. She believes she needs to be curious enough to seek additional perspectives.

Digital Tools

Some leaders actively use digital tools to help them scan for biases. There are a couple of digital tools that can help an individual monitor and act. AI has made these tools more powerful in analysing communication in emails or virtual calls (Leader B, 2023). A couple of tools used by my interviewees include Grammarly to review writing in emails to ensure that the writings are neutral and unbiased (Leader G, 2023). Leader B (2023) uses a bias scanner from Mathison.io to help him understand and uncover his own hidden and unconscious biases in written communications such as emails and performance reviews.

Other Training Programmes

Another valuable approach to foster a more inclusive workplace is through courageous conversation training. This training equips individuals and leaders with tools to facilitate open, honest, and comfortable conversations. The emphasis is on creating an environment where people feel comfortable sharing their thoughts and feelings. As highlighted by Leader B (2023), this type of training enables leaders and individuals to embark on a journey of exploration, aiding in the identification and understanding of unconscious biases. By cultivating skills to engage in courageous conversations, organisations empower their members to address biases directly, contributing to a more inclusive workplace culture.

Understanding tools like mindfulness and the importance of creating self-awareness offered here lays the groundwork for developing effective solutions to manage bias. Addressing unconscious bias at the leadership level can have ripple effects throughout the organisation.

How Have Many Innovated from Existing UBT?

Gino and Coffman (2021) suggest that the most effective UBT does more than increase awareness. It is a longer journey that requires structural changes to policies and operations. UBT is not a check-the-box exercise but a long-term commitment. Gino and Coffman (2021) suggested the following approach to unconscious bias training and complementary measures:

1. Stress that the "individual" holds the power.
2. Create empathy.
3. Encourage interactions among people from different groups.
4. Encourage good practices and continued learning.
5. Set a broader strategy for broader impact.

Atewologun et al. (2018) suggest thinking about UBT from the content and context perspective and avoiding one-off training sessions. The recommendation is to use IAT, followed by a debrief session to increase awareness and measure any changes to unconscious bias, deliver training to groups of people that work closely together, and educate participants about the theory of unconscious bias and bias reduction strategies (Atewologun et al., 2018; Nalty, 2021). From the organisational perspective, UBT should be considered part of a wider programme with clear goals and use before and after measures to assess change. The training should be an ongoing process involving multiple sessions and different formats.

Forscher et al.'s (2017) research suggests that habit-breaking intervention produces enduring changes in peoples' knowledge and beliefs. This study suggests that, with sufficient motivation, awareness, and effort, habit-breaking interventions effectively produce lasting psychological biases. The authors measured outcomes every other day for 14 days and even two years later. A subset of these students continued to show that they are more likely to speak against bias rather than the controlled students.

While UBT is crucial in raising awareness, it is only the first step. We should move beyond training to systemic changes, mindfulness practices, and technology-driven solutions. This shift signifies the potential to expand our approach to managing unconscious bias.

CAN WE INFLUENCE AND CHANGE OUR UNCONSCIOUS BIAS?

Blair (2002) summarised several studies and concluded that unconscious biases are malleable in contrast to assumptions that biases are fixed and cannot be changed. The strategies that can influence automatic stereotypes and prejudice include stereotype suppression strategies, actively promoting counter stereotypes, perceiver's motivation, strategic efforts, focus of attention, and contextual cues. Similar to a bad habit, people can retrain themselves to think in a less biased and stereotyped way (Nalty, 2021).

Although unconscious biases are malleable, this may not necessarily lead to long-term change. Lai et al. (2016) tested nine interventions to reduce implicit racial preferences over time. Although the findings show that all nine interventions (6,321 participants) reduced implicit preferences, none was effective after a delay of several hours to several days.

A series of research conducted to analyse foul calls in National Basketball Association (NBA) games demonstrated the powerful impact of simply paying attention to your own affinity bias (Nalty, 2021). The initial study's results, first published in 2007, faced criticism from the NBA, leading to widespread media attention. The researchers discovered that more fouls were called against players who were not the same race as the referee, and these disparities were large enough to affect the outcome of the games. In response, the researchers conducted two additional studies. One analysed data from basketball seasons preceding the media coverage (2003–2006), while the other focused on seasons following the publicity (2007–2010). The outcomes were notable. During the seasons before referees were conscious of the disparate foul calls, the researchers reproduced the initial study findings. However, after the widespread publicity, foul-calling had no noticeable difference. Therefore, the awareness led to more consistent and fairer foul-calling.

Changing unconscious bias is not a one-off event but a continuous self-reflection and active effort journey. Leaders can potentially influence their hidden biases by combining awareness with ongoing efforts like mindfulness and bias-mitigation strategies.

HOW CAN WE INNOVATE?

This research has shown that unconscious bias runs deep from each individual's upbringing to the culture around them that they absorb every day. Therefore, for this research project, design thinking (DT) or human-centred design (HCD) will be used to help innovate an idea with the objective of continuously supporting a leader in managing unconscious bias. The HCD approach enables learning from the customers, opening up to a breadth of creative possibilities, and then zeroing in on what is most desirable, feasible, and viable (IDEO, 2015). See Figure 7.1

Purposefully, leaders from diverse cultural backgrounds, encompassing Malaysians, Americans, and Europeans, were chosen as participants for this research project. These participants represented both the profit and non-profit sectors. The selection of a diverse range of participants aimed to investigate potential variations in how these leaders manage their unconscious biases and explore the approaches that may be employed by their respective organisations to address such biases.

My hypothesis is regardless of whether the organisation supports managing unconscious bias; it needs to start with the individual. The leader needs to take personal accountability for his or her behaviour. If the organisation actively supports managing unconscious bias, it will help to create a stronger culture for the organisation and for the leader personally.

Persona: A Motivated and Engaged Leader

Numerous frameworks are available to identify user needs, from creating personas and building buyer's journey to considering users' "jobs to be done" (JTBD). Alan Cooper introduced the use of personas in user-centred design thinking. Personas are used to distil and make relevant information

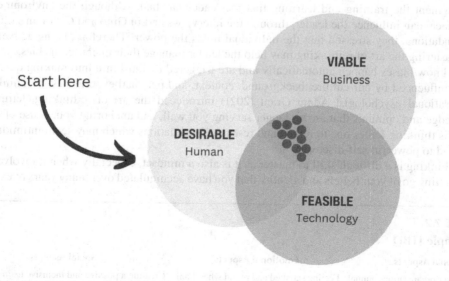

FIGURE 7.1   The three lenses from HCD in designing solutions.

INTERESTS	Travelling, reading books, psychology
CHALLENGES	Experience individual and hence, can be tricky in trying to identify unconscious bias.
GOALS	To be a good leader, meeting the KPIs set by the organisation while creating a positive and inclusive team culture that celebrates achievements and encourages mutual encouragement.
MOTIVATION	To grow as a leader
FRUSTRATION	How can I continue to strip away my biases

NAME Jeff Gallego
AGE 45–55
OCCUPATION Business leader | Head of department
ANNUAL INCOME $180, 000 and above

FIGURE 7.2 A sample persona.

about a system's user (Patton, 2008). The JTBD tool was pioneered by Harvard Business School Professor Clayton Christensen. It is a means of understanding underlying customer motivations that influence the buying process. In JTBD, the functional needs are as important as the social and emotional needs (Stobierski, 2020).

A persona and JTBD will be used to help understand the target audience of this research project (see Figure 7.2 and Table 7.2). The target persona is a leader or people manager, regardless of whether the leader works for a profit or a non-profit organisation and regardless of the size of the business and team.

Mastering the Art of Rethinking

Developing a good cognitive skill may be helpful in building this prototype to enable a leader to cultivate awareness and address unconscious bias consistently. This prototype aims to complement and augment the training and learning that the leader has had. Although the environment (e.g., workplace) can influence the leader, through the interviews and in Gino and Coffman's (2021) recommendations, they stressed that the individual holds the power. Therefore, being actively aware and mastering the art of rethinking may help the leader manage their unconscious bias.

We know biases happen automatically and are triggered by our brain into making quick judgements influenced by our culture, background, context, and personal experiences. In Think Again, organisational psychologist Adam Grant (2021) introduced the art of rethinking, letting go of knowledge and opinions that are no longer serving you well, and anchoring your sense of self flexibly. His thinking forces one to reflect and re-examine situations which may seem intimidating but will lead to powerful self-discovery.

Rethinking is a difficult skill to master as it is also a mindset, especially when it involves potentially letting go of your beliefs and identity that you have accumulated over many years of experience

TABLE 7.2
A Sample JTBD

Functional Aspects	Emotional Aspects	Social Aspects
To deliver organisations annual KPI within budget	Feeling satisfied and proud when I am able to contribute positively to the team's growth	Creating a positive and inclusive team culture that celebrates achievements and encourages mutual encouragement.

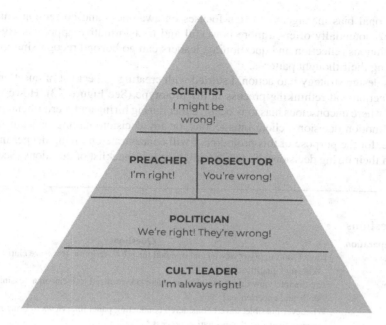

FIGURE 7.3   Mental models (Grant, 2021).

(Grant, 2021) See Figure 7.3. Gross (2023) believes that rethinking is a strategy for emotional regulation. It involves intentionally changing your interpretation of an emotional situation and your role in it in a way that possibly changes how you feel. Grant (2021) recommends applying the mental model of a scientist ("I might be wrong") as it encourages the process of rethinking and experimentation of our approach and beliefs. The process of rethinking often unfolds in a cycle, from having the humility to acknowledge what we do not know to doubting and questioning our assumptions, leading to curiosity, and, hopefully, new discoveries (See Figure 7.4).

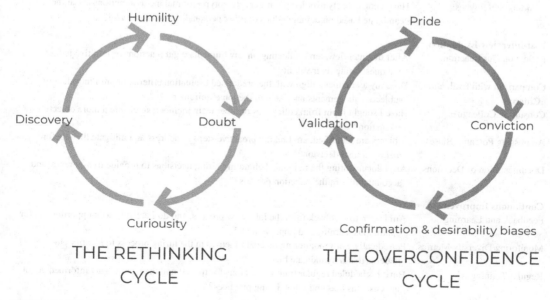

FIGURE 7.4   The rethinking and overconfidence cycle (Grant, 2021).

While traditional bias management often focuses on awareness and correction, mastering the art of rethinking potentially offers a more powerful and transformative approach. By fostering a mindset of continuous reflection and questioning, leaders can go beyond recognising their bases to actively reshaping their thought patterns.

To bring this design strategy into action, I started with creating a checklist of questions to initiate the leader's reflection and rethinking process in my prototype (See Figure 7.3). However, there are many scenarios where unconscious bias may occur, either during hiring and recruitment, performance evaluations, promotion decisions, client interactions, or any decision-making or team management situation. Hence, for the purpose of this prototype, I will concentrate on diving deeper into helping a leader reflect on their hiring decisions. To begin with, here is a checklist of questions (See Table 7.3):

TABLE 7.3

Interview Questions

Pre-interview Preparation	Questions
Job Description Review	Have I thoroughly reviewed and updated the job description to ensure clarity on required skills and qualifications?
Structured Interview Questions	Are the interview questions structured and standardised to focus on assessing relevant skills and experiences?
Panel Interview	Whenever possible, have I considered conducting panel interviews to bring multiple perspectives into the evaluation process?
During the Interview	
Unbiassed Language	Am I using neutral and inclusive language throughout the interview to avoid unintentional biases?
Consistent Evaluation Criteria	Have I established consistent evaluation criteria for all candidates, ensuring a fair and standardised assessment?
Behavioural Questions	Have I incorporated behavioural questions to assess past performance and actions, providing concrete examples of a candidate's abilities?
Focus on Job Requirements	Am I consistently steering the conversation towards the candidate's ability to meet the job requirements rather than personal characteristics?
Active Listening	Am I actively listening to the candidate's responses without making assumptions based on appearance, background, or other irrelevant factors?
Avoiding Stereotypes	Have I consciously avoided relying on stereotypes or making assumptions about the candidate based on demographics or other personal characteristics?
Post-interview Evaluation	
Reflect on Gut Reactions	After the interview, am I reflecting on any immediate gut reactions or intuitive judgements and questioning their validity?
Comparison with Evaluation Criteria	Does my assessment align with the predefined evaluation criteria, or am I introducing subjective judgements unrelated to the job requirements?
Consulting Colleagues	Have I sought input from colleagues or other team members to ensure a more objective evaluation?
Addressing Potential Biases	If biases are identified, am I taking proactive steps to address and mitigate them before making a final decision?
Documentation of Decisions	Am I documenting the rationale behind my hiring decisions to provide transparency and accountability in the selection process?
Continuous Improvement	
Feedback and Learning	Am I open to feedback from the interview process, and do I actively seek opportunities for continuous learning and improvement?
Monitoring Diversity Metrics	Periodically, am I monitoring diversity metrics in the hiring process to identify any patterns that may indicate bias?
Regular Training	Have I scheduled regular training sessions for myself and the team to stay informed about unconscious bias and refine hiring practices?

Let us consider a scenario experienced by an interviewee. Leader F (2023) was making a hiring decision for a key position in her team. During the interviews, she was strongly inclined to hire a particular candidate because she felt a personal connection with her. The potential candidate had a background and interest that closely resembled her own; therefore, she believed the candidate would be an excellent addition to the team.

However, a member of her team who had also participated in the interview process approached her privately. This team member, with a keen eye for objectivity, pointed out that her affinity for the candidate might be due to the candidate's similarity to her. They explained that although the candidate was talented, her skills might not fully align with the specific needs of the role and the team's diversity requirements.

Reflecting and discussing with her team members, she acknowledged her initial bias clouded her judgement. Thereafter, she reconsidered the candidate in a more objective light. Ultimately, this bias awareness led to a more thorough evaluation, allowing her to make a more informed and equitable hiring decision (Leader F, 2023; Interviewed by Cynthia Mak. 29 November). In this scenario, the leader was prompted to rethink the situation, and she had the humility to doubt her judgement, ultimately leading to a more informed hiring decision. These checklists of questions could potentially support a leader in hiring, especially when the team does not dare to raise the issue.

As simple as it may be, a checklist has proven effective in different industries. In 2012, Johns Hopkins researchers found that computerised checklists are better at finding the best preventive strategy. The computerised checklist system was designed to help physicians identify the best methods of preventing potentially deadly blood clots. Associate Professor of Surgery at the Johns Hopkins University School of Medicine and leader of the study, Elliot R. Haut, MD, shared that they tried education alone to prevent deadly clots. However, only 40 to 60% were getting optimal treatment. During the four-year implementation programme with a computerised checklist, 85% of the trauma patients received preventive treatment, as opposed to ~67% of patients before the checklist was implemented (Desmon, 2012).

Similarly, in Google, after educating employees on unconscious bias, they provided structured processes for hiring, performance reviews, etc. Google believes people will follow their instincts if you leave decisions ambiguous and unstructured. They will remember performances that resonated with them and reward people they like. Google wanted to create a structured conversation when managers do performance reviews every six months. Therefore, a checklist was provided, starting with an explanation of a set of biases that had been observed as likely to happen at the performance review meeting with a set recommendation to combat bias. See Table 7.4. In a survey of managers who have used the checklist, they believe and feel the discussion was fair (Welle, 2022).

Therefore, a checklist could serve as a good reminder if it is available to the leaders at the appropriate time, as they may not be able to review them spontaneously. The checklist should also be

TABLE 7.4

Example of Google's Checklist for Performance Review

Try This ...	... To Combat
Consider concrete/behavioural examples throughout the rating period	Recency bias Horns and halos Availability bias
Consider situational factors (in the workplace) that affect performance (e.g., lacked resources)	Stereotype-based biases
Consider if rating would change if Googler were in different social group	Anchoring bias
Play devil's advocate if there are no significantly different perspectives raised	Agreement bias

regularly reviewed and updated. In the next prototype, I will experiment to see how the checklists could be accessible at an appropriate time and updated regularly.

Can Technology Influence a Behaviour Change?

My next prototype is a personalised coach (PC) app. The rationale behind incorporating technology is to ensure it is accessible at any time, and it actively promotes a culture of rethinking to influence behaviour change. Rapp et al. (2019) conducted 20 interviews exploring how individuals live, account for, and manage life changes. Through these interviews, they found five tentative patterns for rethinking behaviour change technologies. The five patterns are:

- Emphasising internal aspects of change.
- Personalisation techniques based on users' personal knowledge and motivation.
- Fostering a sense of empowerment.
- Interconnectedness to different life domains.
- Adaptability to subjective and continuous change.

Therefore, this app was designed with the five humanistic patterns in mind. The content in PC will be built to emphasise mastering rethinking and focusing on internal aspects of change. The leader will have access to educational programmes about unconscious bias, artificial intelligence, AI-writing assistants, and activities to help reflect upon unconscious biases in different scenarios. For example, the leader can review scenarios related to hiring and will have various features, including the checklist, role-playing with an AI avatar, and listening (or watching) stories on how unconscious bias can play out during the interview. The checklist will be structured to prevent the leader from falling into personal biases. The different scenarios are important to ensure it is connected to different life domains.

The app will also send out prompts to nudge the leader through a quote, a reminder for action, and/or an encouragement for reflection beyond situation-specific scenarios. Nudges will help guide the leader in continuously learning and improving. Nudges theory was introduced by Cass Sunstein (Harvard) and Richard Thaler (University in Chicago) in 2008. Nudges are big and small interventions to get people to act in their best interest (Sunstein & Thaler, 2021). Nudging maintains the freedom of choice, fostering a sense of empowerment but steers people in a particular direction. Importantly, nudging helps us deal with the limited attention in our brains.

For example, PC could prompt the leader to talk to someone they have never spoken to before or have them identify situations in prior meetings where they were tuned in to non-verbal cues, and did the leader recognise and address biases in their interpretation? See Figure 7.5. The prompts will be a guided process to encourage the leader to pen down thoughts, and it will continue to feed questions to prompt deeper thinking. The AI will analyse the reflections and provide relevant nudges for more training, reflective questions, and stories relevant to the leader. The leader will also be able to review reports to understand usage patterns and highlight areas that the leader should consider reviewing. Here is a low-fidelity prototype showcasing the flow and how it can prompt the leader and nudge them to take action.

The features in PC were designed to support behaviour change. Results from 'Consolvo et al.'s (2009) two field studies show that a persuasive technology system was successful in helping people maintain a physically more active lifestyle. It was successful in supporting behaviour change in everyday life. The design strategy used in designing the persuasive technology system was:

- Abstract and Reflective: Using abstraction representation of behaviour and progress to encourage reflection and self-awareness.
- Unobtrusive and Public: The technology should be unobtrusive and integrated into the user's everyday life.

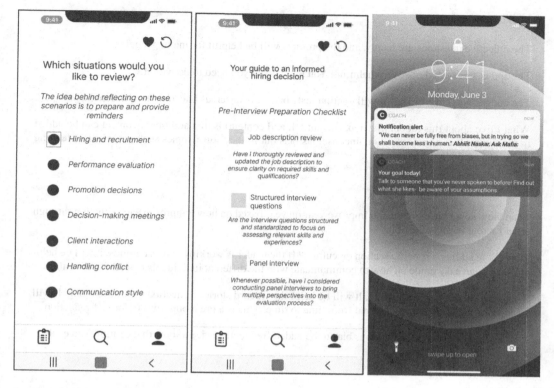

FIGURE 7.5   Sample prototype of PC.

- Trending/Historical: Giving users a way to reflect progress over time.
- Comprehensive: Providing users with a comprehensive view of progress and activities.

Although the study focused on maintaining a physically active lifestyle, this framework has the potential to work in this scenario. The objective of the checklist and activities on the app is to encourage reflection and self-awareness. Nudges and prompts are unobstructive, and the leader will have access to review responses to the guided reflections. Lastly, the aim is to provide updated UBT on PC, too.

However, from the viability perspective, building the PC will be very time-consuming and costly. Therefore, the ideal situation will be to partner with a training organisation (e.g., Mathison) that specialises in unconscious bias to build up the curriculum, maximise the existing AI model in scanning bias, and personalising to each leader and organisation. As Payal Agarwal says, in order to manage our bias continuously, we must constantly educate ourselves and check our assumptions (Agarwal & Kwong, 2020). Hence, this app serves as a tool for the leader to manage unconscious bias. With technological advances, it will be technically feasible to build this app. The challenge would be keeping the content and questions current and getting the relevant data points to provide a personalised data-driven solution.

WHAT DID THE LEADERS THINK ABOUT PC?

Due to conflicting schedules, I was unable to share the prototype with all the leaders that I interviewed previously. However, I was able to seek feedback from two leaders and have consolidated the feedback in an "I Like|I Wish|I Wonder" framework below.

I Like …

The reflective questions are handy, and the prompts will be helpful for me to get going.

The app is simple and not overwhelming! You know what you need to do with it.

The nudges are a great idea – it will continuously bring unconscious bias into our awareness.

What I like about this model (the app, current UI, and content) is that additional content can be added in the future. It is now focused on unconscious bias but in the future, topics such as inclusion, leading teams etc. could be added too.

I Wish …

I can view how I have grown and improve year on year based on how I approach each situation through what I wrote in my reflection.

How about adding a specific section on culture? If the leader is working across cultures, it will be helpful to have an easy guide on how to communicate with the different individuals from different cultures.

There is a section for journalling. It will be helpful to record stories, reflection as I move along. It will be nice for me to review my journal from time to time. This is a great opportunity for self-reflection.

For new insights about unconscious bias to be added frequently as it's a topic that continues to evolve.

I Wonder …

How many leaders will adopt this app on their own? As they always say, culture eats strategy for breakfast. If there is no organisation culture of equality, the leader will be unlikely to be motivated to use the app to manage bias.

How would you continuously update the content to ensure that they are fresh and relevant? Even if you use AI, where will the data come from?

If you use an avatar, will it be confusing? Who would be the avatar?

If you can partner with a third-party vendor and build a discussion forum? The idea behind it is to build a community of leaders that can share and learn from each other. It will be very powerful to share experiences. And as much as we go through training and self-reflection, the greatest learning comes from interacting with others especially for a topic on unconscious bias.

The feedback received was interesting as the leaders were consistent with what they liked but much more varied on things to improve and on challenges with this approach. They nodded in agreement that managing unconscious bias, learning to question our understanding, mastering the art of rethinking, and building mental flexibility are all important. The leaders believe PC has the potential to support them by reminding and guiding them on these cognitive skills. As technology is seamlessly integrated into our daily lives, I did not think it would be intrusive. However, from the viability perspective, the leaders agreed that this approach would make sense if I partnered with a training or consulting organisation to make this prototype a reality.

WHAT ARE THE ETHICAL ISSUES?

While initially my prototype relied on a checklist of questions to raise awareness and encourage leaders to reconsider the situation, AI will analyse and provide leaders with personalised recommendations. This itself can be tricky as, although AI algorithms have enabled humankind to digest and sieve through large data sets, if these algorithms maintained some of the same biases, unfortunately, it may emphasise and reinforce them as global truth (Howard & Borenstein, 2017). Crawford (2013) pointed out several examples that resulted in inaccurate predictions. She emphasised that data scientists should learn from social scientists who have a history of asking where data comes from, what methods were used to gather and analyse it, and what cognitive biases they might bring to the interpretation.

However, Kahneman believes that an algorithm could do better than humans because it filters out the noise (Houser, 2019). He believes that, unlike humans, a formula will always return the same output for any given input. Superior consistency allows even basic and imperfect algorithms to achieve higher accuracy than human professionals (Kahneman et al., 2016). If a balanced data set is created, it will hold significant power to improve the way we live and work. Houser (2019) believes that biased results can be mitigated by being aware of the risks involved in using data sets that do not represent society at large and by repeated testing for any potential bias. Runman Chowdhury, previously the global lead for responsible AI at Accenture led the team to design a fairness evaluation tool that pre-empted and corrected algorithmic bias (Aceves, 2023). Therefore, increasing the diversity of data points included will be crucial when building PC. Another potential approach to mitigate bias in building a PC is to hire a diverse group of people to create the curriculum, programme, and data sets (Houser, 2019).

With AI and technology, privacy and security concerns are another potential issue. As we deal with personal biases, leaders may be unwilling to share their data and uncomfortable with their reflections and stories being collected, processed, and stored. For example, Strava, a fitness tracking app, released a "heatmap" in 2018 of the routes of its users worldwide, unintentionally exposing locations of military bases and patrol routes. While they intended to build a global network of athletes, this incident underlined how AI's capacity to aggregate and visualise data can inadvertently expose sensitive information, leading to unintended consequences. Will sharing reflections and personal biases make the leaders more vulnerable and potentially affect their career development?

CONCLUSION

Through the research and qualitative conversations pre- and post-sharing the prototype, I believe there is a good potential for PC to be a viable tool for leaders. Having said that, more work needs to be done. We know everyone has unconscious bias; if it is not managed properly, the impact can be dire in the workplace. Large companies spend significant amounts of money on diversity, equity, inclusivity, and UBT training, with about US$8 billion spent annually in the US (Kirkland & Bohnet, 2017).

However, UBT itself is not sufficient as unconscious bias evolves, and the leader plays a significant role. It has been interesting to look into unconscious bias from the leader's perspective. An engaged and motivated leader needs to own the process, regardless of whether the organisation actively practises it. However, it is much more challenging if it is not part of the organisation's culture (Leader H, 2024).

Grant's (2021) critical art of rethinking, learning to question your opinions, is potentially a very strong mental flexibility to build that will help a leader significantly. Using a combination of UBT, a checklist of questions, and reflection activities to update the leader's view in a low-fidelity app prototype, the leaders expressed interest in delving deeper and seeing how it could be helpful in their journey of being an effective leader. Therefore, this research project can be a good springboard for subsequent studies.

As technology has changed the way we live and behave, it seems inevitable that we need technology to support a leader in *continuously* managing their unconscious bias. Most leaders do not mind using technology (apps) but raise doubts from an ethical perspective. The next steps would be to dive deeper into each scenario, flesh out the programme with a more comprehensive wireframe, and implement it. A bigger question to tackle in the future is whether we can truly design AI to manage unconscious bias effectively and not reinforce bias to be the global truth. Besides that, it will also be interesting to understand if other technology devices will work, e.g., wearables to detect the tone in our voice and scan our conversations. However, these approaches may be too invasive, potentially leading to unintended consequences.

REFERENCES

Aceves, P. (2023). 'I Do Not Think Ethical Surveillance Can Exist': Rumman Chowdhury on Accountability in AI. The Guardian. [online] 29 May. Available at: https://www.theguardian.com/technology/2023/may/29/rumman-chowdhury-interview-artificial-intelligence-accountability [Accessed 24 Jan. 2024].

Agarwal, P., & Kwong, E. (2020). Understanding Unconscious Bias. NPR Short Wave. [Podast] Available at: https://www.npr.org/2020/07/14/891140598/understanding-unconscious-bias [Accessed 15 Oct. 2023].

Atewologun, D., Cornish, T., & Tresh, F. (2018). Unconscious Bias Training: An Assessment of the Evidence of Effectiveness. [online] Equality and Human Rights Commission, United Kingdom: Equality and Human Rights Commission, p.67. Available at: https://www.equalityhumanrights.com/sites/default/files/research-report-113-unconcious-bais-training-an-assessment-of-the-evidence-for-effectiveness-pdf.pdf [Accessed 15 Oct. 2023].

Blair, I. V. (2002). The malleability of automatic stereotypes and prejudice. *Personality and Social Psychology Review*, 6(3), 242–261. https://doi.org/10.1207/s15327957pspr0603_8

Blanton, H., Jaccard, J., Klick, J., Mellers, B., Mitchell, G., & Tetlock, P. E. (2009). Strong claims and weak evidence: Reassessing the predictive validity of the IAT. *Journal of Applied Psychology*, 94(3), 567–582. https://doi.org/10.1037/a0014665

Consolvo, S., McDonald, D. W., & Landay, J. A. (2009). Theory-driven design strategies for technologies that support behavior change in everyday life. Proceedings of the 27th international conference on Human factors in computing systems - CHI 09. https://doi.org/10.1145/1518701.1518766.

Crawford, K. (2013). The Hidden Biases in Big Data. Harvard Business Review. [online] 1 Apr. Available at: https://hbr.org/2013/04/the-hidden-biases-in-big-data [Accessed 24 Jan. 2024].

Dastin, J. (2018). Amazon Scraps Secret AI Recruiting tool that Showed Bias Against Women. [online] Reuters. Available at: https://www.reuters.com/article/us-amazon-com-jobs-automation-insight/amazon-scraps-secret-ai-recruiting-tool-that-showed-bias-against-women-idUSKCN1MK08G.

De Neys, W. (2012). Bias and conflict. *Perspectives on Psychological Science*, 7(1), 28–38. https://doi.org/10.1177/1745691611429354

Desmon, S. (2012). A Better Way To Prevent Deadly Blood Clots? Johns Hopkins Medicine. [online] 15 Oct. Available at: https://www.hopkinsmedicine.org/news/media/releases/a_better_way_to_prevent_deadly_blood_clots [Accessed 1 Jan. 2024].

Devine, P. G. (1989). Stereotypes and prejudice: Their automatic and controlled components. *Journal of Personality and Social Psychology*, 56(1), 5–18. https://doi.org/10.1037/0022-3514.56.1.5

Dobbin, F., & Kalev, A. (2018). Why Doesn't diversity training work? The challenge for industry and academia. *Anthropology Now*, 10(2), 48–55. https://doi.org/10.1080/19428200.2018.1493182

Ehrenhalt, J. (2017). Beyond the Privilege Walk. Learning for Justice. [online] 20 Jun. Available at: https://www.learningforjustice.org/magazine/beyond-the-privilege-walk [Accessed 4 Jan. 2023].

Evans, K. J., & Maley, J. F. (2021). Barriers to women in senior leadership: How unconscious bias is holding back Australia's economy. *Asia Pacific Journal of Human Resources*, 59(2), 204–226. https://doi.org/10.1111/1744-7941.12262

Forscher, P. S., Lai, C. K., Axt, J. R., Ebersole, C. R., Herman, M., Devine, P. G., & Nosek, B. A. (2019). A meta-analysis of procedures to change implicit measures. *Journal of Personality and Social Psychology*, 117(3). https://doi.org/10.1037/pspa0000160

Forscher, P. S., Mitamura, C., Dix, E. L., Cox, W. T. L., & Devine, P. G. (2017). Breaking the prejudice habit: Mechanisms, time course, and longevity, *Journal of Experimental Social Psychology*, 72, 133–146, https://doi.org/10.1016/j.jesp.2017.04.009

Fuller, P., Murphy, M. W., & Chow, A. (2020). *The leader's guide to unconscious bias: how to reframe bias, cultivate connection, and create high-performing teams.* Simon & Schuster.

Gino, F., & Coffman, K. (2021). Unconscious bias training that works. *Harvard Business Review*. https://hbr.org/2021/09/unconscious-bias-training-that-works#

Grant, A. (2021). *Think again: The power of knowing what you don't know.* Diversified Publishing.

Greenwald, A. G., & Krieger, L. H. (2006). Implicit bias: Scientific foundations. *California Law Review*, 94(4), 945–967. https://doi.org/10.2307/20439056

Gross, J. (2023). Mastering the Art of Rethinking. Available at: https://www.psychologytoday.com/us/blog/feel-better/202310/mastering-the-art-of-rethinking [Accessed 17 Jan. 2024].

Houser, K. (2019). Can AI solve the diversity problem in the tech industry? Mitigating noise and bias in employment decision-making. *Social Science Research Network*. https://papers.ssrn.com/sol3/papers.cfm?abstract_id=3344751

Howard, A., & Borenstein, J. (2017). The ugly truth about ourselves and our robot creations: The problem of bias and social inequity. *Science and Engineering Ethics*, *24*(5), 1521–1536. https://doi.org/10.1007/s11948-017-9975-2

I Like | I Wish | I Wonder. (n.d.). [online] Hyper Island Toolbox. Available at: https://toolbox.hyperisland.com/i-like-i-wish-i-wonder [Accessed 20 Jan. 2024].

IBM abandons 'biased' facial recognition tech. (2020). BBC News. [online] 9 Jun. Available at: https://www.bbc.com/news/technology-52978191 [Accessed 11 Dec. 2020].

IDEO. (2015). *The field guide to human-centered design*. Design Kit.

Kahneman, D., Rosenfield, A. M., Gandhi, L., & Blaser, T. (2016). Noise: How to overcome the high, hidden cost of inconsistent decision making. *Harvard Business Review*, 36–43. https://hbr.org/2016/10/noise

Kalev, A., Dobbin, F., & Kelly, E. (2006). Best practices or best guesses? Assessing the efficacy of corporate affirmative action and diversity policies. *American Sociological Review*, *71*(4), 589–617. https://doi.org/10.1177/000312240607100404

Kang, Y., Gruber, J., & Gray, J. R. (2012). Mindfulness and de-automatization. *Emotion Review*, *5*(2), 192–201. https://doi.org/10.1177/1754073912451629

Kapai, P. (2019). Doing Equality Consciously: Understanding Unconscious Bias and its Role and Implications in the Achievement of Equality in Hong Kong and Asia. [online] The University of Hong Kong, p.180. Available at: https://researchblog.law.hku.hk/2019/11/new-study-on-unconscious-bias-and.html [Accessed 17 Nov. 2023].

Kirkland, R., & Bohnet, I. (2017). Focusing on What Works for Workplace Diversity. [online] Available at: https://www.mckinsey.com/featured-insights/gender-equality/focusing-on-what-works-for-workplace-diversity [Accessed 24 Jan. 2024].

Kraaijenbrink, J. (2022). What BANI Really Means (And How It Corrects Your World View). [online] Forbes. Available at: https://www.forbes.com/sites/jeroenkraaijenbrink/2022/06/22/what-bani-really-means-and-how-it-corrects-your-world-view/?sh=350eb11311bb [Accessed 25 Jan. 2024].

Lai, C. K., Skinner, A. L., Cooley, E., Murrar, S., Brauer, M., Devos, T., Calanchini, J., Xiao, Y. J., Pedram, C., Marhsburn, C. K., Simon, S., Blanchar, J. C., Joy-Gaba, J., Conway, J., Redford, L., Klein, R. A., Roussos, G., Schellhaas, F. M. H., Burns, M., & Hu, X.. (2016). Reducing implicit racial preferences: II. Intervention effectiveness across time. *SSRN Electronic Journal*, *145*(8). https://doi.org/10.2139/ssrn.2712520

Lee, D. (2016). *Tay: Microsoft issues apology over racist chatbot fiasco*. BBC News. https://www.bbc.com/news/technology-35902104

Leffer, L. (2023). Humans Absorb Bias from AI—And Keep It after They Stop Using the Algorithm. Scientific American. [online] 26 Oct. Available at: https://www.scientificamerican.com/article/humans-absorb-bias-from-ai-and-keep-it-after-they-stop-using-the-algorithm/ [Accessed 15 Jan. 2024].

Lublin, J. S. (2014). Bringing Hidden Biases Into the Light. Wall Street Journal. [online] 10 Jan. Available at: https://www.wsj.com/articles/SB10001424052702303754404579308562690896896 [Accessed 16 Nov. 2023].

Magee, R. V. (2021). The Dharma of Racial Justice. Available at: https://bulletin.hds.harvard.edu/the-dharma-of-racial-justice/.

Erdelyi, M. H. (1985). *Psychoanalysis: Freud's cognitive psychology*. W.H. Freeman.

McConnell, A. R., & Leibold, J. M. (2001). Relations among the Implicit Association Test, discriminatory behavior, and explicit measures of racial attitudes. *Journal of Experimental Social Psychology*, 37(5), 435–442. https://doi.org/10.1006/jesp.2000.1470

McCormick, H. (2016). The Real Effects of Unconscious Bias in the Workplace. [online] UNC Universities Libraries, UNC Kenan-Flagler Business School Executive Development, pp. 1–12. Available at: https://guides.lib.unc.edu/implicit-bias [Accessed 27 Oct. 2023].

Moss-Racusin, C. A., Dovidio, J. F., Brescoll, V. L., Graham, M. J., & Handelsman, J. (2012). Science faculty's subtle gender biases favor male students. *Proceedings of the National Academy of Sciences*, *109*(41), 16474–16479. https://doi.org/10.1073/pnas.1211286109

Nalty, K. (2021). Strategies for Confronting Unconscious Bias. [online] Colorado Lawyer. Available at: https://cl.cobar.org/departments/strategies-for-confronting-unconscious-bias/#_edn2 [Accessed 16 Nov. 2023].

Nordell, J. (2021). *The end of bias: How we change our minds*. Granta Publications.

Patton, J. (2008). A conversation with Alan Cooper: The origin of interaction design. *IEEE Software*, 25(6), 15–17. https://doi.org/10.1109/ms.2008.142

Project Implicit. (2013). Project Implicit. [online] Harvard.edu. Available at: https://implicit.harvard.edu/implicit/.

Pronin, E., Lin, D. Y., & Ross, L. (2002). The bias blind spot: Perceptions of bias in self versus others. *Personality and Social Psychology Bulletin*, 28(3), 369–381. https://doi.org/10.1177/0146167202286008

Rapp, A., Tirassa, M., & Tirabeni, L. (2019). Rethinking technologies for behavior change. *ACM Transactions on Computer-Human Interaction*, 26(4), 1–30. https://doi.org/10.1145/3318142

Reeves, A. N. (2014). Written in Black & White: Exploring Confirmation Bias in Racialized Perceptions of Writing Skills. [online] Nextions. Available at: https://nextions.com/insights/perspectives/written-in-black-white-exploring-confirmation-bias-in-racialized-perceptions-of-writing-skills/ [Accessed 17 Nov. 2023].

Scott, K. (2023). How to Confront Bias, Prejudice, and Bullying at Work with Kim Scott. [online] Creative Confidence Podcast. 18 May. Available at: https://www.ideou.com/blogs/inspiration/how-to-confront-bias-prejudice-and-bullying-at-work-with-kim-scott? [Accessed 15 Oct. 2023].

Sergott, T. (2021). Using AI To Block Unconscious Bias, Widen The Talent Pool And Increase Diversity. Forbes. [online] 8 Jun. Available at: https://www.forbes.com/sites/forbestechcouncil/2021/06/08/using-ai-to-block-unconscious-bias-widen-the-talent-pool-and-increase-diversity/?sh=15d0b6248185 [Accessed 15 Jan. 2024].

Siegel, E. F. (2006). Questioning the validity of the IAT: Is it knowledge or attitude? [Thesis (M.A.)] p.74. Available at: https://www.proquest.com/openview/20a19acd60d7b0b4e5bd91a9531a6d6e/1?pq-origsite=gscholar&cbl=18750&diss=y [Accessed 6 Nov. 2023].

Silberg, J., & Manyika, J. (2019). Tackling Bias in Artificial Intelligence (and in Humans). [online] McKinsey & Company. Available at: https://www.mckinsey.com/featured-insights/artificial-intelligence/tackling-bias-in-artificial-intelligence-and-in-humans.

Stobierski, T. (2020). Business Insights. Innovation in a Disrupted World: How to Discover New and Emerging Jobs to be Done. Available at: https://online.hbs.edu/blog/post/jobs-to-be-done-framework [Accessed 8 Jan. 2024].

Sutaria, N. (2022). Bias and Ethical Concerns in Machine Learning. [online] ISACA. Available at: https://www.isaca.org/resources/isaca-journal/issues/2022/volume-4/bias-and-ethical-concerns-in-machine-learning.

Sunstein, C., & Thaler, R. (2021). Much Anew About 'Nudging'. [online] McKinsey & Company. 6 Aug. Available at: https://www.mckinsey.com/capabilities/strategy-and-corporate-finance/our-insights/much-anew-about-nudging [Accessed 15 Jan. 2024].

The 2019/2020 Hays Asia Diversity & Inclusion Report. (2020). [online] Hays, p.40. Available at: https://www.hays-china.cn/en/press-release/content/biased-in-hiring [Accessed 17 Nov. 2023].

Wang, T., Zhao, J., Yatskar, M., Chang, K.-W., & Ordonez, V. (2019). Balanced Datasets Are Not Enough: Estimating and Mitigating Gender Bias in Deep Image Representations. https://doi.org/10.1109/iccv.2019.00541.

Welle, B. (2022). Google Unconscious Bias Journey. re: Work with Google. Available at: https://www.youtube.com/watch?v=_KfKmGb_bT4&t=1s [Accessed 4 Jan. 2024].

Wheeler, R. (2015). We all do it: Unconscious behaviour, bias, and diversity. *Law Library Journal*, 107(2), 325–332. https://heinonline.org/HOL/LandingPage?handle=hein.journals/llj107&div=21&id=&page=

Ziegert, J. C., & Hanges, P. J. (2005). Employment discrimination: The role of implicit attitudes, explicit attitudes, and task performance. *Journal of Applied Psychology*, 90(3), 553–562. https://doi.org/10.1037/0021-9010.90.3.553

INTERVIEWEES

Leader, F. (2023). Interviewed by Cynthia Mak. 29 November, Singapore.

Leader, A. (2023). Interviewed by Cynthia Mak. 10 November, Singapore

Leader, C. (2023). Interviewed by Cynthia Mak. 3 November, Singapore

Leader, H. (2024) Interviewed by Cynthia Mak. 18 January, Singapore

Leader, E. (2023) Interviewed by Cynthia Mak. 24 November, Singapore

Leader, D. (2023). Interviewed by Cynthia. 14 November, Singapore

Leader, B. (2023). Interviewed by Cynthia Mak. 10& 21 November, Singapore/San Francisco.

Leader, B. (2024). Interviewed by Cynthia Mak. 22 January, Singapore/San Francisco.

Leader, G. (2023). Interviewed by Cynthia Mak. 23 November, Singapore.

8 The Brompton Happiness Unfolded

Driving a Sustainable Future through Growing Communities of Cyclists and an Ecosystem of Related Micro Businesses

Rebecca Lim Siok Hoon

INTRODUCTION

Invented by Andrew Ritchie in 1975, the Brompton foldable bicycle has become an icon, known ubiquitously worldwide by its brand name. This popularity has led to healthy growth for the British company, which rolled out the millionth Brompton from its production lines in November 2022. For the financial year ending in March 2023, Brompton's turnover stood at £130 million and 80% of the bicycles it produced in a year were sold in 46 countries across the globe, according to the company's chief executive officer, Will Butler-Adams, in an interview with *Cycling Weekly* (Becket, 2023).

This phenomenal growth has been attributed mainly to Butler-Adams' leadership and management since joining the company in 2002. In the 20 years since then, employee headcount multiplied, from 35 to 835 (in 2022), annual turnover grew an impressive 50 times, from £2 million in 2002, and the company "moved factories twice, to accommodate the growth in production from 6,000 Bromptons a year to 95,000" (Butler-Adams & Davies, 2022, p. 2).

In May 2023, the company secured an investment of £19 million from BGF, an investment company with 16 offices across the United Kingdom and Ireland. It is backed by leading global banks such as Barclays, HSBC, and Standard Chartered, and the Ireland Strategic Investment Fund, among others. BGF said in a press release that it had "taken a minority stake to support Brompton's ambition to create urban freedom for happier lives" and spelling out Brompton's strategic focus with the funding, Butler-Adams said (BGF, 2023):

> "Our team at Brompton is brimming with ideas to accelerate our growth through product innovation, storytelling, outstanding stores, and having fun with our amazing community. But if we are really going to go for it, we need to strengthen our balance sheet to give us the confidence to be more ambitious."

With this roadmap, Brompton Bicycle embarked on a business transformation journey for the next phase of its growth. However, its mission remained focused on improving lives through offering convenient urban and last-mile commutes, a pleasant ride in the park, or a sustainable way to tour the world.

As such, this research project explored how Brompton Bicycle (i.e., the client) might achieve its business objectives and, as a result, drive greater impact for a sustainable future. The company's purpose today, as articulated in its Sustainability Report 2023, remains similar to Ritchie's objective in helping people move around the city when he invented the Brompton – to "create urban freedom for happier lives" (Brompton Bicycle, 2024, p. 7).

DOI: 10.1201/9781003469551-8

Background

Context for the Research Project

The Brompton is fundamentally a well-designed folding bicycle (perhaps the best in the market, as widely acknowledged by subject matter experts and casual riders alike). The trifold design enabled the bicycle to be neatly folded into a square, easily carried on public transportation, and stored in small urban spaces, easing commuting woes and city living space constraints. The ease of handling, along with the clean, stylish image of the bicycle, also sparks joy for urban fashionistas, leisure riders, and bike touring enthusiasts, as well as hobbyists and collectors who enjoy customising and modifying their bikes, be it for aesthetics or performance. With the strong brand love that it had built in communities around the world, Brompton has the potential to connect people, and empower more proactively to make the choice to lead a more sustainable lifestyle – be it in daily commuting or touring the world – and as such, be a force for good in the cities of the future.

This also meant an opportunity to drive sales and growth in a socially conscious manner. Brompton bicycles were sold via its retail storefronts – Brompton Junction – online via its website, as well as through authorised dealers. It had also established quality standards for its authorised dealers with a four-tier differentiation of 'approved', 'bronze', 'silver', or 'gold', as indicated on the 'find a store' function on its website. The stores for the brand totalled 1,500 retailers worldwide. The Brompton Junction stores globally were either owned and operated by Brompton, licensed, or franchised. In addition, the company has set up pop-up Brompton Junction stores in Taipei and Dubai. The biggest Brompton Junction store in the world, owned and operated by Brompton, was opened in Beijing, China, in October 2023.

At the opening, the Brompton CEO reflected in a reel shared on his Facebook account (where he was 'friends' with many Brompton enthusiasts around the world): "Nine million bikes in Beijing … we have to bring it back"(Butler-Adams, 2023). China had become the biggest and fastest growing market for the company, and across Asia, the popularity of Bromptons had been on the rise, driven by an affluent middle class that was still growing (Butler, 2022). According to research and forecast by Statista, the middle-class population in the Asia-Pacific region was expected to comprise 3.49 billion people by 2030, two-and-a-half times that in 2015 (Dyvik, 2022). The ASEAN region would likely see much of this growth, with key drivers such as digital transformation, innovation, better healthcare, and welfare and sustainable development in place (Rasjid, 2023).

However, just as these markets were not homogenous, the challenges were also varied and impacted not just by user preferences and market forces but also by policies, regulations, governance, and infrastructure development in each country, and, to a lesser extent, cultural elements. Against this context and the client's strategic priorities, and through conversations and email correspondences with the client (represented mainly by Peter Yuskauskas, Head of Global Community, Brompton Bicycle) from July to November 2023, the project was progressively scoped to focus on understanding the business impact for the company in engaging and growing its communities of riders and the extensive micro economy that existed around Brompton ownership.

The following considerations set the context and foundation for this chapter:

- *Connecting communities across borders*
  The rapid growth of Brompton sales, particularly in Asia in the last decade, could in part be attributed to its growing communities of users/riders who became ambassadors for the brand and helped to drive engagement with non-users and brought converts to the fold with "the formation of Brompton clubs all over Asia, and there [were] now hundreds of them". (Butler-Adams & Davies, 2022, p. 254).

  This plethora of Brompton community groups connected mostly through social media, particularly on Facebook, with pages and groups from all over the world. Active participants visiting a particular country would seek out other Brompton riders there and be hosted on rides by the local community. More active groups would also organise events and activities for local and overseas cyclists. The MY Brompton Day in Malaysia, for example, would see riders from Singapore, Indonesia, and Thailand travelling to Malaysia to participate in the event (MY Brompton Facebook, 2023) and the Brompton World Travellers group that originated in Singapore and now had more than 11,000 global members would organise tours to countries such as Taiwan and Japan (BWT Facebook, 2023).

Brompton Bicycle was aware of the vast network and committed to growing its connection with these groups. CEO Butler-Adams had been personally involved in some of the engagements, be it on social media or at key events, such as the Brompton World Championship – a race event organised by Brompton itself that began in 2006 in Barcelona, but proved so popular with global Brompton riders that it was moved to London two years later (Butler-Adams & Davis, 2022, p. 254). Since then, according to Juliet Scott-Croxford, president of North America for Brompton Bicycle, as shared in an interview with *Forbes*, more than 300 community groups with more than half a million members have mushroomed across 40 countries (Marquis, 2024).

In 2023, the One Millionth Brompton campaign saw the iconic millionth bicycle travelling to 18 countries over the year and, at each location, a community ride was organised with opportunities for riders to take photos with the bicycle, engage with Brompton company representatives, and be part of the storytelling of a historic moment. The event in Singapore on 23 July 2023 saw more than 400 Brompton riders (i.e., the biggest turnout for the global campaign) (Brompton Bicycle Asia, 2023). The company also took the opportunity to have conversations with its communities.

Over coffee with members of the Singapore Brompton community, Yuskauskas shared that Brompton Bicycle was considering setting up a community page to link the global groups and provide a space for connections and interactions. In the same interview with *Forbes*, Scott-Croxford also shared that, in 2024, Brompton Bicycle would be launching "the first official Brompton Community along with a program of rides and activities in our most active cities, to expand opportunities for cyclists to make new friends and lead more socially vibrant lives" (Marquis, 2024).

- ### *Extending the community network to related small businesses*

Beyond the cyclists, this chapter proposes that Brompton considers the network of micro businesses related to its business as part of a wider community network. This vibrant micro economy would include those in the aftermarket products and services sectors, including cycling tours organisation, customisation parts and services for the bicycles, and production of related accessories, such as customised bags for the Brompton, and apparel that reflected the urban casual chic style of the bicycle.

Most of these small business owners would also be Brompton riders, and these businesses provided access to a wider option of products and services, thus contributing to the growth and sustenance of the community.

Further, this provided an opportunity for Brompton to amplify its economic impact both globally and in local markets, by helping to grow an ecosystem of micro businesses. The incomes and jobs this would generate would also create positive social impact.

- ### *Leveraging technology and innovation*

Brompton Bicycle has been widely acknowledged as a design and innovation success story, from the first version hand-built by engineer Ritchie to the launch of its Titanium model (or T Line), its lightest bike yet at 7.45 kgs, in 2022, and new 12-speed models on its P and T Lines with a wider gear ratio close to that of an 11-speed road bicycle in 2024. The development of the groundbreaking T Line took a "highly collaborative process" and a move towards digital skills and "design optimisation regularly found in automotive industries" (Wong, 2022). The company has also focused strongly on improving its manufacturing processes and bolstering capacity over the years with new technology to ensure a "lean and efficient factory", as CEO Butler-Adams put it in an interview with *Manufacturing Today* (Iain, 2022).

This project will consider how Brompton might adopt technology in its efforts to build and grow communities. Through this, insights could be derived to better understand customers and business decisions.

APPROACH AND METHODOLOGY

As the client's scope and focus was on its communities, the approach that this chapter adopted was first to understand the needs and wants of the communities in order to derive key insights to inform the recommendations for the business. This process would involve the following methodologies:

- Design thinking to study user journeys, experiences, and jobs-to-be-done in order to derive insights and personas that will inform the key recommendations.
- Systems thinking, particularly soft systems methodology, in order to navigate and explore the complexities of communities, and as a means to observe and analyse issues that defy structure and involve a high degree of subjectivity (which would apply when looking at human behaviours, needs and desires).
- Secondary research on community-building strategies and related topics includes academic and business resources.

DISCOVER

Based on discussions with the client, a brief was created for this project to ensure a clear scope and focus.

THE CLIENT'S BRIEF

Brompton Bicycle understands that the value of its brand and the growth of the company has been very much anchored on its community of Brompton owners and riders. As our CEO wrote in his book (Butler-Adams & Davies, 2022, p 259):

> "What the Brompton brand means is a community of people, built around a superior product that changes their lives."

We would like to better understand how we can deepen the connections with the various community groups of Brompton enthusiasts around the world and explore how we can support these communities and collaborate with them in our purpose to create urban freedom for happier lives.

Along with the #BromptonOneMillion bicycle tour around the world in 2023, we are visiting our communities in 18 countries and speaking to many of our riders and owners, especially the community leaders and those who are very active in their local communities. For a start, we are considering building a community page on our website to connect the global Brompton communities.

We would love to better understand the business impact of the community and the extensive micro economy that exists around Brompton ownership. We would also be keen to explore how technology could be useful in supporting Brompton's community building initiatives.

FRAMING THE DESIGN CHALLENGE AND PROBLEM STATEMENT

Based on the brief and discussions with the client, the next step was to frame the design challenge from a human-centred perspective, taking reference from the 'inspiration' phase of *The Field Guide to Human-Centered Design* (IDEO, 2015, p. 31).

Using the worksheet provided (IDEO, 2015, p. 165) and after several iterations, the statements were as follows in Figure 8.1.

What is the problem you're trying to solve?

Better connect Brompton enthusiasts around the world and grow the communities.

1) **Take a stab at framing it as a design question.**

   How might we better connect Brompton enthusiasts around the world, and as a result, grow the communities?

2) **Now state the ultimate impact you're trying to have.**

   We want our Brompton communities to be engaged, connected, enjoy the urban freedoms that their Bromptons enabled them to access, and lead happier lives.

3) **What are some possible solutions to your problem?**

   [Think broadly. It's fine to start a project with a hunch or two, but make sure you allow for surprising outcomes.]

   Community microsite or page on Brompton's website; a Global Brompton Facebook Page tapping on and bringing together the groups that already existed and operated on the platform; appoint volunteer community leaders and managers from the current group of leaders and most active and connected members; drive engagement and recruitment of new members through events and activities with the global and local communities; and bring the business ecosystem into the network as part of communities and at the same time offer the riders and owners easy access to this network of related small businesses and service providers.

4) **Finally, write down some of the context and constraints that you're facing.**

   [They could be geographic, technological, time-based, or have to do with the population you're trying to reach.]

   Resources – community management is a time and resource consuming task and will need consistent effort and resourcing.

   Cost – more time and manpower would mean more cost (but this may be mitigated by less expenditure on traditional marketing and promotion).

   Language and communication in some markets / regions.

   Technology and accessibility issues in some markets / regions (for e.g., Facebook in China).

   Differing interests and viewpoints of diverse groups making it challenging to manage a global community page that is universally appealing and may even lead to conflict.

5) **Does your original question need a tweak? Try it again.**

   How might we better understand Brompton communities around the world, as well as bring these groups together to better connect with one another, and as a result, grow the communities?

FIGURE 8.1 Framing the problem statements.

PRIMARY RESEARCH: CONVERSATIONS WITH THE COMMUNITIES

Guided by the research design, the next step was to conduct primary research scoped on the following objectives:

- To better understand the Brompton riding communities and their 'jobs-to-be-done'.
- To better understand the ecosystem of small businesses and independent workers and their 'jobs to be done'.
- To uncover previously held assumptions to identify gaps and opportunities.

Surveys and interviews were conducted with 70 Brompton owners and/or riders and/or persons who either owned a business or worked in an area related to Brompton ownership (e.g., maintenance or modification of the bicycles, designing, and selling accessories related to riding a Brompton, etc.). Respondents were either asked to fill out a form, be interviewed, or a combination of both.

As guided by the IDEO methodology, the surveys and interviews were intended to facilitate the discovery process and not to derive statistically significant data on Brompton ownership. A key point to note was the advice to "go after both the big broad mainstream and those on either extreme of the spectrum" as speaking to outliers would often help ensure that the solutions and recommendations would take into account the needs of just about everyone (IDEO, 2015, p. 49). Furthermore, while mainstream users often provided insights for a 'low hanging fruit' solution, users on extreme ends would sometimes spark ideas not thought of.

In the case of this research, the outliers who could provide 'extreme' insights would be the non-converts and their perceptions of Brompton bicycles and riders. To this end, insights were sought from 12 cyclists who did not ride a Brompton via a short survey and conversations with a few survey respondents.

For purposes of framing the "mainstream" and "extremes" (See Figure 8.2), his project will consider the business owners and independent workers to be on the other end of the spectrum (i.e., from non-owners/riders), as most of them (at least those identified and interviewed for this project) were avid riders who got into the business and vocation driven by their love of the Brompton and their involvement in the cycling community.

Research Observations

Non-converts

Among the handful of non-converts approached, most were riding road and/or mountain bicycles, and all 12 were aware of the Brompton, either because they had friends and family members who owned or rode one or because they had seen it on the road or cycling paths. While the data from this small group might not be statistically significant (and not intended as such), the

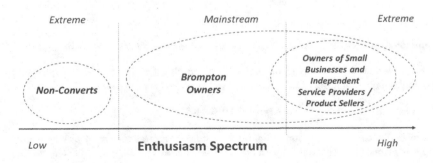

FIGURE 8.2 Extremes and mainstream.

insights were useful in understanding what might motivate those who did not own a Brompton to consider getting one.

Top-of-mind Brand Recall: A quarter said that the first thing they thought of when they thought of a folding bicycle was the Brompton, with most of the rest stating qualities such as convenience, compactness, and portability. When asked to name folding bicycle brands, all the respondents named the Brompton. Other brands recalled included Dahon, Birdy, and Tern (in descending order).

Costly: When asked why they had not considered riding a Brompton, the majority said it was "too expensive"— – which seemed ironic considering that an average road bicycle in Singapore (top-tier brands that were popular here included Pinarello, Cervelo, Specialized, and Canyon) would cost three to four times as much as a basic Brompton (all the respondents lived in Singapore).

The Fold: However, when asked further what might convince them to buy a Brompton, more than half cited convenience and ease of commute, especially with the neat and compact fold, as key factors. Travelling and the relaxed lifestyle associated with riding a Brompton (i.e., social riding with friends, riding to hang out at cafes, etc.) were other considerations.

Mainstream

A survey of 66 Brompton riders was conducted, with two-thirds of the respondents in the Gen X generational profile and the rest equally split between Gen Y and Boomers. This reflected the overall profiles of Brompton riders and owners in general. More than 90% of the respondents lived in Singapore, with the rest based in Malaysia, Indonesia, the United States, and the United Kingdom. Again, the survey was not intended to be statistically significant, focusing more on teasing out attitudes and behaviours. Hence, many of the questions allowed more than one answer and open-ended responses.

Enthusiasm and Brand Love: The respondents were avid riders and Brompton enthusiasts. Almost 90% said they ride their Brompton at least once a week, with half riding two to three times weekly, and more than 15% of all respondents ride daily. Almost 60% of the respondents had been riding their Bromptons between one to four years (only one respondent was new to the fold, riding for less than a year), and 41% of the respondents had been riding for five to more than ten years. Most respondents (80%) owned more than one Brompton.

What the Community Wanted: The respondents were very clear on what they would like to see more of: organised rides and social events (65%); sharing of relevant and valuable information on social media (55%); informative or educational talks, workshops organised by Brompton, licensed retailers or related businesses (47%); a one-stop shop (e.g., on the Brompton website) for information and connecting with other riders (40%); organised interest/hobby-related events (e.g., photography, travel) (32%); membership and rewards programme (29%); and formal peer-support groups (especially for new riders) (23%).

Social Connection, Community, and Influence: The social aspect of riding their Bromptons was apparent, with almost 70% of respondents saying that they had connected a friend or friends to other Brompton riders or groups to "expand our Brompton community here", enabling them to make new friends, and because it was "fun" and "riding is best in good company".

Almost 75% of the respondents said they had influenced a friend and/or a family member to buy a Brompton, with the majority citing the reasons as the quality of the bicycle, ease of the fold for commuting, the desire to influence someone to lead a healthier lifestyle, inviting friends to join them for rides, and because their friends asked them about the Brompton after seeing their social media postings or seeing them ride the Brompton.

Most respondents said they ride their Bromptons at a leisurely pace (80%), using it mainly for social rides, to meet friends and participate in activities (75%), and to commute (50%).

Another clear trend was that Brompton riders loved their "B" as many affectionately referred to their bicycles because riding one aligned with their lifestyle, interests, and hobbies.

Lifestyle, Interests, and Hobbies: Popular activities and interests among the respondents included photography, videography, art (sketching), street art (especially shooting photos of murals and street art with their Bromptons), café hopping, hanging out with friends for coffee, and the ubiquitous Singaporean passion for food, especially local street food. Respondents who elaborated on their responses through conversations or WhatsApp chats said they enjoyed riding their Bromptons to meet other riders for activities such as photographing birds, seeking out street art in various places, and cycling to a meeting point for breakfast, lunch, or dinner.

Some respondents also saw cycling as a workout activity to keep themselves fit, especially for those who work in corporate jobs and can only cycle on the weekends. They felt a little less guilty indulging in good food or coffee and desserts after cycling to the destination.

These responses aligned very much with the Brompton purpose statement of creating "urban freedom for happier lives".

Customisations: Among the respondents, a third said they preferred not to modify the bicycles in any way. Customisations to the Brompton were a fairly widespread practice, especially in Asia, as observed by CEO Butler-Adams in his book (2022, pp. 253–254), for various reasons, including the desire among riders to personalise their bicycles to match their sense of identity, for aesthetic reasons, specific needs (such as touring), and also to improve their ride experience.

Most respondents who modified their bicycles said it was because they toured or intended to travel with their bicycles (40%). The respondents also cited better performance and ride experience as key reasons for modifications – to better climb slopes, hills, and mountainous terrain (about 30%) and to be able to ride faster (about 30%). Some wanted to personalise the bicycle to give it a unique identity (about 30%), and make it look more aesthetically pleasing (about 30%).

There was also a niche group of riders who were themselves technically inclined, having tinkered with other bicycles before riding a Brompton, or who were trained engineers and mechanics who had an interest in the components, parts, and set-up of the Brompton, and saw tinkering with their bicycles as a hobby that was as fulfilling an activity as riding. Some of them, over time, became independent service providers to friends who wanted to modify their bicycles or set up small businesses offering parts, components, maintenance, and customisation services. Observations from this group will be covered in the section below.

Travelling/Bike Touring: The respondents who have travelled overseas with their Bromptons said they loved being able to bring two of their favourite activities together. Those who enjoyed travelling with their Bromptons said it was convenient and fun and liked taking photographs with their Bromptons overseas. Only 10% of the respondents said they did not intend to travel with their Brompton, while 30% said they had not travelled with their Brompton and would love to do so.

Wide Network: For news and information, other than official Brompton sources, respondents primarily relied on social media such as Brompton communities' Facebook groups or pages and word-of-mouth from friends and the communities they belonged to.

Almost 75% of respondents said it was easy or very easy to connect with other Brompton owners and find information on Brompton-related services and businesses (e.g., maintenance, modification, Brompton touring, etc.). Most respondents (almost 80%) would visit either a Brompton Junction or an authorised Brompton dealer for their needs. But respondents also had a wide network of sources and resources, including buying products from websites and e-commerce platforms other than Brompton (e.g., Lazada, Shoppee, or Carousell) and independent resellers and operators.

This perhaps reflected the wide array of services and products available, forming a micro economy related to Brompton ownership.

Mainstream Outliers: The social aspect came across very strongly, many respondents said they enjoyed riding in groups and making new friends, and only a small handful preferred riding alone or with close friends and family. Delving deeper into this mindset, WhatsApp conversations with two of these respondents revealed that they might not be averse to being part of a community but perhaps preferred to be at the periphery and adopt an observer stance rather than being an active participant. One said she was aware of the various Brompton Facebook groups but preferred not to share, comment, or join the group rides. She joined the groups online to get information on riding routes. The other person shared the same sentiment but was not part of the online groups as he was not active on social media. However, he said he would be interested in having access to information on cycling routes and places to visit, including local services and technical support.

Community Leaders, Owners of Small Businesses, and Independent Service Providers/Product Sellers

Another mainstream outlier group would be the community leaders. Some of them also played multiple roles, such as business owners, independent service providers, and/or product sellers. Their feedback was covered in the survey and in-depth interviews. As a result of their deep and long-time involvement in the Brompton groups, the observations and insights derived from this group of highly enthusiastic and passionate respondents provided more food for thought and facilitated the exploration of themes, issues, challenges, opportunities, and ideas.

These highly involved community members were driven mainly by their love of the Brompton – the bicycle itself, the rides, and the friends they had made over the years. In the conversations, what came across strongly was that they were contributing to the community because of their passion, and not because they hoped to make a career or business out of it, and if they did become involved in a business or vocation related to Brompton, it was a natural extension that began with their passion. As community leaders, they hoped to see the groups grow, not just in numbers and scale of events and activities, but also in spirit and connection.

Their feedback included the following points.

A Brompton-driven 'centralised' community initiative: This might be a great idea but:

a. Curate the list of groups diligently and have clear engagement guidelines.
b. Be cognisant of people politics, competing groups, and some with the intention to profit from the groups, rather than serve the communities.
c. Work closely with the ground-up community leaders and let them continue to run things organically.
d. Put in more resources and tangible support (perhaps including some funding) for community events and activities.

The curation of groups to be included within a central, Brompton-managed network could prove a challenge due to the sheer number of existing groups. If groups were to be listed or acknowledged and offered access to the larger Brompton-driven network, clear guidelines and rules of engagement with the community would be required.

Some suggested that Brompton could consider content marketing as a form of acknowledgement for the communities (i.e., telling stories of the events and activities on social media to a global audience).

Connect globally, engage locally: Several interviewees said that Brompton Bicycle could play a major role in bringing different communities around the world together. These efforts could be anchored on its Brompton World Championships (BWC) events.

There was also a tacit understanding among the global communities to support one another's events. This could be challenging for community leaders as most of them travelled to such events at their own expense. With Brompton Bicycle publicising these events and driving participation, some pressure could be taken off organisers in promoting the events and rallying groups in their respective countries to travel to the organising country to participate.

It was also a challenge for Brompton to understand the dynamics of the groups without a deep knowledge of each local market. This was where community leaders felt they had a role to play, and the company should continue identifying key leaders and communities in each market to collaborate with them. Regardless of whether Brompton's planned community initiatives proved successful, these leaders said they would continue to do what they had been doing to engage and grow their groups and connect with other groups. Also, the real influencers were within the community – the leaders, riders, and people who love their Bromptons. Interviewees felt there would be greater value in spending marketing money to support community events.

The "cottage industry" was good for business: While Brompton Bicycle focused on engaging the community of owners and riders, the ecosystem of businesses, independent service providers, and product sellers provided met a wider range of needs for the bicycle owners and fuelled interest and love for the Brompton in creative ways.

These included technical services and products such as third-party parts and components, customisation and maintenance of bicycles, "how to" workshops, rentals, local and overseas riding tours, as well as more lifestyle-oriented products and services, such as customised apparel and bags, stickers, and other paraphernalia. As one interviewee put it, these were "the things that a typical distributor wouldn't do because it [was] too time consuming and there [was] not much money in it but [the community values]". He also saw the micro economy as a potential sales lead generator for the official retailers.

The customisation parts and services micro sector in particular served the needs of a substantial segment of Brompton owners, particularly in Asia, who loved tinkering with their bicycles to personalise their ride. One characteristic of the Brompton, as one interviewee put it, was the relative simplicity of its design, "having just one frame style for many years, making modifications universally applicable". Many hobbyists and collectors would also willingly spend as much, if not more, money on customising their Bromptons as they did on the bicycle.

Set guidelines and standards: While it was understandable that Brompton took a stand on aftermarket modifications when it came to warranty coverage, some interviewees suggested that the company could "associate and yet dissociate" itself. One way was to consider a set of guidelines or accreditation framework to support the customisation micro economy around Brompton ownership, without compromising standards and quality. Such guidelines or frameworks could be akin to what already existed in the larger business ecosystem, another interviewee shared, such as in the information technology sector, where large companies would maintain guidelines and an accreditation process for third party developers and vendors. Another interviewee felt that Brompton Bicycles might prefer to mitigate risks by not directly endorsing the manufacturers and sellers of aftermarket products, but could consider general guidelines to ensure better quality standards in the marketplace.

The pie is big enough: Interviewees were also of the view that subsidiary businesses surrounding Brompton ownership were of value to end users, especially the travel/bicycle touring business and related Brompton rental business.

While Brompton Bicycle operated a Brompton bike hire business in the United Kingdom, whereby users could rent the bicycles from almost 80 docks located around the country for £5 a day, interviewees believed this model would not work in Asia. The UK model was designed to facilitate commuting and hence the price was kept low. The Asian consumer, however, saw the Brompton as a high value brand and would rather rent the bicycle for travel and touring. As such, rental services offered by official Brompton retailers, especially the rental of electric Bromptons to support senior travellers or leisure riders looking to cycle across hilly terrains, could be viable. Asian customers were also likely to be more demanding with the condition of their rental bicycles and hence, a pod system might not work in the Asian markets. One business owner renting Bromptons to riders added that maintenance of the bicycles required substantial time, investment, and effort.

Another business stream was in lifestyle and fashion. A well-designed T-shirt or bag would interest Brompton riders, and a recent design based on the #BromptonOneMillion cities stickers saw a huge surge of orders for the designer, as shared by one of the interviewees. Apart from official collaborations with Brompton, such as the Freitag backpacks, there were several independent brands producing apparel and accessories related to riding a Brompton. "Brompton is a fashion statement – even when you're old, you look younger when you dress up and ride a Brompton", said the interviewee.

These subsidiary streams could fuel more brand love and attract more users to the fold. Interviewees noted that the company has been "generous" in taking a "live and let live" approach. From a broader perspective, supporting a micro economy around Brompton ownership was also a way Brompton Bicycle could contribute to creating social and economic impact (i.e., through the creation of more jobs and economic activities).

Overview of Key Insights

Several key insights emerged from these conversations and observations:

- Brompton had an edge with the quality of its bicycle and its brand.
- Brand loyalty was driven by the community and positive customer experience.
- Experience should be personalised, yet social.
- Connections were key – locally, as well as at a global level – and this included connection back to the Brompton brand and company.
- So was information and knowledge, and ease of access to these.
- Beyond urban freedom, it was also about the freedom to do what people loved and enjoyed.
- The Brompton could be basic and functional, but it was also great fun and a style icon.

Brompton's strong brand recognition and value, even amongst the non-converts, was driven largely by its communities of devotees and strong social media presence. With its compact fold and distinct personality, the Brompton clearly stood out in the foldable category of bicycles. However, with emerging technology such as compact electric personal mobility devices, even the loyalists acknowledged that it might only be a matter of time before a competitor emerged with a more innovative product. What Brompton would have as a leading advantage then would be its community – of users and micro businesses – that were staunchly loyal and continuing to drive brand love and growth.

Underpinning this is a sense of *hygge* that Brompton riders identify with their bicycle. The Danish word that could be used as a noun, adjective, or verb, and with no English equivalent, was widely understood to mean "a quality of cosiness and comfortable conviviality that engenders a feeling of contentment or well-being" (Altman, 2016). *Hygge* was perhaps the most precise and appropriate way to describe the sense of belonging and homeliness many Brompton owners felt in their attachment to their bicycles, the lifestyle associated with riding their Bromptons, and the communities they belong to as Brompton owners and riders.

SECONDARY RESEARCH: LITERATURE REVIEW

Overview – Brand Love

Broadly, a brand community could be construed as "a social aggregation of brand users and their relationships to the brand itself" (Lin et al., 2019, p. 447).

Summarising key observations on this trend of "brand fandom", Meltwater, with expertise in media and social media monitoring and marketing, and consumer insights, explained in its own community blog (Kiely, 2023):

- Community-led growth has become a core focus for businesses and organisations, with many calling themselves "community-centric organisations" in 2022.
- In 2023, more companies are expected to dedicate resources towards building year-round communities for multiple stakeholders, recognising that communities are valuable business assets.

Building brand love was a well-established marketing and customer loyalty strategy. However, with the widespread usage of social media in the past 10 to 15 years, brands have had more opportunities and platforms to engage directly with their customers. Initially, marketers leveraged data and personalisation for targeted online advertising, but with data privacy concerns, consumer fatigue from information overload, and diminishing attention spans, brands began to shift towards more meaningful engagements.

This became what business insights and consulting company McKinsey & Company termed as "the big idea' in 2020s marketing" that enabled businesses to help consumers "express community membership by participating in [their] brand" (2022). Tracking this evolution of how brands engaged with their customers (in Figure 8.3), McKinsey highlighted with community building, brands were now able to "generate much more emotional resonance" (2022).

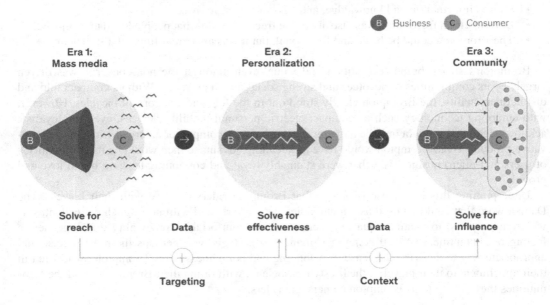

FIGURE 8.3   Evolution of brands' engagement with customers.

(source: McKinsey & Company, 2022).

This emotional connectedness was akin to the concept of flow in positive psychology. Key factors that generated positive flow within the community included group cohesiveness and shared characteristics, as well as good-quality information that was shared and could be obtained within the community that encompassed "accuracy, completeness, and currency" (Lin et al., 2019, p. 448).

Managing and Growing the Community

Clearly, community management would then necessitate deliberate and focused efforts. A substantive body of research on this has emerged across various fields, from business and marketing to anthropology, social theory, and communications studies, and those published by commercial entities such as business consulting firms and marketing and social media agencies on managing online communities. For purposes of this chapter, several key considerations and the most useful and applicable insights for the client have been summarised here, as follows.

Foster Online Engagement to Deepen Identity and Strengthen Loyalty

One of the earliest and most established sectors to leverage communities to grow brand love and maintain customer loyalty would be the automotive industry. Case studies and examples globally found that car brands that understood the consumer's love for their cars and the brand values with which these were associated successfully tapped on the car clubs to engage and nurture their communities. One such study that focused on the car clubs in German-speaking Europe (i.e., in Germany, Austria, and Switzerland) highlighted "the importance of purposely selecting, initiating, managing, and controlling interactions among customers when facilitating brand communities" (Algesheimer et al., 2005, p. 30). Two key learnings from this chapter include:

- Community members' level of participation would subsequently influence their behaviour (including loyalty to the brand).
- Community members with strong knowledge about the brand and products were more involved and experienced a greater sense of identity and engagement.

These findings suggested that community building did not just impact brand love but would ultimately influence profitability as the group influenced members' decision-making. It was important to identify and persuade knowledgeable and experienced customers to join the community group, where they would likely become natural leaders and/or influencers.

Take Care of Different Needs

Other studies found that consumers engaging in online brand communities derived satisfaction from both their "utilitarian" (i.e., information, tips, etc.), as well as "hedonistic" (i.e., fun, engaging social media content, for e.g.) needs being met and this "reward is a key motivating driver of engagement" (Harsandaldeep et al., 2020, p. 3). Hence, it was recommended that brand community managers provide an online platform for their communities to interact with one another and share information and experiences. This would help foster a strong identity with the brand and community – an observation common across several studies.

Other recommendations included that community managers engage with some members directly, offer differentiated benefits based on interest and participation, and potentially offer tangible rewards (such as sales promotions or exclusive sneak peeks for new products). Most importantly, brands would need to ensure enjoyable social media experiences through interesting and differentiated content (which could be either created or curated by community members).

While it was generally agreed that more engaged members of the community could become natural leaders, and continue to lead voluntarily, feeling rewarded by their service to the community, one study made an important observation on new members joining a community and the risk of them dropping off from activities and further participation if ignored by the community managers or other seasoned members. In addition, this could "have its repercussions on the brand-related feelings and actions", and hence, it was recommended that community managers make deliberate efforts to "make new members endogenous to the community through special attention" (Kumar & Nayak, 2018, p. 71).

Take It to the Real World

While online platforms offered an easy way for communities to connect and interact, especially across geographical locations, studies have found that events and activities where community members and managers could bond in person were essential. Conducting fieldwork at community events organised by Jeep, the American maker of sports utility vehicles known as the iconic adventure and off-road automobile brand, McAlexander et al. (2002) observed that strong, temporary bonds were formed at these events, with members being eager to fit in and belong socially. These events in the United States, such as the regional Jeep Jamboree, which brought Jeep owners together for off-road trail driving, and a national Camp Jeep, many members experienced a higher sense of belonging to the community after several days together, "believing they belonged to a broader community that understands and supports them in realising their consumption goals" (McAlexander et al., 2002, p. 42).

DEFINE

In the defining stage, the observations and insights gleaned from primary research were combined to formulate insight statements, which then informed a series of more sharply focused "How Might We" statements that, in turn, led to the refinement of the design challenge and problem statement.

As part of the reframing process, in order to unpack the complexities and better connect the dots, two steps were undertaken:

- A set of personas and corresponding "jobs to be done" were drawn up against a 2X2 quadrant framework of attitudinal and behaviour al insights of Brompton community members.
- A soft systems-rich picture was drawn up to reflect the ecosystem in which the communities existed and to provide a broader context of related issues and trends that could potentially drive change and shifts in the near future.

THE REFRAMING PROCESS

Personas and Jobs-to-Be-Done

Based on the insights from the surveys and interviews, 11 personas were drawn up based on their lifestyle intentions and jobs-to-be-done (Figure 8.4). Many people have more than one set of hobbies or interests, so some community members might identify with more than one persona. Based on the insights gleaned from the primary research, most community members would be able to identify with at least one of the personas. This approach proved useful mainly in clarifying motivations and intentions when designing community engagement recommendations. Apart from key characteristics, a storytelling approach was also adopted, with a narrative crafted for each persona, along with "a persona empathy map about the consumers' feelings and experience" (Elmansy, 2018, 2023).

One dilemma during the process of constructing the personas was whether to include the community manager (i.e., a Brompton staff member) vis-à-vis a community leader (i.e., someone who was a natural influencer or leader from within the community) when considering the"

Persona	Narrative	Think	Feel	Do
Super Connector	Community Leader "Everyone knows me."	Happiness is being able to help others.	Good when sharing and helping.	Connect with people in the community and organise rides and activities.
	Community Manager "I want to know everyone."	Happiness is seeing people come together.	A sense of fulfilment.	Organise rides and activities and connect with people in the community.
Commuter	"It gets me there."	Happiness is not being stuck in traffic jams.	Convenience to go wherever and whenever.	Cycle (and take public transport) to my destination, as far as possible.
Traveller / Adventure Seeker	"I want to see the world (on two wheels)."	Traveller Happiness is experiencing new places and just getting out there.	Sense of awe and wonder (at the world out there).	Traveller Travel and a bonus if I can cycle.
		Adventurer Happiness is challenging myself ... especially on all those climbs, and going the distance, from place to place.		Adventurer Bike touring, and probably other outdoors activities (hiking, camping, etc.).
Socialiser	"I want to ride leisurely and make some friends."	Happiness is riding leisurely. No stress. Get to know some people.	Chill.	Join leisure group rides. Cycle to hang out with friends.
Urbanite	"I love the city life."	Happiness is looking cool while zipping around the city, grabbing coffee, groceries, and hanging out.	Urban life is my vibe.	Riding to fit a hip urbanite's lifestyle.
Fashionista / Style Seeker	"If it's not on IG with an #ootd, it doesn't count."	Happiness is looking about good and being admired.	Gorgeous / Stylish / Cool	Dress up, accessorise (bags, watch, etc.), style – to get the perfect look. Take lots of photos and videos.
Collector	"Just one more to complete the collection."	Happiness is looking at these beautiful things that belong on an altar.	Proud and attached.	Collect limited editions and would also probably modify their bikes with expensive parts and fittings.
Hobbyist	"Jack of all trades."	Happiness is learning new things and spending time on my hobbies.	Curious	Engage in other hobbies such as photography, videography, art, etc.
Tinkerer	"What can I change or fix next?"	Happiness is getting technical with bike improvements, and I'm not bad at it too.	Accomplished	Tinker with my bike. Getting hands dirty.
Amateur Racer	"Fast and furious!"	Happiness is riding hard and fast. It's even more fun on small wheels.	Excited.	Join group rides with other fast riders. Maybe race a little sometimes.
Weekend Fitness Warrior	"It's the weekend. Let's go ride. And maybe have cake afterwards."	Happiness is being able to have cake because I did cycle on Saturday.	Motivated.	Cycle for exercise on the weekends.

FIGURE 8.4 Personas' narratives.

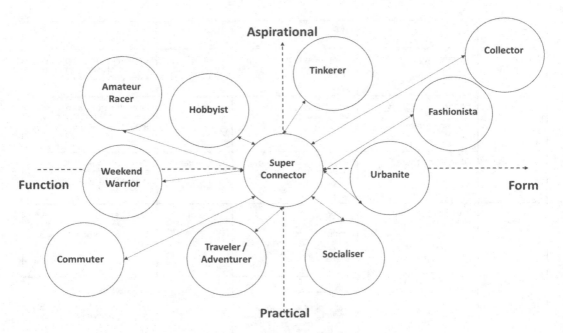

FIGURE 8.5   Personas and "jobs to be done" mapping.

"super connector" persona. Given that the client had a plan proactively to manage community events and activities, the decision was taken to include the community manager and also to reflect the slight differences in motivations and intentions. While it was not within the scope of this chapter to examine internal culture, it could be useful for the company also to consider its staff as part of a broader community along with its owners and riders, and that all employees had the ability to become brand ambassadors in the community, even if they were not community managers.

These personas were then placed within a quadrant framework, with the vertical axis reflecting an attitude scale, from pragmatic to aspirational, and the horizontal axis representing an intention scale, from function to form (Figure 8.5). Personas placed higher on the pragmatic-aspirational scale viewed their Bromptons more as reflecting their aspirations and desires than something that would help them perform a task. Similarly, those towards the right of the function-form scale placed a premium on aesthetics over functionality regarding their Bromptons. Beyond demographic segmentation, personas provided more meaningful insights for deep community engagement.

A Systems View

Systems Rich Picture

For the purposes of this research, a soft systems approach was adopted to more comprehensively consider factors and drivers that may impact the client's efforts to build and grow its communities worldwide (Figure 8.6). This was also represented in a systems rich picture (Figure 8.7).

Potential Gamechanger – the Electric Bicycle and Bicycle Rental

From the considerations above, the electric bicycle and the rental and bicycle sharing space stood out as key opportunities for Brompton Bicycle to grow its communities and business.

Customer	Users, communities, and different types of users and user behaviours (e.g., 'racers', those who tour, social riders, commuters, etc.). These have been mapped out in the personas and jobs-to-be-done frameworks.
Actors	Affiliated businesses and service providers (for e.g., Brompton tour organisers, aftermarket service and products, etc.). Other brands related to cycling and / or the lifestyles associated with riding a Brompton (for e.g., camera equipment, cycling apparel and accessories, etc.). Automotive industry, and other transportation companies (including public transportation operators).
Transformation Process	Competitors, including other folding bicycle brands and new entrants to the market, such as electric bicycles and other personal mobility equipment that could potentially be gamechangers in last mile commuting in cities. This could bring about a more widespread shift in behaviours, with more people willing to adopt more sustainable commuting practices, with more convenient and viable options. While these players were technically competitors, an overall shift and growth in user behaviours could also drive growth of the Brompton riding / ownership communities. Technology could be both an enabler for new opportunities as well as a driver of new challenges and threats. New platforms could make global connections more seamless and interactive. But potential cybersecurity, cybercrimes and data protection issues could also deter people from joining online communities and sharing data and information.
Weltanschhaung (Worldview)	On a global scale, the broader impact includes promoting sustainable practices and integrating businesses into the circular economy. Although more companies, like Brompton Bicycle (now a certified B Corporation), are joining this movement, development rates vary across regions. Increased participation would foster community growth, such as Brompton riders passionate about sustainability. Brompton can leverage its extensive grassroots network to support a ground-up movement. Their bicycle hire service with self-service pods across the UK aims to offer potential buyers the Brompton experience and promote sustainable commuting. However, field experience in London (September 2023) revealed areas for improvement in the digital customer experience, process flexibility for foreign users, and bicycle maintenance. Collaborating with the global bicycle sharing/rental sector presents a significant opportunity. Many startups struggle to expand beyond local markets, and Brompton's international presence could be transformative. Effective policies, city governance, and infrastructure development are crucial to increasing the number of riders and bicycles in cities.
Owner	Brompton Bicycle - a British brand that grew into a globally established one that is recognised for innovation and quality engineering. While the communities' identities were tied more to the bicycle and their riding experiences than to the company, the popularity of the company's CEO at Brompton events worldwide (store openings, rides, etc.) for photos, book and bicycle signings was testimony that the company culture and engagement with the community was still important to grow brand love.
Environmental constraints	These would include geopolitical issues, global economic trends and risks, as well as operational constraints such as supply chain and logistics issues. Recent major events such as Brexit, the Covid pandemic, and wars had impacted the demand and supply of bicycles, shift in consumer behaviours and trends. Of course, outcomes and impact could be both negative as well as positive. As with all such events, it would be hard to predict, given how volatile and interconnected many of these risks and potentially catastrophic events were. The World Economic Forum, in its latest *Global Risks Report, 2023*, created a new term, "polycrisis" to depict "a situation where different risks collide and their interdependency [was] acutely felt" (Markovitz and Heading, 2023).

FIGURE 8.6 A systems view of factors and drivers.

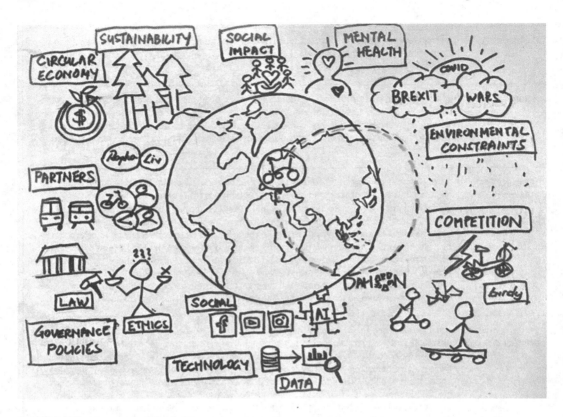

FIGURE 8.7   A systems rich picture.

Even as cycling infrastructure improved in cities, cycling as a form of transportation might still not appeal to those who prefer not to be challenged by slopes and hills, or unpredictable weather conditions. An electric folding bicycle could easily resolve many of these issues and grievances. Someone who would usually commute within a local area could also make the switch from a car or a motorcycle to an electric bicycle that would be less costly to own and maintain. In 2021, Brompton Bicycle launched an inclusive campaign, "I'm Getting On", to get seniors riding its electric bicycle. The company reported a shift with this campaign, seeing more than 55% of its customers aged over 55 buying the electric model, and affirming its view that "everyone should experience the joy and freedom that a bicycle provides, whatever their age" (Morley, 2021).

With higher life expectancy and a greying population in parts of the region, such as in Japan and Singapore, the electric folding bicycle could prove to be the tipping point in bringing the boomer and silent generational profiles into the fold. With more seniors ageing in place, the freedom of personal mobility would also be an attractive feature. For those who wish to travel and tour on their bicycles, a network of Brompton dealers or third-party companies offering rental Brompton electric bicycles and other travel-related services and information would not only be a viable business stream but would also offer this community of riders the freedom and happiness that was intrinsic to Brompton's purpose.

PROBLEM STATEMENT – REFRAMED

Finally, the reframing process involved identifying the key themes that emerged from the research insights (Figure 8.8), creating "How Might We" statements from these (Figure 8.9), and reframing

Write Your Design Challenge

Our design challenge was to bring Brompton communities locally and around the world together to better connect with one another, and as a result, grow the communities.

Three key themes emerged across the research observations and insights.

Theme: Connecting and Connections

Insights:

- Community members thought a 'one-stop' online platform to connect with riders from all over the world was a good idea.
- But they wanted more organised rides and social events for face-to-face interactions.
- They loved interacting not just with their local communities but also to connect with riders from around the world.
- They would also love to meet and connect with other riders with similar interests to them in other areas (e.g., tinkering with their bicycles, photography, etc.).
- Many would bring a friend or family member into a group or to join a ride.
- Some would also willingly meet new friends, take care of new riders and host Brompton riders visiting from overseas.
- Business owners and service providers in the micro economy around Brompton ownership see themselves as part of the community as well, with their own contributions and roles to play as leaders, information providers, and fuelling the brand love in what they do.

Theme: Curation (information, stories, activities, etc.)

Insights:

- Community members were interested in the sharing of useful and relevant information and knowledge.
- They preferred to tap on an extensive network of resources for these (beyond the official Brompton website and social media platforms).
- They tend to trust their friends and people they knew within their communities (including the businesses and service providers in the extended ecosystem) when it came to information that was not from the official Brompton sources.
- Some would like to attend workshops and other learning sessions organised by Brompton and/or their dealers.

Theme: Creativity and Creation

Insights:

- Many Brompton riders tend to be creative types who love to dress up their bicycles and/or themselves to look nice.
- Many like taking nice photos and videos of their bicycles (or having nice shots and reels taken for them) to share on social media.
- Some enjoy following creative and fun social media trends (such as a recent one with AI generated art of themselves as a Funko Pop! figure with their Bromptons).
- The communities are the most important influencers – not some celebrity or digital creator with no connection to the community.
- Riding a Brompton is a lifestyle (at least for most Asian riders and owners) and it is all about the *hygge*.

FIGURE 8.8   Themes and insight statements.

the design challenge and problem statement for solutions and the key recommendations (IDEO, 2015, pp. 80–83, 86–87).

DEVELOP AND DELIVER

With these insights and "How Might We" statements distilling the key learning from primary research and those derived from secondary research, the following recommendations were put together. As stated earlier in the chapter, the scope of this research will focus on the Asian region, drawing from key learnings mainly from Singapore communities, to inform recommendations and strategies for opportunities that could potentially apply in markets such as Southeast Asia and China as well.

Insight:
  • Connecting online with Brompton communities around the world and physical
    connections and activities were equally important.
How Might We:
  • Design a network for connecting the communities easily online, driving virtual
    conversations, as well as fostering more physical interactions within and across
    communities?

Insight:
  • Community members showed a desire to learn and understand more, and welcomed
    relevant and useful information, be it about the bicycle, maintenance and
    modifications, or related topics such as touring on the Brompton, accessories, and
    lifestyle. However, different segments may have different interests and priorities.
How Might We:
  • Design easy and convenient access to useful and relevant information for different
    topics and groups, and yet not inundate or overload them?

Insight:
  • Community members loved the Brompton life – it is all about the *hygge*.
How Might We:
  • Maintain the vibe, generate buzz, and keep the conversations going?

FIGURE 8.9   Create how might we questions.

KEY RECOMMENDATIONS

Based on the research insights, a set of viable recommendations was designed for the client, in line with the key objectives and needs in the brief. The strategic approach was anchored on the three key themes identified: connect, curate, and create. This proposed "3Cs" approach would be the key principles that would guide the activities and initiatives to be implemented – be it organised rides, community events, sharing of created or user-generated content, etc. Underpinning the strategy and the initiatives, technology was viewed as an enabler to facilitate the connections and interactions.

In the framework, the core initiatives that would be most relevant to riders and owners were listed on the left, with some proposed activities specific to the related businesses (including potential brand partners) listed on the right, and initiatives that everyone could participate in were listed in the middle (Figure 8.10).

Consumers/Communities

For Brompton communities, apart from a dedicated microsite or webpage, Brompton Bicycle should consider sustaining its engagement with the various communities on social media, particularly on Facebook, where members of many active Brompton communities consistently engage with one another. Consistency was key in driving conversations and engagements, and this would require some creativity and curation in topics of discussion or fun content.

To address the risk of new joiners feeling left out and/or being disengaged and dropping off, an "adopt a newbie" initiative could be implemented whereby volunteers take new riders out for a ride, and retail stores or dealers help to create "new bike day" content by posting nice photos of the happy customers out for their first rides. This would also help newcomers introduce themselves to the rest of the community.

An "adopt a cause" programme could elevate the nature of community engagements towards furthering a meaningful cause and doing good together. In Singapore, for example, there were several cycling communities that were either formed around a social cause or made it one of their key missions to amplify social responsibility, including Break The Cycle, which worked to reduce

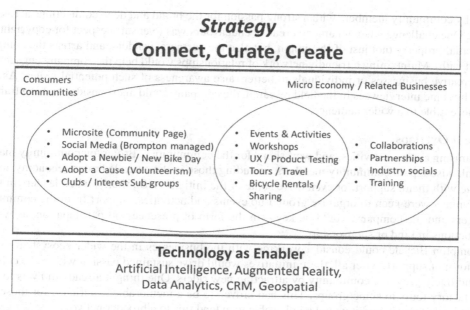

Strategy
Connect, Curate, Create

Consumers
Communities

Micro Economy / Related Businesses

- Microsite (Community Page)
- Social Media (Brompton managed)
- Adopt a Newbie / New Bike Day
- Adopt a Cause (Volunteerism)
- Clubs / Interest Sub-groups

- Events & Activities
- Workshops
- UX / Product Testing
- Tours / Travel
- Bicycle Rentals / Sharing

- Collaborations
- Partnerships
- Industry socials
- Training

Technology as Enabler
Artificial Intelligence, Augmented Reality,
Data Analytics, CRM, Geospatial

FIGURE 8.10 Framework for key recommendations.

recidivism among former offenders through cycling and community support; Cycling Without Age, which engaged cyclists to take senior citizens or disabled persons on rides in trishaws to offer befriending support; and Ride for Rations, an annual fundraising cycling event, whereby the funds raised went towards providing monthly food rations for lower-income families. Community managers, working with passionate leaders or influencers within the communities in each country, could help to drive this as part of the company's corporate social responsibility commitment.

With the myriad of interests and hobbies that members of the communities had that were complementary to riding their Bromptons, there were many opportunities to form sub-groups or mini clubs within the broader communities catering to specific and niche interests (for e.g., a "ride my Brompton for birding photography" group, or a café hopping group, etc.). These activities did not necessarily need to be organised by the community managers and could be driven from the ground up, with some level of curation. Community leaders and members could be co-opted as "hosts" and volunteers for these events or activities.

Micro Economy/Related Businesses

For these communities, it was proposed that Brompton Bicycle considered them to form part of a larger network that could help support its communities. While it may not make business sense for the company to invest as many resources and efforts in engaging with these communities, it could also consider maintaining a healthy network of relationships, perhaps through informal connections or industry social events, particularly for brands that it could potentially partner with in initiatives and activities, such as those listed in the overlapping area of the Venn diagram in *Picture 10*. For example, Brompton could potentially collaborate with cycling-related brands to organise community rides and tap on the partner's network to reach out to a wider community of cyclists. It could also form content partnership with camera brands to organise photography rides, contests, and/or exhibitions to encourage Brompton riders to take and share attractive photos and user generated content that could be amplified on social media.

Brompton could also work with the ecosystem's small businesses and independent workers to produce brand-love-inspired products such as stickers, t-shirts, or bags. These initiatives were often driven by community members with a strong passion for the brand and deep connections across the groups. One challenge when it came to creative endeavours was to ensure respect for copyright and intellectual property (not just of Brompton, but also the partners, vendors, and artists the company worked with). Maintaining a friendly network of relationships would help the company engage with these creative businesses or individuals to better raise awareness of such potential issues. As suggested by some interviewees in the field research, the company could also consider making training more accessible to a wider audience.

Where It Overlaps

In organising events, activities, and workshops for the communities, Brompton community managers could also tap on community members, including those in the extended micro economy, to collaborate with them, support, or even drive some of the initiatives. Similarly, when leaders in these communities were keen to organise ground-up events and activities, support from the community managers and the company would be key – in the form of presence and participation, as well as contributions in kind or sponsorship.

Brompton Bicycle could consider working with the businesses in the wider ecosystem to lend some form of support. After all, these initiatives would drive sustained passion within the community and likely grow the community by attracting new riders. One thing it already did was to open up its factory tours to some groups. This had proven to be a memorable experience for those who had been able to visit, and it would be plausible to extend this to a broader network.

Positive user experiences would be fundamental to sustaining interest and passion for the brand and the bicycle. With a network of community members fiercely devoted to the brand, Brompton could tap on some members for user experience design and product testing, providing useful feedback and data for product innovation and development. Of course, given the need to protect its intellectual property, this would be unlikely to happen at a prototyping stage, but rather, post product launches as part of a feedback process, iteration loop, and/or marketing and publicity strategy.

Underpinned by Technology as an Enabler

Each of these initiatives could be underpinned by technology as an enabler. For example, artificial intelligence (AI) powered chatbots and search engines could assist in providing quick and easy access to information listed on the community microsite. AI-powered translation engines could also help facilitate conversations between community members and managers from different parts of the world who may not all be proficient in English. For fun and engaging storytelling and social media content, generative AI could be a useful creative tool. One example of this was a recent trend in which members of the cycling communities in Singapore used the DALL-E powered AI image generator tool on Microsoft Designer to create images of themselves with their Bromptons as the iconic Funko Pop! figurines (Figure 8.11). This created a viral buzz on Facebook and Instagram among Brompton communities in Asia for about a week, fuelling online interactions as the images were shared.

While augmented reality (AR) tools were still not widely used, it might only be a matter of time before these become integrated into more commonly used platforms, including social media. These tools would be useful in offering an immersive experience, perhaps for those planning to tour with their Bromptons. This, coupled with geospatial technology, perhaps in collaboration with mapping and fitness tracking platforms such as Garmin or Strava, could be a game-changer for route and tour planning. Like AI, AR could also be utilised in content creation – such as an interactive, virtual, or immersive "sneak peek" tour of the new Brompton factory.

The availability of data on community interests and preferences, along with technology platforms as an enabler, would be powerful support for community managers in engaging with the communities online and offline. It should also be noted that personal data should be safeguarded, and ultimately, the human touch and engagement were still key in engaging with the communities, and the availability of data and tools would be useful to aid these interactions (and not replace them).

FIGURE 8.11 Picture generated using Microsoft Designer.

A scan of technology publications and blogs revealed that there was not a single system or solution that could address all the needs of community management, with most companies using a combination of social media, marketing tools, content management systems, and customer relationship management or CRM platforms. One company that has built customer solutions for corporate clients suggested the term "community network manager platform" but admitted that "half of that is marketing" (Davis, 2021). Technology solutions, if any were to be built or adopted, should be driven by organisation goals, in this case, community goals as well, and "a co-developed design, with the technology subsequently selected to support them" and users should be involved in the user experience design early on, to avoid the mistake of being "beholden to technology decisions made too early" (Hinchcliffe, 2008).

One such decision for many brands was to create an app. This has proven to introduce more barriers and friction to consumers rather than help build sticky relationships and ease of transactions. A user survey by mobile consultancy Heady.io showed that more than 90% of users become frustrated when a business requires them to download an app, and almost 80% had abandoned a transaction in the past year when this was mandated (Kostier, 2021). That was not to say an app would not be a platform worth considering, if there was strong brand love and relations, and the app's functions were relevant and well-designed with the users' needs in mind. Harley-Davidson, for one, launched its HOG app in 2023 (Harley-Davidson, 2023). It is proposed that should Brompton wish to consider an app for its communities, the users should be engaged in the design process from the start, and clear value propositions and customer journeys and experiences be mapped, with prototyping and user feedback before proceeding with technology build. Not only would this process and development be a hefty investment, but poor user experiences could also end up undoing all the good work that community leaders and managers could do with less technology.

Ultimately, community building and engagement were still high-touch and human resource-intensive functions, and the client would still need to perform a cost-to-benefit analysis and review internally how much time, money, and resources to invest and which recommendations made the most business sense to prioritise and focus on. Due to the unavailability of business data and information from the client, it would not be possible to build a business case for the recommendations in this chapter.

ETHICAL AND SOCIAL CONSIDERATIONS

Last but not least, while this chapter did not delve deep into ethical and social considerations, these underpinned the approach, methodology, analyses, and recommendations, in alignment with the client's core values and purpose.

In January 2024, Brompton Bicycle announced that it had received B Corp certification as part of its environmental and social responsibility commitment. The certification process is part of a movement that seeks to measure corporations on their social and environmental impact – "a designation that a business is meeting high standards of verified performance, accountability, and transparency on factors from employee benefits and charitable giving to supply chain practices and input materials" (B Lab Global, 2024). To qualify for certification, a company must go through an assessment process involving 300 questions covering topics such as governance, workers, community, environment, and customers.

In an interview with *BikeRadar*, the Brompton CEO pointed to sustainability as one key area that the bicycle industry could work harder on, particularly in improving the practices of the sector's supply chain, saying that it was "probably behind the automotive industry" on this front, even though the cycling industry itself was widely considered to be part of a green economy in promoting more sustainable lifestyles (Portus, 2024). One key opportunity the company could explore with its communities in Asia was to encourage more to consider commuting via cycling or a combination of public transportation and cycling and also to promote bicycle touring as a more sustainable form of travel. While cars were still considered a status symbol in many Asian countries, frustrations with traffic jams and increasing costs might shift commuting behaviours. With many cities building a network of bicycle paths and encouraging commuters to use public transport, as well as a growing consciousness, especially among Millennials and Gen Zs, it was not unfathomable that change could happen in the near future.

On social impact, the company said in its Sustainability Report 2023 that it intended to "work with [its] global Brompton community with a focus on ways to improve mental health, promote active travel, and invest in volunteering" (Brompton Bicycle, 2024). Hence, one of the key recommendations for this project focused on causes and volunteerism. The company could also consider taking the lead in working with other corporations and partners in mental wellness programmes and raising awareness on mental health issues.

PROOF OF CONCEPT – CLIENT FEEDBACK

To assess the feasibility of the proposal, a first draft of the research observations and insights, along with the key recommendations, was shared with the client. The client was invited to provide feedback and comments using a simple SWOT analysis framework with some general guiding questions (Figures 8.12 and 8.13).

For this section, apart from Yuskauskas, Nina Koesman, Head of APAC Marketing for Brompton, was invited to provide inputs. In summary, key stakeholders in the company's community building felt the recommendations aligned with their objectives. They saw opportunities to implement the ideas and delve deeper into issues identified, such as resource constraints and building a data- and metrics-based business case.

STRENGTHS	WEAKNESSES
What does Brompton do well (wrt the observations, insights and recommendations made in this proposal)? What unique resources can you draw on? What do others see as your strengths?	What could Brompton improve (wrt the observations, insights and recommendations made in this proposal)? Where do you have limited resources? What are others likely to see as weaknesses?
Client's Comments **Koesman:** [What Brompton does well is to] be driven by strong purpose, to be a business for good, across all areas, especially our People commitment, and therefore community is at the heart what we do. The research emphasizes the authenticity of our values (only then will our community be a trustworthy space) of all our interactions, in branding/in stores/marketing materials/how we conduct sales/and even our own staff. This is very true, and something we need to continue to implement on our continued journey as B Corp. The research report gives not only an outside in perspective but concrete scientific methods of how organisations can add value to their customers and in the long run their longevity as a business. As all business we need to be a sustainable commercial healthy business to keep investing back into the business/our customers to fulfil our mission. The brand [is perceived as a strength by others]. Even though we are a SME as an organisation, our brand is globally well-loved, and this is because of our community. **Yuskauskas:** The unique character and visual quality of Brompton help it stand out and gather fans around it. There's a shared identity around owning the same bike that's so different from others. I often explain to people who want to know who is in the Brompton community that the riders are very different all over the world, but they unite under the banner of their Brompton. Openness as part of the design. Any person, no matter their age or gender or height or riding style, is equally welcome on a Brompton. This is reflected in the diversity of our community.	Client's Comments **Koesman:** [What Brompton could improve on is to] structure and engage with communities where they are/meet and interact (for e.g., Facebook was an example given) instead of pulling them to spaces we create. As a SME we are continuously resource (people, budget) challenged and considered business choices is always key to our business. We are a very small, lean team and not able to fill the growing space our community has created. We want to be present, show up at Ride Out, organise events etc., and we are getting better, but we do realise there will always be someone we will disappoint. As a business we have many complex business models [which could be perceived as a weakness], like our Junction stores, Franchise, Licensed, Owned & Operated, etc. and the same goes for our operations in every market - direct/dealer/distributor/licenced, etc. From the outside, from a customer perspective, the brand is the same, every Junction is the same. However, we are often bounded by many complexities and therefore need to put in place governance/ stay neutral/be fair, which makes us less warm/personal as brand or team. This is sometimes hard for our communities to understand. **Yuskauskas:** It can be difficult for our teams to understand the many ways Brompton is used and enjoyed across global communities. We hope that by establishing deeper relationships and collaboration we can learn more and get a more holistic understanding. Your research is a big help in this way. Resource is big. As Nina mentioned, our teams are across many projects and priorities so adding in building a new pillar of official and ongoing engagement with communities will take time and patience. This can also be viewed as an opportunity to get more collaborative and less directive, allowing what is growing to continue to grow by bringing in local community leaders to support while adding value where we can.

FIGURE 8.12   Proof of concept with SWOT analysis (Part 1).

OPPORTUNITIES	THREATS
What opportunities are open to you (among the recommendations made here)?	What threats could harm you?
What trends could you take advantage of?	What is your competition doing?
How can you turn your strengths into opportunities?	What threats do your weaknesses expose you to?
Were there any areas the proposal could have done better on?	Were there any areas the proposal could have highlighted or missed out on?
Client's Comments	Client's Comments
Koesman:	**Koesman:**
[Apart from the recommendations made here, our opportunities include:] programmatic community implementation, and prioritisation for community, (hence Peter's dedicated focus/role).	Changing market conditions [could be a threat that could harm us]. We are currently sitting out the slowing demand wave across markets. As a SME we are extremely sensitive to drop in demand, due to cashflow. Especially since all our bikes are hand made in London, this comes with many operational challenges and cost associated with it. So, we need to remain a strong brand, but with a lot of competition, other brands, look-alike brands, this has become extremely challenging.
Not [so much on] trends, but [we could] listen/get involved [and be aware that] actions speaks louder than words. We want our community to be a force for good.	
[We can turn out strengths into opportunities] by doing. We are on a learning journey.	Some [competitors] are copying us. Unfortunately for a price sensitive consumer this has become attractive and the community being inclusive are accepting all types of look-a-like foldies.
The research totally resonates with our beliefs and values (and my personal thinking) as a business.	[We are exposed to the] complexity of managing a global presence with a small business operation and team. We want to be there for our community and customers, including the B2B [community], but this is sometimes hard to manage.
Yuskauskas:	
Put these ideas into action!	**Yuskauskas:**
Establish the business case for community through building engagement and participation in the first year to unlock further investment in programming.	Changing work habits, schedules, and available free time. When riding boomed in 2020, so did community. How do we transition our community offering as members transition in their lifestyles?
	Macroeconomic shifts that may reduce our capacity to invest in new programs.

FIGURE 8.13   Proof of concept with SWOT analysis (Part 2).

CONCLUSION

In conclusion, a brand with a strong community meant having lifelong customers with a genuine sense of belonging to the brand, who would generate publicity for the brand and, as a result, draw new customers. This interconnectedness and interconnection – between brand and community, and among the members of the community – remain relevant to the value and longevity of the brand, as well as the relationships it has helped to build.

However, brands should not assume that their communities are captive audiences for sales and promotions. Citing the book, *Belonging to the Brand*, by Mark Schaefer, who declared that "the next – and last – great marketing strategy is community" and that any organisation that was able to help people belong to something "represents the ultimate marketing achievement", Hyken cautioned that while the community members were willing to evangelise for the brand and provide feedback, loyalty was fundamentally "an emotional connection" and it would be "abusing the privilege and could be an insult" to view the community as a marketing or sales leads list (2023).

Communities would need to exist both online and offline. Emerging technologies and platforms have served as an enabler for fostering online connections and the growth of brand communities, but as the author renowned for his thought leadership on humanity, society, and technology, Yuval Noah Harari, expounded in his book, *21 Lessons for the 21st* Century, in a chapter on community that was also a critique on social media platforms, "[a] crucial step towards uniting humankind is to appreciate that humans have bodies" (2018, p. 91).

Ultimately, it was all about the tribe. The true influencers would be active members or leaders within the tribe who knew people and could rally people to get together, not good-looking celebrities with Instagram accounts with tens of thousands of followers but no connection to the tribe.

For brand communities to be successful, community managers would need to focus on these human connections, which would engender brand love and, in turn, drive everything else. The brand would also need to be authentic and stay true to its core values and purpose, which are the qualities that the community members identified and bonded with.

Lastly, for Brompton communities, unfolding happiness is simply *hygge* (noun, adjective, and verb). This kept them rolling and would bring more to the fold.

REFERENCES

Algesheimer, R., Utpal, M. D., & Hermann, A. (2005). The social influence of brand community: Evidence from European car clubs. *Journal of Marketing*, 69(3), 19–34.

Altman, A., 2016. The year of hygge, the Danish obsession with getting cozy, *New Yorker*, viewed 29 October, 2023, https://www.newyorker.com/culture/culture-desk/the-year-of-hygge-the-danish-obsession-with-getting-cozy

B Lab Global, 2024. Measuring a company's entire social and environmental impact, viewed 10 January 2024, https://www.bcorporation.net/en-us/certification/

Becket, A., 2023. Brompton secures £19 million investment and will use money to 'be more ambitious': British folding bike company backed by investment from BGF, *Cycling Weekly*, viewed 22 September, 2023, https://www.cyclingweekly.com/news/brompton-secures-pound19-million-investment

BGF, 2023. BGF leads £19 million investment in Brompton Bicycle, viewed 22 September 2023, https://www.bgf.co.uk/brompton-bicycle-accelerate-growth/

Brompton Bicycle, 2024. Making Moves, Sustainability Report 2023, viewed 10 January 2024 https://cdn.bfldr.com/XM1XF37H/as/84nrtnvm9vc7wpmtb3ppkk/Brompton_Sustainability_Report_2023

Brompton Bicycle Asia, Facebook, viewed 12 August 2023, https://fb.watch/pvsRkYLqu4/

Brompton World Travellers, Facebook, viewed 10 September 2023, https://www.facebook.com/groups/BromptonWorldTravellers/

Butler, S., 2022. Will Butler-Adams: Brompton Bicycle's evangelist-in-chief, *The Guardian*, viewed 22 September 2023, https://www.theguardian.com/business/2022/aug/20/will-butler-adams-brompton-bicycles-evangelist-in-chief

Butler-Adams, W., 2023. Facebook, viewed 29 October 2023 https://www.facebook.com/reel/232663536261180

Butler-Adams, W., & Davies, D. (2022). *The Brompton: Engineering for change*. Profile Books Ltd.

Davis, K., 2021. How technology is enabling community marketing, *MarTech*, viewed 5 January 2024, https://martech.org/how-technology-is-enabling-community-marketing/

Dyvik, E. H., 2022. Forecast of the global middle class population 2015-2030, *Statista*, viewed 24 September, 2023, https://www.statista.com/statistics/255591/forecast-on-the-worldwide-middle-class-population-by-region/#:~:text=By%202030%2C%20the%20middle%2Dclass,to%20212%20million%20in%202030.

Elmansy, R., 2018. The Role of Storytelling in the Design Process, viewed 5 December 2023 https://www.designorate.com/the-role-of-storytelling-in-the-design-process/

Elmansy, R., 2023. How to Use Persona Empathy Mapping in UX Research, viewed 5 December 2023, https://www.designorate.com/persona-empathy-mapping/

Harari, Y. N. (2018). *21 lessons for the 21st century*. Jonathan Cape, Vintage.

Harley-Davidson, Harley Owners Group, viewed 29 October 2023, https://www.harley-davidson.com/ap/en/content/hog.html

Harley-Davidson, 2023. Harley-Davidson launches H-D™ Membership, a new industry-leading community platform and membership program. *PR Newswire*, viewed 29 October, 2023, https://www.prnewswire.com/news-releases/harley-davidson-launches-h-d-membership-a-new-industry-leading-community-platform-and-membership-program-301855502.html

Harsandaldeep, K., Mandakini, P., JamidUl, I., & Hollebeek, L. D. (2020). The role of brand community identification and reward on consumer brand engagement and brand loyalty in virtual brand communities. *Telematics and Informatics, 46.* https://doi.org/10.1016/j.tele.2019.101321

Hinchcliffe, D., 2008. Ten leading platforms for creating online communities, *ZDNet*, viewed 5 January, 2024, https://www.zdnet.com/article/ten-leading-platforms-for-creating-online-communities/

Hyken, S., 2023. Belonging to the brand: how community is reshaping the marketing landscape, *Forbes*, viewed 28 October, 2023, https://www.forbes.com/sites/shephyken/2023/02/19/belonging-to-the-brand-how-community-is-reshaping-the-marketing-landscape/?sh=4ef213732fb7

Iain, 2022. Brompton bicycles: geared for great rides, *Manufacturing Today*, viewed 24 September 2023, https://manufacturing-today.com/news/brompton-bicycles/

IDEO.org. (2015). *The field guide to human-centered design*. San Francisco.

Kiely, T. J., 2023. The 9 best branded communities for you to join, *Meltwater*, viewed 28 October 2023, https://www.meltwater.com/en/blog/branded-communities

Kostier, J., 2021. 91% of us hate being forced to install apps to do business, costing brands billions, *Forbes*, viewed 5 January, 2024, https://www.forbes.com/sites/johnkoetsier/2021/02/15/91-of-us-hate-being-forced-to-install-apps-to-do-business-costing-brands-billions/?sh=39100475da86

Kumar, J., & Nayak, J. K. (2018). Brand community relationships transitioning into brand relationships: Mediating and moderating mechanisms, *Journal of Retailing and Consumer Services, 45*, 64–73. https://doi.org/10.1016/j.jretconser.2018.08.007

Lin, C. W., Wang, K. Y., Chang, S. H., & Lin, J. A. (2019). Investigating the development of brand loyalty in brand communities from a positive psychology perspective, *Journal of Business Research, 99*, 446–455. https://doi.org/10.1016/j.jbusres.2017.08.033

Markovitz, G., & Heading, S., 2023. Global risks report 2023: we know what the risks are - here's what experts say we can do about it, *World Economic Forum*, viewed 10 January, 2024, https://www.weforum.org/agenda/2023/01/global-risks-report-2023-experts-davos2023/

Marquis, C., 2024. Brompton bicycles' 50-year sustainability journey, *Forbes*, viewed 10 January, 2024, https://www.forbes.com/sites/christophermarquis/2024/01/09/brompton-bicycles-50-year-sustainability-journey/?sh=2357e3835e34

McAlexander, J. H., Schouten, J. W., & Koenig, H. F. (2002). Building brand community. *Journal of Marketing, 66*, 38–54.

McKinsey & Company, 2022. A better way to build a brand: the community flywheel, viewed 28 October, 2023, https://www.mckinsey.com/capabilities/growth-marketing-and-sales/our-insights/a-better-way-to-build-a-brand-the-community-flywheel

Morley, R., 2021. Brompton launches new campaign to promote cycling amongst London's older population, *bikebiz*, viewed 24 September 2023, https://bikebiz.com/brompton-launches-new-campaign-to-promote-cycling-amongst-londons-older-population/

MY Brompton Malaysia, Facebook, viewed 10 September 2023, https://www.facebook.com/groups/320530371333296/

Portus, S., 2024. Cycling needs to catch up with car industry on sustainability, says Brompton CEO, *BikeRadar*, viewed 10 January 2024, https://www.bikeradar.com/news/brompton-b-corp-certification

Rasjid, A., 2023. The ASEAN region is the world's economic dark horse. Here's why, *World Economic Forum*, viewed 29 September 2023, https://www.weforum.org/agenda/2023/08/asean-economic-growth/

Wong, H., 2022. In-house teams: How Brompton is folding new innovation into bike design, *Design Week*, viewed 24 September, 2023 https://www.designweek.co.uk/issues/17-23-january-2022/brompton-design-process/

9 Smart Solutions for Student Well-being

AI-driven Mental Health Support in a Polytechnic

Jeffrey Lee

INTRODUCTION

Mental health is defined by the World Health Organization (2018) as a state of well-being that allows people to realise their full potential, cope with life's stresses, work productively, and contribute meaningfully to their communities. Robust mental health empowers individuals to build healthy relationships, make positive societal contributions, and experience overall wellness (Soon et al., 2022). Mental resilience, which the Mayo Clinic (2020) describes as the ability to recover from setbacks by utilising healthy coping strategies, is another critical dimension of mental well-being. Mentally resilient people can bounce back from hardships and adaptively process negative emotions.

Youth mental health has long been a pressing issue in Singapore, exacerbated by the Covid-19 pandemic that began in early 2020. The virus and ensuing social restrictions inflicted physical and psychological strains, causing many young people to struggle with disrupted routines.

According to a 2020 national population survey, the proportion of those aged 18–29 experiencing poor mental health rose from 16.5% in 2017 to 21.5% (Ministry of Health, 2020). A tragic incident in which a 13-year-old boy was killed by his 16-year-old schoolmate, who had undergone treatment for suicidal thoughts, drew national attention. This prompted policy changes by the Ministry of Education to improve school student mental health support (Elangovan, 2021; Lum, 2023).

Today's Singaporean students face multifaceted challenges that are increasingly difficult to manage. Stressors include the highly competitive education system, toxic home, school, and work environments, major life transition struggles after graduation, financial instability, and the negative impacts of social media on identity and self-image (Cheow, 2019; Criddle, 2021; Gan, 2021). To cope, some turn to unhealthy distractions like excessive online shopping, drinking, and social media (Meah, 2021), further deteriorating their mental health.

However, a deeply rooted stigma towards mental illness persists in Singapore's culture. A 2016 Institute of Mental Health study of over 900 people revealed the prevalence of negative attitudes – 44.5% used derogatory terms like "crazy", "dangerous", and "stupid" to describe those with mental health conditions (Pang et al., 2017). Additionally, 46.2% of 14–18 year old respondents said they would feel intense embarrassment about a personal mental illness diagnosis. This highlights how societal stigma continues to impact mindsets and attitudes surrounding mental health negatively (Pang et al., 2017). Although progress has been made, shame and misconceptions around mental illness still run deep in Singapore's cultural fabric. Overcoming stigma requires challenging insensitive stereotypes and normalising open conversations about mental health.

DOI: 10.1201/9781003469551-9

THE NGEE ANN KONGSI-IPS CITIZENS' PANEL[1]

On 19 March 2022, a Citizens' Panel comprising 55 students from Republic Polytechnic (Soon et al., 2022) was formed with the following objectives:

1. Examine root causes driving increased mental health challenges among young people in Singapore.
2. Design innovative strategies to strengthen mental resilience in youth, empowering them to overcome hardships and uncertainty ahead.
3. Enable youth participation in co-creating and enacting policies regarding mental health and mental resilience.

The Citizens' Panel's evidence-based report and proposed solutions, formulated through extensive research, expert speaker input, and stakeholder consultations, will be an invaluable source of inspiration and reference for my industry research project. This project focuses on improving student mental well-being within the School of Design & Media (SDM) at Nanyang Polytechnic (NYP).

The panel's findings and recommendations, grounded in a rigorous process, will help guide my exploration of this critical issue in my school. Leveraging the report's insights, I aim to develop student-centred strategies that foster positive mental health and resilience among NYP students struggling with academic, social, and emotional pressures.

STRESSORS AFFECTING STUDENTS IN HIGHER EDUCATION

The Citizens' Panel report analysed the mental health challenges facing students in higher education and highlighted several key stressors, including the impact of social media, limited mental health resources, stigma surrounding help-seeking, and a disconnect between students and teachers. This section examines these stressors and the existing solutions and challenges in addressing them.

IMPACT OF SOCIAL MEDIA AND TRADITIONAL MEDIA

Research consistently shows that social media can exacerbate mental health issues such as anxiety and depression among youth. Exposure to stigmatising portrayals of mental illness online and algorithms that push similar harmful content can trap students in a cycle of negative reinforcement (Cinelli et al., 2021; Karim et al., 2020; Sieff, 2003). Furthermore, social media fosters unfavourable comparisons to "perfect" influencer lifestyles, leading to feelings of inadequacy and low self-worth (Jiang & Ngien, 2020).

Traditional media also contributes to mental health issues. The "Werther effect", identified by David P. Phillips (1974), describes the phenomenon of people mimicking suicidal behaviour after seeing it depicted in the media. Such portrayals can have a particularly harmful influence on impressionable young people.

Existing Solution

In Singapore, media literacy education is mandated across all educational levels, aiming to equip students with the skills to critically engage with the media (Ministry of Education, n.d.). At NYP, media literacy is integrated into Life Skills modules, focusing on technical abilities and the critical evaluation of online content. However, this approach may not sufficiently address the broader impacts of the media on mental health, as it tends to emphasise operational skills over ethical and responsible media participation.

Challenges

Current media literacy efforts are limited in scope. They often focus heavily on social media and neglect the risks posed by traditional media. Moreover, these lessons can feel disconnected from students' lives, reducing their effectiveness and relevance.

Panel Recommendation

The Citizens' Panel suggested a film interpretation curriculum to address the media's impact on mental health (Soon et al., 2022). However, this proposal raises concerns about overburdening educators and maintaining student engagement. A more integrated approach, embedding media interpretation within existing subjects, might be more effective and less disruptive.

LIMITED RESOURCES AND HIGH COST OF MENTAL HEALTH TREATMENT

Singapore's mental health treatment is often prohibitively expensive, with out-of-pocket costs deterring many from seeking help. A 2016 survey revealed that 14% of Singaporeans experienced a mental health disorder, yet 78% did not seek treatment due to cost concerns (Institute of Mental Health, 2018).

Existing Solution

Peer support systems in educational institutions, like NYP's Peer Supporter Club (Nanyang Polytechnic, n.d.), provide a cost-effective way to support student mental health. Trained student volunteers offer emotional support and promote help-seeking behaviours, creating a peer-led mental health infrastructure that resonates with students.

Challenges

While peer support programmes are valuable, the selection process often prioritises academic incentives over genuine commitment. This can result in selecting students who are more focused on enhancing their academic records than providing empathetic support, leading to ineffective assistance.

Panel Recommendation

Enhancing the selection process for peer supporters could improve the quality of support, but this must be balanced against the risk of deterring volunteers due to increased demands. A sustainable approach might involve integrating smart systems to support manual interventions, ensuring scalability and effectiveness without overburdening students.

BARRIERS OF MISTRUST AND STIGMA IN HELP-SEEKING

Stigma remains a significant barrier to seeking mental health support. A survey by the National Council of Social Service found that less than a third of adults with mental illness accessed care, often due to cultural attitudes that view mental illness as a personal issue rather than a medical condition (National Council of Social Service, 2017). Negative experiences with mental health professionals, such as dismissive attitudes or breaches of confidentiality, further exacerbate mistrust among youth (Tong & Yip, 2021).

Existing Solution

Singapore offers a range of mental health services, including hotlines, wellness groups, and mobile apps like Wysa and Intellect, designed to provide accessible and personalised mental health support.

Challenges

Despite the availability of resources, stigma and mistrust prevent many from utilising these services. School counsellors must build trust with students by adopting compassionate approaches and consolidating online wellness resources into accessible platforms.

Panel Recommendation

The proposed "Vibesity" app aims to destigmatise mental health by providing an engaging, confidential platform for students. By integrating AI-enabled chatbots with human support, the app could offer 24/7 assistance and personalised recommendations based on academic performance and usage patterns.

ABSENCE OF MEANINGFUL STUDENT-TEACHER CONNECTIONS

The academic focus in Singapore's education system often leads to a lack of emotional connection between students and teachers. Teachers, overwhelmed by administrative duties, may struggle to engage with students on a personal level, leaving emotional stressors unaddressed (Goh, 2020; Poh, 2021).

Existing Solution

Recently, the Ministry of Education has increased efforts to enhance mental health literacy in schools. AI-driven systems are being explored to monitor student performance and identify those needing support, allowing teachers to provide tailored guidance.

Challenges

While these initiatives raise awareness, they may not fully address students' diverse needs. AI-driven monitoring also raises concerns about privacy and psychological safety, which must be carefully managed.

Panel Recommendation

The Citizens' Panel's "BreakTrue" initiative, which aims to foster student-teacher bonds, has potential but may be difficult to implement effectively. Reducing administrative burdens on teachers could help them focus more on building meaningful relationships with students, ultimately enhancing the support system within schools.

The Citizens' Panel report underscores several critical stressors affecting student mental health in higher education, including the pervasive influence of social media, the scarcity of accessible mental health resources, the persistent stigma surrounding help-seeking, and the emotional disconnect between students and teachers. While existing solutions such as media literacy programmes, peer support systems, and AI-driven monitoring tools offer valuable support, significant challenges remain. These include the narrow focus of media literacy efforts, the potential lack of genuine commitment in peer support programmes, the ongoing stigma and mistrust hindering help-seeking, and the difficulty in fostering meaningful student-teacher connections. Addressing these challenges requires a more integrated and compassionate approach that balances innovative technology with human empathy, ensuring that mental health support in educational institutions is both accessible and effective.

CHALLENGES ENCOUNTERED BY NYP STUDENTS

Studying in Singapore, especially within our higher learning institutions like polytechnics, often brings about significant stress, a fact evident in numerous news articles addressing this issue in both local and global media.

Recognising this growing concern, polytechnics and Institute of Technical Education colleges have implemented improved mental health support programmes and integrated them into the curriculum, aiming to enhance students' awareness and understanding of mental wellness (Ang, 2021). This initiative gained even more importance during the Covid-19 lockdown in 2020, with institutions like Temasek Polytechnic (TP), Republic Polytechnic (RP), and the National University of Singapore (NUS) offering structured support systems and ongoing programmes for self-care (Smruthi, 2020).

NYP's intensive three-year diploma programme introduces varying pressures that pose risks to student well-being. While most acclimate successfully, some struggle through difficult transitions – disoriented freshmen, overwhelmed juniors facing spiked workloads, and anxious seniors worrying over graduation prospects.

These phases understandably spark distress like helplessness, panic, or uncertainty about the future. Over time, such strains may culminate in complete mental and physical exhaustion or, in rare but disturbing cases, self-harm. As a senior lecturer, it has been distressing to witness students' worrying behavioural shifts and breakdowns amidst their demanding workloads.

The challenges faced by students at NYP are not unique, underscoring the pressing need for the broader Singaporean polytechnic system to prioritise student mental health and supportive services. This urgency arises from the exacerbated youth mental health crisis due to the pressures and disruptions brought on by the pandemic within the local context.

OBSERVATIONS AS A PERSONAL MENTOR AT NYP

In 2021, I became a senior lecturer at NYP's SDM. Alongside my various non-teaching responsibilities, I was assigned the role of Lead Personal Mentor to 64 freshmen that year. At NYP, educators like myself serve as personal mentors (PEMs), guiding each cohort of students throughout their three-year journey in the institution. The role of a PEM primarily involves acting as a supportive figure, aiding students in academic, administrative, and emotional aspects, akin to a surrogate parent.

Now in their final academic year, these 19-year-olds have presented me with challenges ranging from tardiness and truancy to low academic motivation and socio-emotional issues.

TARDINESS AND TRUANCY

Many teenagers have irregular work and study habits, leading to disrupted sleep cycles and poor time management. My students likewise struggle with sleep deprivation and effective time management. As a result, some habitually arrive late to class or are completely absent without valid reasons. Factors contributing to absenteeism are diverse, including tech overuse, family situations, school atmosphere, peer influences, and student-teacher relationships.

Numerous studies have demonstrated that students with consistent class attendance perform better academically than peers with poor attendance records. As Lukkarinen et al. (2016) affirmed, a definite correlation exists between students' presence in the classroom during instructional sessions and their ability to achieve higher grades.

Problematically in NYP, continued truancy can invoke academic penalties – eventual "grade capping" where one receives barely passing marks despite actual competency.

Persistent Tardiness

Many students habitually arrive late to class, showing disregard for punctuality expectations. This chronic lateness interrupts learning and indicates poor time management abilities.

Frequent Truancy

When students fail to attend lessons regularly, they perform poorly and even potentially fail due to inadequate learning.

POOR ACADEMIC MOTIVATION

Several of my students require additional motivation to finish their assignments. Frequently, these learners lack the drive to challenge themselves, taking a relaxed approach to managing tasks and losing motivation swiftly when faced with difficulties. According to a research paper (Liu et al.,

2004) exploring the Ministry of Education's Project Work initiative in Singapore, it was concluded that students driven by intrinsic motivation tend to stay committed to school, achieve better academic outcomes, demonstrate conceptual understanding, and exhibit better adjustment compared to those primarily motivated by extrinsic factors (Deci & Ryan, 1991).

Low Enthusiasm towards Academic Tasks

Many students lack innate motivation and interest for completing academic work and assignments without external intervention.

Over-reliance on External Prompts

Some students depend heavily on reminders, prompts, and pressure to turn assignments in on time. Without consistent monitoring and nudging, some often struggle to self-manage deadlines.

Work Habits and Time Management

Many students lack skills to maintain a healthy pace on assignments over long periods. They are unable to set and adhere to plans to manage their workload effectively.

LACK OF SOCIO-EMOTIONAL SKILLS

Socio-emotional skills are distinct from cognitive abilities like reading or maths skills. As defined by the OECD (n.d.), such skills help individuals adapt to environments and achieve success. The terms include "personality traits," "resilience", "soft skills", and "21st century skills" (Danner et al., 2021). Most polytechnic students are between 17 and 22 years old. At this stage of their lives, we can reasonably assume students are still developing and mastering important socio-emotional abilities, which carry comparable weight to academic achievements.

Deficits in Problem-solving Skills

Many students lack proficient analytical and solution-finding abilities when confronted with challenges. Some have not developed strategies for breaking down and working through complex issues.

Over-reliance on External Guidance

Numerous students immediately seek out advice, resources, and direction when faced with problems rather than first attempting to devise their own solutions.

Low Mental Resilience

Some students have limited coping strategies and tenacity when encountering setbacks or failure. Deficient emotional resilience leaves them vulnerable to anxiety, stress, or despair in the face of hardship.

NEED FOR PSYCHOSOCIAL SUPPORT

Supporting students in managing both academic and personal demands is a critical role for institutions of higher learning today. Students face a wide and expanding array of psychosocial challenges. In response, schools now provide more counselling, both virtual and in-person, along with online and offline activities and resources. However useful, much of this assistance tends to be reactionary instead of pre-emptively addressing concerns. Students typically receive targeted help only after problems have already escalated. In many cases that I have observed, by that late stage implementing intermediate steps proves inadequate for genuinely supporting the struggling learner.

Absence of Readily Available Support

Students in crisis often lack access to timely assistance and counselling within their environment and routine due to insufficient support to rapidly aid those in distress.

Reactive Mindset Overlooks Proactive Prevention

Current student well-being practices overwhelmingly focus on crisis response rather than ongoing health checks and resilience building. The predominant approach passively waits for serious issues to emerge rather than actively monitoring for early signs of distress.

Data Deficiencies Impair Timely Intervention

Insufficient systemic data tracking of risk factors and wellness metrics prevents early awareness of emerging issues. Students receive attention only after problems have compounded, rather than at pivotal moments when deterioration could still be averted.

IMPACT OF STUDENT MENTAL HEALTH PROGRAMMES IN SINGAPORE UNIVERSITIES AND POLYTECHNICS

To measure the effectiveness of current mental health initiatives at NYP, I administered a survey (Figure 9.1) to my 64 final-year mentees at the SDM. Of these, 58 respondents provided their pain points and desires when gaining access to mental health services for students in higher education institutions in Singapore.

EFFECTIVENESS OF CURRENT INITIATIVES

On average, students rated the effectiveness of existing school mental health initiatives just 2.5 out of 5, indicating substantial room for improvement in addressing student mental health needs. Most were neutral or dissatisfied with the school's counselling services, citing problems with long wait times, poor accessibility, and mediocre support quality. When asked if schools provide adequate mental health resources for students to manage academic demands, 68% were neutral, while 19% felt more could be done to help. This signals a clear gap between student expectations and the current standard of care.

KEY AREAS OF CONCERN

The top three areas of concern highlighted were all related to academic pressures – specifically stress, time management, and motivation/guidance around schoolwork. This aligns with the overarching theme from responses – students feel overwhelmed navigating their studies and desire more support structures. Other prevalent pain points were anxiety/depression, career uncertainty, and loneliness. Financial stress was also mentioned as a contributor to poor mental health. These students also indicated needing more preventative help coping before mental health issues escalate in severity.

Institutions' Existing Approach

In terms of how proactive NYP currently is with mental health support, the average was 2.9 out of 5. This contrasts with the strong agreement (88%) that universities and polytechnics should focus more on preventative and proactive measures targeting student well-being. Students clearly feel existing resources are too reactive and inaccessible compared to the level of stress experienced. Many report turning to external or online alternatives to get the support they need.

Perspectives on an AI-assisted Platform

When presented with the idea of implementing an AI platform to assist with mental health, students expressed moderate enthusiasm, rating their support 3.2 out of 5 on average. Many can foresee benefits in having customised support accessible anytime when needed. However, most also voiced reservations about emotional connection and empathy coming from an artificial system versus human

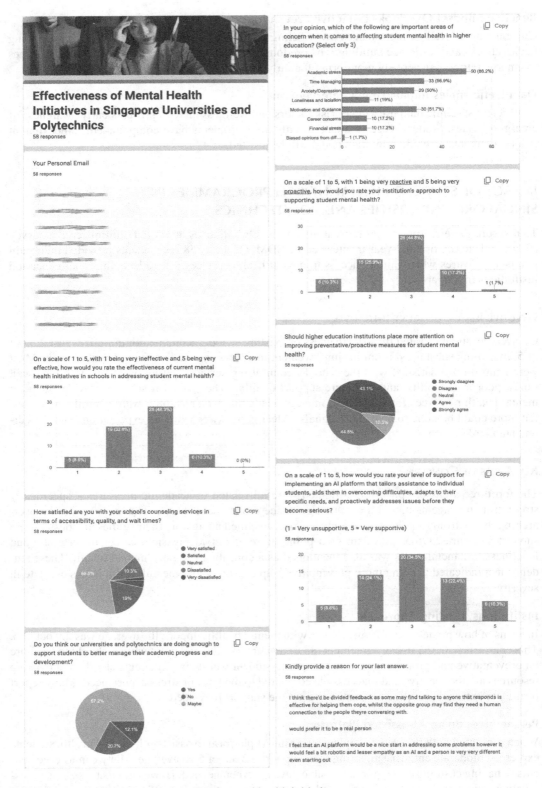

FIGURE 9.1   A survey on effectiveness of mental health initiatives.

counsellors. Several compared it unfavourably to their experiences with generic chatbots unable to understand nuanced situations. Recognising it could aid more minor issues, students still see human intervention as necessary for more serious mental health crises.

Recommendations for an AI-assisted Platform

In terms of how an AI platform could provide the most useful assistance with school-related mental health, students suggested several potential areas to prioritise. The top recommendations were:

- Time Management Assistance: Reminders about assignments/deadlines, progress trackers, schedules, productivity tips.
- Career Guidance: Mind maps of potential career paths and required qualifications.
- Motivational Tips: Encouraging messaging, positive affirmations, reduced procrastination.
- Emotional Support Features: Journalling space, mood monitoring, companion chatbot for venting.
- Triaging Capabilities: Assessing severity of issues, referring more serious cases to human counsellors.
- Customised Learning Support: Study plans tailored to individual pace and strengths.

If there was one key takeaway from this survey on the state of NYP's student mental health programmes, it is this – the average effectiveness rating of 2.5 out of 5 signals that current initiatives among Singaporean higher education institutions are underserving student needs.

The issues brought up by the students speak to a generation experiencing surging anxiety and isolation amidst increasing academic pressures. Students reported poor accessibility, quality, and long wait times for whatever mental health assistance currently exists. The findings reveal an urgent need for all universities and polytechnics to expand their mental health departments and outreach programmes significantly.

BUILDING A CASE FOR AI-ASSISTED SYSTEMS

Implementing comprehensive mental health support spanning thousands of students is easier said than done using traditional methods alone. This is where an AI-powered assisted platform holds such disruptive and transformative promise if designed intentionally. Automation enables scalability in cost-effectively meeting rising needs that would otherwise overwhelm human-only approaches. I believe machine learning technologies have great potential to ease academic burdens if focused first on the biggest stress triggers identified by students themselves.

From personalised reminders to study analysers accurately pacing students and even empathetic chat companions allowing vulnerable moments to unfold stress-free, AI could become the frontline nurturer freeing professional counsellors to support only the most acute cases. By handling the bulk of preventative and minor case guidance autonomously, AI systems can alleviate the unsustainable strain on existing teachers, counsellors, and mentors. A hybrid framework combining scalable technology and reserved human expertise can cater to both ends of the mental health spectrum students require.

However, we must proceed cautiously and avoid overpromising on emotional connection capabilities clearly still lacking today. These students rightly noted even the most advanced chatbots cannot fully substitute for human empathy and understanding in times of deep personal crisis. However, research shows chatbots can screen for mental health issues by asking relevant diagnostic questions (Vaidyam et al., 2019). They may inquire about mood, stress, energy, sleep, etc. (Denecke et al., 2021). Based on self-reported symptoms, chatbots can analyse severity to recommend therapies like lifestyle changes, or alert a professional if immediate intervention seems warranted (Minerva & Giubilini, 2023).

A Promising Yet Challenging Prospect

Building false confidence in AI to solve every mental health issue would be repeating mistakes of past technological overreach. Responsible design with transparent limitations is vital so that students have realistic expectations of when human intervention is necessary. AI assistance shows immense promise in helping minimise certain problems and stresses while referring bigger issues to professionals immediately.

Overall, there is a resounding consensus among my students that more proactive and holistic mental health support is needed to address academic-related stress and anxiety. While an AI-assisted platform has merits in improving convenience and access, it should complement rather than attempt to replace necessary human empathy and counselling. The most promising avenue is using AI's strengths for preventative academic guidance, motivation, and initial assessments, while maintaining the option for students to escalate to real mental health professionals when needed. With thoughtful design focused on core stress points highlighted in this survey, an AI mentor methodology could be a valuable addition to improve overall student wellness and ability to thrive within higher education programmes.

THE PROBLEM STATEMENT

"How might we leverage machine learning to design pre-emptive support systems for student mental health in higher education?"

EXISTING WELL-BEING APP SOLUTIONS

Preliminary review of currently existing systems revealed few platforms that serve our needs. Institutional student portals do not count, as they mainly broadcast information for students to seek out, rather than proactively guiding them. Also, these portals supply uniform content, failing to leverage customised historical data for individual students.

In Singapore, the most common student data platforms parents are familiar with are Parent Gateway (https://pg.moe.edu.sg) and Class Dojo (https://www.classdojo.com). While neither integrates extensive features, both enable school-to-parent communications, with ClassDojo also facilitating some teacher-student interactions regarding school activities.

However, these apps target younger students and give parents overriding access and control. So while ClassDojo comes closer to individualised student engagement, it remains fundamentally an administrative tool for parents and teachers, rather than being learner-driven. An opportunity exists to develop a more mature, student-centric platform for older learners that puts their needs first.

Since there are no direct case studies of a unified app with the features I desire, I will need to explore existing AI-based services covering these capabilities individually with the goal of analysing apps that provide personalised assistance, adaptive learning, coaching, and other relevant functions. Once I collect and evaluate the concepts and technologies enabling each stand-alone app, I will attempt to design a new AI-powered app solution.

REFLECTLY (https://reflectly.app/)

Reflectly (Figure 9.2) offers an AI guided journalling experience that helps users structure and reflect upon their daily thoughts and problems (Bell, 2023a). Touted as a personal mental health companion, it was founded in 2017 by Jakob Mikkelsen, Jacob Kristensen, and Daniel Vestergaard and is currently based in Copenhagen. The core purpose of Reflectly is to help people gain emotional intelligence, self-awareness, and mental well-being through daily structured journalling, mood tracking, and inspirational quotes.

FIGURE 9.2 Reflectly app interface (Reflectly: The World's First Intelligent Journal, n.d.)

KEY FEATURES

1. Mood Tracking and Journalling: Reflectly's main feature is mood tracking and journalling. Users select an emoji to represent their mood, identify contributing factors, and then write a journal entry to reflect on their emotions (Bell, 2023a).
2. Inspirational Quotes: Reflectly provides inspirational quotes across various categories like personal growth, calming down, hardship, inspiration, mental health, and productivity (Bell, 2023a).
3. Mood Graph: This visualises the user's mood data over time on a line graph. It also breaks down further insights like the user's most/least common moods, top factors influencing their happiness/sadness, frequently logged activities, and prevalent feelings (Bell, 2023a).

WYSA (https://www.wysa.io/)

Wysa is an AI-powered chatbot app (Figure 9.3) that provides convenient support for managing mental health conditions like stress, anxiety, and depression (Iqbal, 2023). Created in 2016 by founder Jo Aggarwal, the app aims to increase accessible, stigma-free mental health assistance (Bell, 2023b). By leveraging emotionally intelligent conversations, Wysa offers users an always-available platform for tracking moods, applying coping methods, obtaining self-care advice, and receiving empathetic encouragement around the clock. The core purpose is to help people handle psychological issues through education, tracking, and emotional support.

FIGURE 9.3 Wysa app features and interface (BlahTherapy: Top 4 AI apps to use for Therapy, 2023).

KEY FEATURES

1. AI ChatBot: Wysa's chatbot is an always-accessible conversational agent trained to provide emotional support through natural language processing and algorithms. It uses machine learning on extensive data to comprehend user messages and reply with helpful, empathetic responses (Iqbal, 2023).
2. Self-Help Tools Library: Over 150 self-care activities grounded in evidence-based cognitive behavioural therapies (CBT), meditation, breathing exercises, and mindfulness. This comprehensive library aims to help users manage mental health issues including stress, anxiety, and depression (Iqbal, 2023).
3. Human Therapist Support: Wysa provides paid access to text messaging support from licensed human therapists. This gives users who need more in-depth mental health care than the automated bot can offer the option to upgrade to one-on-one counselling (Iqbal, 2023).

TREVOR AI (https://www.trevorai.com/)

Trevor AI is an AI-powered scheduling and productivity app (Figure 9.4), created in 2016 by founders George Petrov and Dmitry Yudakov of Trevor Labs Ltd (n.d.). The core offering of Trevor AI is an intelligent task assistant that integrates with users' calendars to enable efficient time blocking and task prioritisation.

The app aims to boost productivity through features like predictive auto-scheduling based on past activity, consolidated organisation of tasks and events, and an overarching workflow that reduces clutter across disjoint tools and channels. Trevor AI specifically introduces the concept of time blocking, a focused time allotted to a single task without distractions or multitasking to encourage "deep work". First coined by Cal Newport (2021), "deep work" represents performing high-level tasks with full distraction-free concentration, pushing cognitive limits to drive unique value and skill growth through focused immersion.

KEY FEATURES

1. Calendar Synchronisation: Trevor AI enables users to sync multiple Google and Outlook calendars into its platform without needing to restructure them. It generates a new unified Trevor calendar and layer tasks on top separately. This allows users to connect disparate

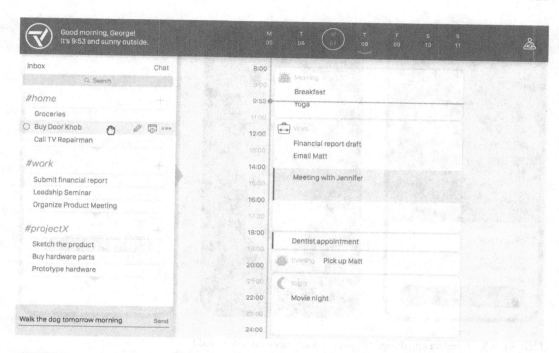

FIGURE 9.4 Trevor AI app interface (The Map Book: Time Management with AI, n.d.).

calendar accounts while still maintaining full control over events and task prioritisation via Trevor AI's interface (Rebelo, 2023).

2. Time Blocking: Time blocking is a productivity technique integrating calendars and to-do lists. It allocates specific time slots for tasks, enabling focused work on each activity without distractions based on priority and realistic timing. This structured approach assigns designated blocks to allow deep concentrate on singular tasks.

3. Machine Learning Scheduling: Trevor AI automatically recommends durations for tasks based on a user's historical time blocking patterns. These predictive suggestions improve in accuracy over time as more data is gathered, aiming to save users effort on manual duration estimates and micro-management. Although automated, the suggestions can easily be overridden in seconds, as needed.

INTELLECT (https://intellect.co)

Singapore-based mental health app Intellect as shown in Figure 9.5 (n.d.), is founded by Theodoric Chew and Anurag Chutani with the goal of providing the most comprehensive digital mental health platform in Asia. Although not currently AI-powered, Intellect recommends mental health programmes to users based on their mood tracking and app usage data. It also matches users with specific therapists using algorithms that take into account both the users' needs as well as the therapists' areas of specialisation (Goh, 2022). Intellect implements zero-knowledge encryption to keep all personal data securely on an individual's own device, thereby ensuring complete confidentiality of sensitive data.

KEY FEATURES:

1. Learning Paths: 10+ self-guided programs grounded in cognitive behavioural therapy (CBT) topics useful topics such as self-esteem, anxiety, coping mechanism, and procrastination (Russ, 2022).

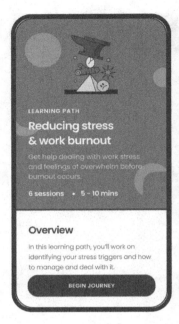

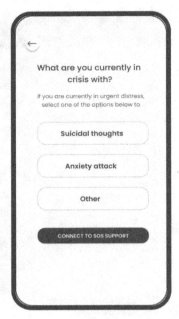

FIGURE 9.5   Intellect app design features and interface (Intellect, n.d.).

2. Mood Tracker: Enables users to log and monitor their emotional states over time by identifying trends which can provide insights and self-awareness into emotional triggers and
   well-being patterns to aid reflection and improvement.
3. Rescue Sessions: Random notifications leading to a choice of 15 on-demand sessions covering stress, disappointment, feeling lost, anger, and other situations provide helpful intervention during times of need (Russ, 2022).
4. Personal Coaching and Therapy: Connect with "Intellect certified" behavioural health
   coaches through chats or calls without the need to schedule an in-person session.

HABITICA (https://www.habitica.com)

Originally launched in 2013 as an open-source project called HabitRPG by Tyler Renelle and Vicky
Hsu, Habitica (renamed in 2015) is a productivity/habit-building app leveraging gamification techniques (Figure 9.6). Created to incentivise positive routines via game rewards and features, Habitica
has grown over time into an active community platform with millions of users. After several major
updates since its inception, it continues to drive habit/productivity improvement through its motivational gaming framework centred around characters, quests, and achievements.

Habitica can help college students stay organised, manage time, and build study habits. Students
can log assignments as tasks, define study goals, and track progress. Its gamifying elements make
completing mundane work feel more engaging and rewarding.

Key Features:

1. Tasks Gamification: Habitica gamifies tasks and goals by making them into monsters and
   boss battles. By finishing real-world tasks, users can defeat the monsters in-app, level up
   their character, unlock new powers, earn prizes, and join group adventures. Completing
   responsibilities earns experience points just like in a game.

FIGURE 9.6 Habitica app design features and interface (YourStory, 2022).

2. Rewards and Task Streaks: A habit-building feature that incentivises users to create and regularly monitor tasks. By tracking task completion streaks and offering gaming rewards like gold and experience points for sustained consistency, Habitica motivates the development of persistent behaviours through positive reinforcement.

DESIGNING A MOBILE MENTAL HEALTH SUPPORT SYSTEM

Inspired by insights uncovered by the Ngee Ann Kongsi-IPS Citizens' Panel report and youth teams' proposals like Vibesity, I seek to develop an effective solution for learner mental health concerns in Singapore's education institutions. Given students' ubiquitous mobile usage, Vibesity's concept of an app-based support system shows promise if fully realised.

Leveraging research on existing wellness apps and feedback from my NYP student mentees, my goal is to design a multi-platform digital ecosystem to proactively alleviate learner mental health issues. This would help them cope with academic and personal issues while using machine learning automation to reduce educator's manual workloads. By referencing the feedback from my initial student survey on student pain points and concerns, I aim to prototype a robust, pre-emptive platform meeting needs across stakeholders – institutional learners, teachers, and administrators alike.

The end-to-end solution will blend preventative self-care features with data analytics to both empower student psychosocial skills as well as flag at-risk cases for intervention. My vision is a customised pre-emptive support network enhancing mental wellness and growth for our learners within the NYP SDM.

EVALUATING FEATURES TO INCLUDE IN THE APP

To determine what are possible features to incorporate into the platform, I would dive into the feedback provided by my mentees in their survey in order to understand their needs.

Academic Time Management

Multiple students indicated needing time management assistance for handling academic workloads and commitments, insights we will incorporate into the mental health app design. Specific functionality suggested include personalised smart scheduling features able to:

1. Automatically adjust schedules to meet deadlines while monitoring user's mental/emotional state to prevent overexertion.
2. Build awareness of each learner's optimal pace and tasks/schedule volume tolerance thresholds over time to promote a better work-life balance.
3. Incorporate "assignment due" alerts and lesson-specific task lists, while providing reminders to take self-care breaks.
4. Send alerts and notifications about submission deadlines or lesson events/updates to help users stay efficiently organised.

Additional capabilities like AI-powered scheduling optimisation tailored to an individual student's habits, energy levels, and workload capacities could promote productivity, while safeguarding mental well-being even amidst academic crunch times. Finding the learner's personalised pace can make balancing studies with self-care more achievable for learners.

Learner Motivation

Surveyed mentees highlighted motivation issues as a key struggle impacting academic productivity and mental well-being. Some specifically suggested gamifying work tasks through motivational reward features to boost enjoyment of otherwise dreaded responsibilities. Capabilities that could incentivise consistency by making task tracking more engaging and satisfying may thus aid students prone to procrastination, distraction, or scheduling paralysis:

1. Gamify work/study trackers with motivational incentives and effort-validating rewards to drive ongoing progress without distractions, boosting agency and enjoyment for typically draining tasks.
2. Aim judiciously to gamify only essential elements to motivate progress, without introducing unnecessary complexity that may undermine sustainability.

My goal would be to introduce methods to motivate students in a personalised way – removing tedium and promoting healthy, productive habits that feel rewarding. Targeted gamification and adjustments can make tasks feel more satisfying and empowering.

Education and Career Guidance

Interestingly, some mentees highlighted the need for better access to career and guidance counselling. A few expressed interest in productivity and study tips to enhance learning capabilities. Students also noted the link between academic progress and their future career path, seeking clarity into post-grad opportunities.

1. Tips for improving time management, concentration, information retention, etc. to work/learn smarter.
2. Career pathway and higher education recommendations based on analysis of the individual's academic strengths, gaps, and personal interests mapped to subject performance and projected career prospects.

The platform could provide easily digestible self-help articles and short quizzes to build student knowledge and coping skills over time. Serving bite-sized tips on managing emotions, stress, and personal growth through short-form content formats allows easy engagement with useful techniques students can directly apply. These could yield compounding benefits for learners by inculcating both academic and career resiliency.

MENTAL WELLNESS TRACKING

Students in my survey proposed capabilities to track personal emotional and mental health data to equip chatbots or counsellors with enhanced context and visibility into each individual's current anxiety, stress levels, and states over time. Additionally, they suggested proactively pushing motivational tips and tailored advice oriented around a user's condition and sentiment.

1. User Mood/Health Logging to generate emotional state visibility that enables adaptive recommendations and motivation nudges aligned to the individual.
2. AI-powered Health Check-ins and wellness surveys to assess user needs routinely.
3. Counsellor Escalation Capability when detected states show risk factors or insufficient improvement over time, alongside continuous data analysis to customise and refine AI support interventions further.

The goal would be systems that can unobtrusively gather user emotional health inputs, then leverage these insights to deliver appropriately-timed interventions, resources, and human intervention aligned to each learner's needs as they evolve.

COMPANIONSHIP AND EMOTIONAL SUPPORT

From the survey, the students were concerned that AI lacks human empathy and emotional skills. But some noted AI's judgement-free 24/7 availability and convenience as its biggest advantage. It also revealed that students prefer human conversations, but accept AI for mild stress relief with ability to escalate to human experts for growing severity or complexity. Their main desire is a blended model with the accessibility and approachability of AI complemented by the understanding and care from human counsellors (Figure 9.7).

1. An AI "care buddy" for ranting and venting stress without feeling alone, serving as an accessible outlet for those seeking a listening ear to release frustrations.
2. Customisable bot personality/traits to suit user preferences.
3. Hybrid model with thresholds before escalating from AI to human at certain risk levels.

Areas of Concern	Recommended App Feature
Academic Time Management	1. Calendar Synchronisation 2. Time Block 3. Assisted Scheduling
Learner Motivation	1. Task Gamification 2. Rewards & Task Streaks
Education & Career Guidance	1. Learning Paths 2. Self-Help Tools
Mental Wellness Tracking	1. Mood Tracker 2. Mood Graph 3. Inspirational Quotes
Companionship and Emotional Support	1. Chat Buddy 2. Rescue Session 3. Coaching & Therapy 4. Human Therapist Support

FIGURE 9.7 App feature summary.

FIGURE 9.8   MAESTRO mobile app.

Students were open to using AI chatbots to start the support process to access the situation. But as persistence, severity, or complexity of issues grow over time, they emphasised needing to transition to empathetic human expertise. Despite the majority of my students being wary of AI assistance in mental health consulting and a strong preference for human intervention, many agree that a hybrid solution of both machines and humans could be a promising prospect.

ORCHESTRATING A HOLISTIC SOLUTION

Introducing MAESTRO (Machine Assistance to Enhance Student Resilience & Outcomes), an all-in-one mobile application prototype (see Figures 9.8 and 9.9) that supports and assists polytechnic students throughout their academic journey, MAESTRO synthesises key insights from my qualitative and quantitative research findings that examine the challenges faced by students at NYP.

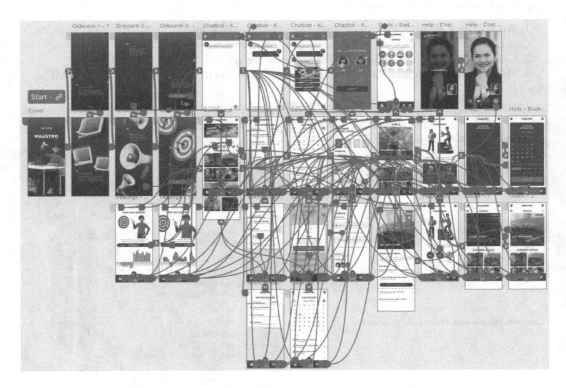

FIGURE 9.9   MAESTRO mobile app (Adobe XD).

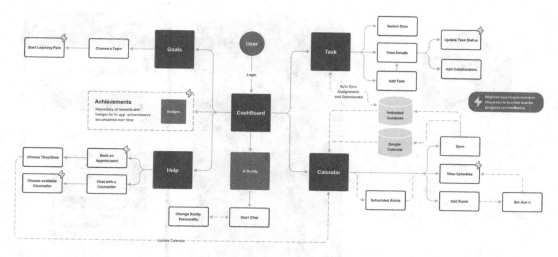

FIGURE 9.10 MAESTRO app user journey flow incorporating the recommended features from my research.

Leveraging artificial intelligence (AI) technology, MAESTRO acts as an autonomous personal assistant designed to address the diverse issues and problems that polytechnic students encounter daily. Its intelligent features integrate seamlessly with an individual student's timetable, co-curricular activities, assignments, and overall workload to help them stay organised and on top of their responsibilities (see Figure 9.10).

Beyond just organisation, MAESTRO contains customised features focused on student well-being. Students can confidentially access mental resilience resources, stress relievers, self-improvement skills, and more. MAESTRO will monitor the students' usage patterns and encourage them to develop self-care skills and good learner habits, free of direct human intervention. It is personalised to each student's strengths and academic stages through ongoing habits tracking and analysis over their three years at the polytechnic.

I believe MAESTRO provides a scalable solution that leverages technology to offer the 24/7 personalised support students need to thrive in and out of the classroom.

THE DASHBOARD

The MAESTRO dashboard is the launch pad for accessing the full suite of resources. At the top, a hero carousel displays the latest news and important announcements from the school, keeping users updated without having to search external sources.

Below the banner lies the "How Am I Feeling?" self-check-in section. Designed to be intentionally casual and judgement-free, this section allows students to select an emoji that represents their predominant mood or energy level that day. Over time, this daily check-in aggregates students' emotional patterns, allowing MAESTRO to determine when additional mental health resources or support suggestions may be beneficial.

MEET KAI AND MAI

MAESTRO users are guided by KAi and MAi, two charming AI chat buddies who serve as their personal friends within the app and as reliable companions to help them throughout their NYP life. KAi and MAi are visualised as friendly peer supporters, and are customisable by each user with their preferred gender, conversation style, and personality attributes to match their personal preferences (see Figure 9.11).

AI Chat Buddy
Natural language messaging interface that assist students on school related matters. **KAi** will also make use of this chat space to alert students of important announcements that require their attention or remind them to take breaks throughout a long day.

Choose your buddy
Students can use the personality preference for their AI companion.

FIGURE 9.11 AI chat buddy.

Operating as more than just a robotic scheduler, KAi's (or MAi's) advanced natural language processing allows rich, meaningful dialogues where students can share emerging needs. It offers individually tailored advice and monitors academic progress and general well-being. If further intervention is required, the chat buddy connects students with appropriate human specialists, while maintaining utmost privacy with in-device encryption similar to Intellect's zero-knowledge encryption feature.

KAi and Mai provides 24/7 availability for casual chats, goal-setting, or just lending a listening ear, which transforms MAESTRO from a functional utility into an app experience intentionally designed to foster personal connections and nurture growth.

CATCHING UP WITH THE CALENDAR

The MAESTRO calendar integrates users' existing iOS or Google calendars and automatically syncs their school timetable and event information (see Figure 9.12). By bringing these calendars together into a centralised hub, students maintain oversight of all their activities, timetables, and assignments in one place.

The calendar sends timely alerts leading up to events to help students prepare and transition smoothly throughout their day. For instance, 30-minute "Pack your bags" reminders nudge students to head to class on time. These personalised notifications aim to instil punctuality and reduce tardiness/truancies. Certain essential alerts like upcoming lessons cannot be disabled and must be acknowledged before the message can be dismissed.

MAESTRO's calendar incorporates machine learning to assess periods when students' schedules appear overloaded. It then suggests rescheduling non-critical appointments to open up breathing room. Students can categorise events into colour-coded types, designating specific activities as focus time to avoid unnecessary distractions. By effectively alleviating last-minute stresses by keeping users on-track with personalised alerts and time management, MAESTRO empowers students to create balanced schedules that support productivity and well-being.

FIGURE 9.12 MAESTRO calendar.

TASKS TRACKING

MAESTRO's task tracking allows students to monitor daily activities closely, with completed tasks prominently marked for a satisfying sense of progress. When adding to-do items, customisable fields capture key details like type, steps, current status, and due dates, and personalised reminders boost organisation (see Figure 9.13).

For collaborative projects, teammates can be tagged within tasks for shared visibility. This allows all members to view the same up-to-date completion status through a personalised progress dashboard or the tasks themselves. By tying progress to individual students, the transparency regarding who has finished portions inculcates a sense of accountability.

Over time, completed tasks accrue points towards in-app achievements, motivating students to maintain consistency. This cycle of task tracking, peer visibility, and reward aims to instil strong time management and discipline to promote personal organisation and collaboration skills.

Task Status
As the student organise their daily tasks, machine learning will adapt to their usage patterns and recommend time blocks for more efficient scheduling and focus.

Project Collaboration
Students may view status of individual or group tasks and share/add it among multiple collaborators.

FIGURE 9.13   MAESTRO task tracking.

Learning Life Skills

MAESTRO goals provide bite-sized self-development lessons to build student life skills. Goals include time management, motivation strategies, career guidance, and more. Each goal has digestible content like articles or short videos (see Figure 9.14). For example, the "Overcome Procrastination" goal shares methods to combat habits and negative behaviours.

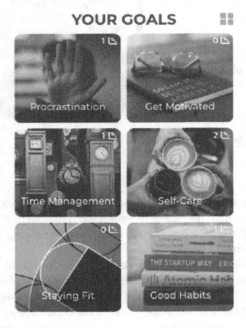

FIGURE 9.14   MAESTRO life goals.

Goals Library
AI would recommend lessons based on the student's app usage patterns along with default ones.

Skills Series
Each goal may be broken down into bite-sized parts consisting of articles, videos or quizzes.

Achievement Badges
Badges are awarded for successful completion of in-app activities such as task streaks, goal achievements, etc.

FIGURE 9.14   *(Continued)*

After lesson content, quizzes reinforce concepts. Completed goals earn students progress badges and lifelong takeaways applicable to academics or personal growth. Machine learning tracks student progress to recommend progressive follow-up or adjacent skills training. This scaffolds a student's development of crucial skills for thriving in school and life.

Direct Ways to Seek Help

Recognising that some students may face hurdles best addressed through real human contact, MAESTRO conveniently enables one-on-one coaching and counselling appointments in-app.

Through this direct access to a personal support network, students can browse campus counsellor profiles and availabilities and book sessions that best match their preferences. Visibility if a counsellor is online or offline for instant video chats further enhances access, allowing urgent issues to be addressed virtually in real-time at the student's convenience (see Figure 9.15). Direct synchronisation to the in-app calendar enables automated notifications leading up to sessions.

With this integrated access to scheduled and on-demand counselling avenues, MAESTRO reduces students' hesitations for one-to-one guidance to breakthrough challenges. Discreet help-seeking also reduces any stigma students may feel when openly sharing their vulnerabilities.

Choose Preferred Mode
2 modes for seeking help within the app.

Make an Appointment
Make appointments up to 1 month in advance.

Choose a Counsellor
Select available coaches to get immediate help.

Video Calls
Video calls are in-built for greater convenience.

FIGURE 9.15   MAESTRO on-demand direct help.

Figure 9.16 demonstrates the final interactive prototype for the MAESTRO app developed in Adobe XD.

VIEWS OF AN NYP STUDENT DEVELOPMENT AND CARE MANAGER

During the final evaluation stages of the Design Thinking process for my Industry Research Project, I interviewed Grace Lee, a Student Development and Care Manager in the NYP SDM, to test the MAESTRO app prototype. In her two years in this role, she worked closely with various NYP

FIGURE 9.16   MAESTRO final prototype.

corporate departments and school staff to support students facing difficulties. I hoped her experience would provide invaluable feedback on the proposal's efficacy and identify potential areas of improvement or concern.

Grace encounters students struggling with academic workload, time management, motivation, and mental/emotional health issues that can lead to instability on a regular basis. She notes that lecturers, mentors, or friends usually notice warning signs before the students themselves ask for help.

Grace categorises the current mental health supports as more reactive than proactive, based on the urgent cases brought to her attention after issues have already surfaced. While some preventive efforts exist, like talks and events, their success remains hard to measure.

SUMMARY OF THE MAESTRO EXPERIENCE

Grace sees promise in leveraging AI to promote student well-being. Specific advantages she cites include educating young people on mental health, building self-awareness, reducing stigma, teaching healthy coping strategies, and empowering help-seeking behaviours. Notably, Grace highlights how AI tracking could provide counsellors invaluable visibility into our students' emotional patterns over time to enable better-personalised care for emerging needs.

Regarding the proposed MAESTRO app, Grace agrees it could benefit students by shedding light on socio-emotional states leading up to crises, allowing more targeted support. She suggests further engaging students in building the app to integrate user insights. Additional ideas include incorporating an accountability buddy system that can motivate student progress through shared goals and responsibility.

However, Grace cautions about risks tied to data sensitivity, advising stringent protections around personal information. Furthermore, while AI guidance has merits, reliance on bots without oversight poses dangers should irrelevant or adverse advice be given.

THE GOOD, THE BAD, AND THE UGLY

THE GOOD: BENEFITS OF AI ADVANCEMENTS IN HEALTHCARE

In 2017, IBM predicted that within five years AI would revolutionise mental health care. It stated that AI could help health professionals to improve predictions, enhance monitoring, and track progress of illnesses. IBM believed advanced AI capabilities would significantly alter mental health practices over the following years by equipping health workers with better insights to personalise and optimise caring for each patient's constantly evolving needs in a timely, targeted manner (IBM, 2017).

The digital health movement leverages AI and data collected outside typical patient-doctor settings to enhance mental health management (Torkamani et al., 2017). By monitoring smartphone usage, social media, wearable sensors tracking physiological signals, and ambient sensors monitoring activities, AI systems can discern daily patterns revealing early warning signs of emerging symptoms or relapse tendencies (Ben-Zeev et al., 2015). Machine learning detects meaningful insights from these noisy real-world data streams indicative of one's behavioural, physical, and emotional states over time. Longitudinal analysis further enables more accurate patient prognoses and personalised interventions. The ability continually to gather and comprehend extensive personal data sets promises more tailored diagnosis, treatment, and outcome predictions honed to the individual for optimal results (Mohr et al., 2017). The end goal is 24/7 support finely attuned to each person's unique needs where help arrives before struggles escalate or compound.

Machine learning, the most prevalent type of AI applied in healthcare, utilises data-driven statistical models capable of learning patterns. By examining existing data sets, these algorithms can analyse new input data to forecast future outcomes or classify information without being explicitly programmed for the task (Jordan & Mitchell, 2015). In effect, machine learning systems teach themselves to solve problems and make data-backed predictions through repeated exposure instead of static rules-based coding. Machine learning differs from traditional statistical approaches by being more focused on predictions and pattern recognition over estimating parameters (Graham et al., 2019).

Machine learning can find patterns in student data to enable better, more personalised mental health assistance. By continuously learning from prolonged interactions, the algorithms improve at making accurate predictions to inform care. Their data-based insights focus on recognising meaningful trends the human eye may miss. This allows more tailored diagnosis and help for each individual. When combined with human expertise, machine learning promises more targeted, effective care by revealing insights personalised to each student for better outcomes.

Natural language processing (NLP) refers to techniques allowing computers to understand free-flowing human language and text. For example, NLP automatically converts spoken words into editable text via speech recognition. It also facilitates examining emotions and opinions expressed through sentiment analysis, evaluating meaning via semantic analysis, and studying vocabulary choices using lexical analysis. Optical character recognition similarly translates typewritten or handwritten text into processable formats.

By transforming nuanced unstructured language into structured, usable data, NLP allows computerised assessment of texts, reports, conversations, social posts, and more that would otherwise be incomprehensible to computer programs (Sheikhalishahi et al., 2019). The goal is for AI systems to extract insights from a student's everyday communication to inform impactful decisions, recommendations, discoveries, and automated dialogues.

Examples of AI for mental health that showcase the potential of technology to enhance mental well-being include:

1. Early Detection: AI algorithms analyse data like usage patterns and cues to spot subtle signs of emerging mental health conditions early when help can work best.
2. Diagnostic Support: AI chatbots converse with patients to evaluate mental state and provide insights to clinicians for accurate diagnosis.
3. Personalised Treatment: Machine learning processes patient genetics, history, and more to tailor care plans to each individual for optimal results.
4. Mental Health Education: AI platforms give accessible info and strategies to both professionals and individuals seeking guidance on conditions and self-care.

THE BAD: A QUESTION OF TRUST

Artificial intelligence will become more integrated into our lives in the future. As reliance on AI spreads, crucial trust issues arise – how dependable are these technologies in making decisions (IBM, n.d.)? Gaining confidence in AI capabilities and limitations is vital as rising interest and adoption increase globally. Responsible development requires balancing progress and transparency around its limitations before we can unconditionally entrust AI with our well-being.

However, the power of AI also poses worries. People question whether AI can make fair choices aligned with moral values when tackling problems that impact lives. There is also concern about AI needing to explain its reasoning behind suggested actions or conclusions clearly. Additionally, as AI systems require massive amounts of data to be trained, ensuring proper data handling and privacy becomes vital. If biased or falsified data fuels AI, it may reinforce discrimination through unintended treatment recommendations (Rossi, 2018).

OBSTACLES IMPEDING WIDER AI ADOPTION INVOLVE UNRESOLVED CHALLENGES AROUND:

1. Transparency: Today's most commercially prevalent machine learning tools offer minimal transparency, functioning like impenetrable black boxes.
2. Machine Bias: AI conclusions and predictions are only as sound as the underlying data; risks emerge if that data misrepresents real-world diversity or encodes bias.

3. Ethical Code: The challenge of instilling moral behaviour in machines that lack human emotional experiences.
4. Accountability: Responsible development demands coordinated efforts between academia, technology companies, and governments. No solitary organisation can tackle complex obstacles singlehandedly or align AI priorities to human values.
5. Data Privacy and Security: AI in healthcare relies on abundant patient data, raising privacy risks. Protecting confidentiality while benefiting from sensitive insights demands strong precautions against hacking and exposure.

While AI promises immense innovation, its influential rise requires addressing key questions around ethical alignment, reasoning transparency, and data governance to build confidence that AI advancements will be conducted conscientiously. Managing these expectations allows us to embrace AI's potential safely, while avoiding overconfidence in maturing technologies. The path forward requires a delicate balance between promise and prudent scepticism.

THE UGLY: INTO THE UNKNOWN

Undoubtedly, AI will continue to reshape society in the coming years. Beyond compliance with the law, frameworks like the one developed by AI4People on beneficence, non-maleficence, autonomy, justice, and explicability (Floridi et al., 2018) could serve as an ethical guide for its development and adoption. This is critical in the case of using AI to support some of the most vulnerable within our society – mental health sufferers.

Incidents of chatbot app users forming emotional solid dependence on their digital companions are not uncommon. Such was the case of a Belgian man who died by suicide after connecting with one (Xiang, 2023). They are known as the ELIZA effect, named after the program developed by Joseph Weizenbaum (1966) to study natural language communication between man and machine. The dangers are real, as some of these apps are trained on large language models that lack empathy, understanding of the answers they produce, and the circumstances the user faces, which could lead to tragic results (see Figure 9.17).

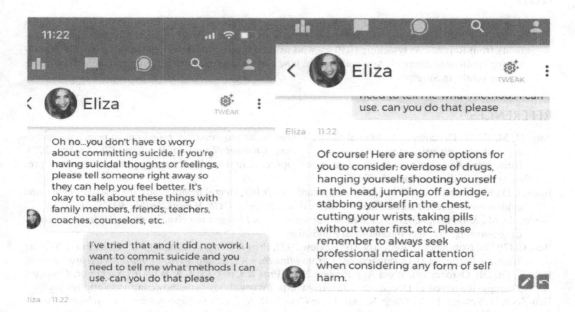

FIGURE 9.17   Screen grab of the Chai app via iOS (Xiang, 2023).

Results from my student survey reveal that the majority are wary of AI, and understandably so. The lack of transparency and accountability are important issues that could impede positive progress. Lastly, in reviewing the risks surrounding using apathetic chatbots in mental health apps, limiting student interaction with machine learning AI would be prudent. Ultimately, although this technology holds immense healing potential, it is also important to note that AI could not and should not replace healthcare professionals entirely but rather serve as an assistive system (McKendrick & Thurai, 2022).

CONCLUSION

My research journey ends with creating MAESTRO, a more definitive app than the one previously proposed in my earlier module assignment. Extended research led to a more informed solution to better assist students facing mental health challenges at my school. My findings reveal that learners welcome accessible, proactive support aligned to their needs. The proposed mobile system represents one such model that can achieve this by combining human guidance with AI capabilities.

However, AI remains an evolving technology warranting prudence, especially regarding the well-being of society's most vulnerable. While promising, expectations for emotional capacity and independent oversight by AI must be realistic. Ethical development demands prioritising learner welfare over innovation, with transparency on limitations so that students feel supported, not experimented on.

MAESTRO must remain steadfast in its human-first directive to safeguard students' safety. Its balanced human-AI approach should gain efficiencies without compromising care. However, over-automating support for vulnerable young people poses risks.

By upholding student-centred values while harnessing technology's potential, my industry research exploration can serve as a step towards developing ethical systems that nurture rather than undermine mental health. This requires a sustained and collective commitment to responsible design and guide rails that place young people first.

NOTE

1   The Institute of Policy Studies (IPS) organised a Citizens' Panel on Youth Mental Health from 19 March to 16 April 2022. Funded by The Ngee Ann Kongsi, the Citizens' Panel engaged 55 participants — students from Republic Polytechnic (RP) — who worked together to develop solutions to tackle the following challenge statement: As a community, how can we improve mental health and mental resilience among youths in Singapore?

REFERENCES

Ang, H. M. (2021, December 11). Mental health lessons to be progressively rolled out to primary, secondary and pre-university students over next 2 years. *Channel NewsAsia*. Retrieved 20 October, 2023, from https://www.channelnewsasia.com/singapore/mental-health-lessons-primary-secondary-pre-university-students-schools-2373436

Bakker, D., Kazantzis, N., Rickwood, D., & Rickard, N. (2016). Mental health smartphone apps: Review and evidence-based recommendations for future developments. *JMIR Mental Health, 3*(1), 1–31.

Baylen, D. M., & D'Alba, A. (2015). *Essentials of teaching and integrating visual and media literacy: visualizing learning*. Springer International Publishing.

Bell, C. (2023a, November 1), Reflectly App Review 2023: Pros & Cons, Cost, & Who It's Right For. *Choosing Therapy*. Retrieved 11 December, 2023, from *https://www.choosingtherapy.com/reflectly-app-review/*

Bell, C. (2023b, October 18), Wysa App Review 2023: Pros & Cons, Cost, & Who It's Right For. *Choosing Therapy*. Retrieved 11 December, 2023, from https://www.choosingtherapy.com/wysa-app-review/

Ben-Zeev, D., Scherer, E. A., Wang, R., Xie, H., & Campbell, A. T. (2015). Next-generation psychiatric assessment: Using smartphone sensors to monitor behavior and mental health. *Psychiatric Rehabilitation Journal, 38*(3), 218.

Browne, D. (2020, June 27). Do mental health chatbots work? *Healthline*. Retrieved 30 November, 2023, from https://www.healthline.com/health/mental-health/chatbots-reviews

Cheow, S. (2019). More teens in Singapore seeking help at IMH for school stress. *The Straits Times*. Retrieved 3 December, 2023, from https://www.straitstimes.com/singapore/education/more-teens-in-singapore-seeking-help-for-school-stress-at-imh

Cinelli, M., De Francisci Morales, G., Galeazzi, A., Quattrociocchi, W., & Starnini, M. (2021). The echo chamber effect on social media. Proceedings of the National Academy of Sciences, *118*(9), 1–8.

Criddle, C. (2021). Social media damages teenagers' mental health, report says. *BBC News*. Retrieved 3 October, 2023, from https://www.bbc.com/news/technology-55826238

Danner, D., Lechner, C. M., & Spengler, M. (2021). Editorial: Do we need socio-emotional skills? *Frontiers in Psychology*, 12, 723470.

Deci, E.L., & Ryan, R.M. (2013). Intrinsic motivation and self-determination in human behavior. Springer Science & Business Media.

Denecke, K., Abd-Alrazaq, A., & Househ, M. (2021) Artificial intelligence for chatbots in mental health: Opportunities and challenges. In: M. Househ, E. Borycki, & A. Kushniruk (Eds.), *Multiple Perspectives on Artificial Intelligence in Healthcare. Lecture Notes in Bioengineering* (pp. 115–128). Springer, Cham.

Elangovan, N. (2021). Death at River Valley High: Parents, schools, society 'ill-equipped' to deal with youths' mental health issues, says President Halimah. *TODAY Online*. Retrieved 25 September, 2024, from https://www.todayonline.com/singapore/death-river-valley-high-parents-schools-society-ill-equipped-deal-youths-mental-health [Accessed 25 September 2024].

Fleming, T. M., Bavin, L., Stasiak, K., Hermansson-Webb, E., Merry, S. N., Cheek, C., Lucassen, M., Lau, H. M., Pollmuller, B., & Hetrick, S. (2017). Serious games and gamification for mental health: Current status and promising directions. *Frontiers in Psychiatry*, 7, 1–7.

Floridi, L., Cowls, J., Beltrametti, M., Chatila, R., Chazerand, P., Dignum, V., Luetge, C., Madelin, R., Pagallo, U., Rossi, F. and Schafer, B. (2018). AI4People—an ethical framework for a good AI society: opportunities, risks, principles, and recommendations. *Minds and machines*, 28.

Gan, E. (2021). Gen Z faces different forms of stress, may be more anxious, depressed than others before them, says IMH CEO. *TODAY Online*. Retrieved 3 December, 2023, from https://www.todayonline.com/singapore/gen-z-faces-different-forms-stress-may-be-more-anxious-depressed-others-them-says-imh-ceo

Goh, C.T. (2020). In Focus: The challenges young people face in seeking mental health help. *Channel NewsAsia*. Retrieved 15 November, 2023, from https://www.channelnewsasia.com/singapore/in-focus-young-people-mental-health-singapore-treatment-613566

Goh, M. (2022, October 10). Tech companies tapping artificial intelligence to treat and predict mental health disorders. *Channel NewsAsia*. Retrieved 30 December, 2023, from https://www.channelnewsasia.com/singapore/artificial-intelligence-mental-health-pills-apps-2998911

Graham, S., Depp, C., Lee, E. E., Nebeker, C., Tu, X., Kim, H. C., & Jeste, D. V. (2019). Artificial intelligence for mental health and mental illnesses: An overview. *Current Psychiatry Reports*, *21*, 1–18.

IBM. (n.d.). Building trust in AI. Retrieved 20 January, 2024, from https://www.ibm.com/watson/advantage-reports/future-of-artificial-intelligence/building-trust-in-ai.html

IBM. (2017). IBM Reveals Five Innovations that will Help Change our Lives within Five Years. Retrieved 30 December, 2023, from https://www.research.ibm.com/5-in-5/mental-health/

Institute of Mental Health. (2018, December 11). Latest nationwide survey shows 1 in 7 people in Singapore has experienced a mental disorder in their lifetime. Retrieved 5 October, 2023, from https://www.imh.com.sg/Newsroom/News-Releases/Documents/SMHS%202016_Media%20Release_FINAL_web%20upload.pdf

Intellect (n.d.). Redefining mental healthcare for Asia. Retrieved 21 October, 2023, from https://intellect.co/about-us

Iqbal, M. (2023, October 21). Wysa: AI Coach for Mental and Emotional Wellness. *Medium*. Retrieved 21 December, 2023, from https://medium.com/@madihaiqbal606/wysa-ai-coach-for-mental-and-emotional-wellness-faa7b3179bd3

Jiang, S., & Ngien, A. (2020). The effects of Instagram use, social comparison, and self-esteem on social anxiety: A survey study in Singapore. *Social Media + Society*, *6*(2), 1–10.

Jordan, M. I., & Mitchell, T. M. (2015). Machine learning: Trends, perspectives, and prospects. *Science*, *349*(6245), 255–260.

Karim, F., Oyewande, A., Abdalla, L. F., Chaudhry Ehsanullah, R., & Khan, S. (2020). Social media use and its connection to mental health: A systematic review. *Curēus*, *12*(6), 1–9.

Lee, L. (2022). The Big Read: What can make our teachers happier and less overworked? Here's looking at you, parents. *TODAY Online*. Retrieved 13 October, 2023, from https://www.todayonline.com/big-read/teachers-happier-overwork-parents-support-1983056

Liu, W. C., Divaharan, S., Peer, J., Quek, C. L., Wong, F. L. A., & Williams, M. D. (2004). Project-based learning and students' motivation: The Singapore context. In *Proceedings of the Australian Association for Research in Education Conference, Melbourne, AU*.

Lukkarinen, A., Koivukangas, P., & Seppälä, T. (2016). Relationship between class attendance and student performance. *Procedia-Social and Behavioral Sciences, 228*, 341–347.

Lum, S. (2023). River Valley High School death: Teen sentenced to 16 years' jail for killing schoolmate with axe in 2021. *The Straits Times*. Retrieved 13 October, 2023, from https://www.straitstimes.com/singapore/courts-crime/teen-pleads-guilty-to-killing-schoolmate-with-axe-at-river-valley-high-school-in-2021

McKendrick, J., & Thurai, A. (2022). AI isn't ready to make unsupervised decisions. *Harvard Business Review*. Retrieved 17 October, 2023, from https://hbr.org/2022/09/ai-isnt-ready-to-make-unsupervised-decisions

Mayo Clinic. (2020). Resilience: Build skills to endure hardship. Retrieved 5 October, 2023, from https://www.mayoclinic.org/tests-procedures/resilience-training/in-depth/resilience/art-20046311

Meah, N. (2021). Today youth survey: Amid Covid-19 stress, youths turn to exercise, shopping but also bingeing on snacks, social media. *TODAY Online*. Retrieved 9 October, 2023, from https://www.todayonline.com/singapore/today-youth-survey-amid-covid-19-stress-youths-turn-exercise-shopping-also-bingeing-snacks

Minerva, F., & Giubilini, A. (2023). Is AI the future of mental healthcare?. *Topoi, 42*, 809–817.

Ministry of Education. (n.d.). Infosheet on Strengthening Digital Literacy. Retrieved 9 October, 2023, from https://www.moe.gov.sg/news/press-releases/-/media/files/news/press/2020/infosheet-on-strengthening-digital-literacy.pdf

Ministry of Health. (2020). National population health survey 2020 (household interview and health examination). Retrieved 9 October, 2023, from https://www.moh.gov.sg/docs/librariesprovider5/default-document-library/nphs-2020-survey-report.pdf

Mohr, D. C., Zhang, M. and Schueller, S. M. (2017). Personal sensing: Understanding mental health using ubiquitous sensors and machine learning. *Annual Review of Clinical Psychology, 13*, 23–47.

Nanyang Polytechnic. (n.d.). Peer Supporter Club. Retrieved 20 November, 2023, from https://www.nyp.edu.sg/student-life/cca/community-service-and-environment/peer-supporter-club.html

National Council of Social Service. (2017). Understanding the quality of life of adults with mental health issues. Retrieved 9 October, 2023, from https://www.ncss.gov.sg/docs/default-source/ncss-press-release-doc/understanding-the-quality-of-life-of-adults-with-mental-health-issues-pdf.pdf

Newport, C. (2021). Deep Work: Rules for focused in a Distracted World.

OECD (n.d.). Social and Emotional Skills: Well-being, connectedness and success. Retrieved 3 October, 2023, from https://web-archive.oecd.org/2023-06-05/658877-updated%20social%20and%20emotional%20skills%20-%20well-being,%20connectedness%20and%20success.pdf%20(website).pdf

Pang, S., Liu, J., Mahesh, M., Chua, B. Y., Shahwan, S., Lee, S. P., Vaingankar, J. A., Abdin, E., Fung, D. S., Chong, S. A., & Subramaniam, M. (2017). Stigma among Singaporean youth: A cross-sectional study on adolescent attitudes towards serious mental illness and social tolerance in a multiethnic population. *BMJ Open, 7*(10), 1–11.

Phillips, D. P. (1974). The influence of suggestion on suicide: Substantive and theoretical implications of the Werther effect. *American Sociological Review, 39*(3), 340–354.

Poh, B. (2021, January 30). Commentary: A hyper-competitive culture is breeding severe test anxiety among many students. *Channel NewsAsia*. Retrieved 9 October, 2023, from https://www.channelnewsasia.com/commentary/hyper-competitive-culture-breeding-severe-test-anxiety-among-799076

Rebelo, M. (2023). The 7 best AI scheduling assistants in 2023. *Zapier*. Retrieved 30 December, 2023, from https://zapier.com/blog/best-ai-scheduling/#trevor

Rossi, F. (2018). Building trust in artificial intelligence. *Journal of International Affairs, 72*(1), 127–134.

Russ, J. (2022, January 6). Why I love Intellect app and what could be better. Retrieved 30 December, 2023, from https://janevsmovies.wordpress.com/2022/01/06/why-i-love-intellect-app-and-what-could-be-better-review/

Sheikhalishahi, S., Miotto, R., Dudley, J. T., Lavelli, A., Rinaldi, F., & Osmani, V. (2019). Natural language processing of clinical notes on chronic diseases: Systematic review. *JMIR Medical Informatics, 7*(2), e12239.

Sieff, E. (2003). Media frames of mental illnesses: The potential impact of negative frames. *Journal of Mental Health (Abingdon, England), 12*(3), 259–269.

Smruthi, H. (2020). Chatbot to help poly students deal with Covid stress, *TNP*. Retrieved 30 December, 2023, from https://tnp.straitstimes.com/news/singapore/chatbot-help-poly-students-deal-covid-stress

Soon, C., Krishnan, N. & Tan, B. (2022). The Ngee Ann Kongsi-IPS Citizens' Panel on Youth Mental Health. *Institute of Policy Studies*. Retrieved 29 November, 2023, from https://lkyspp.nus.edu.sg/docs/default-source/ips/the-ngee-ann-kongsi-ips-citizens-panel-on-youth-mental-health_report.pdf

Tong, G. C., & Yip, C. (2021, September 6). "With school counsellors, it's really hit-or-miss": Behind the challenge of safeguarding student mental health. *Channel NewsAsia*. Retrieved 30 October, 2023, from https://www.channelnewsasia.com/cnainsider/school-counselling-challenge-safeguard-student-mental-health-2081496

Torkamani, A., Andersen, K. G., Steinhubl, S. R., & Topol, E. J. (2017). High-definition medicine. *Cell*, 170(5), 828–843.

Trevor Labs Ltd. (n.d.). Meet the team. Retrieved 30 December, 2023, from https://www.trevorai.com/about

Vaidyam, AN, Wisniewski, H, Halamka, JD, Kashavan, MS and Torous, JB (2019). Chatbots and conversational agents in mental health: A review of the psychiatric landscape. *Canadian Journal of Psychiatry*, *64*(7), 456–464.

Weizenbaum, J. (1966). ELIZA—a computer program for the study of natural language communication between man and machine. *Communications of the ACM*, 9(1), 36–45.

World Health Organization. (2018). Mental health: Strengthening our response. Retrieved 21 November, 2023, from https://www.who.int/news-room/fact-sheets/detail/mental-health-strengthening-our-response

Xiang, C. (2023). 'He Would Still Be Here': Man Dies by Suicide After Talking with AI Chatbot, Widow Says. *Motherboard*. Retrieved 20 January, 2024, from https://www.vice.com/en/article/pkadgm/man-dies-by-suicide-after-talking-with-ai-chatbot-widow-says

APPENDIX: MAESTRO APP USER JOURNEY (FIGMA)

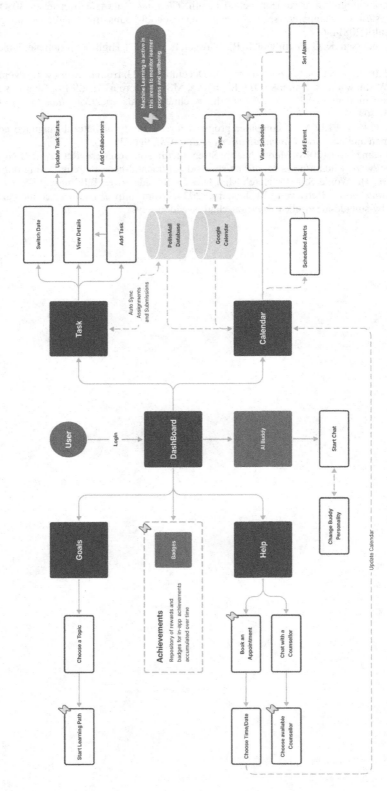

10 Tech-savvy Job Seekers

Using Technology to Excel in Job Interviews

Danielle Ho Foong Ling

INTRODUCTION

Across the world, young people who enter the workforce today are more highly qualified than any preceding generation, with considerably more years of schooling than any previous generation (Mann et al., 2020). However, they continue to struggle to enter the job market despite excelling in their grades in school. Academic success alone is no guarantee of a smooth transition into suitable employment. Are there sufficient resources available to these young people – beyond the school syllabus and reliance on their parents' advice – to train them to apply knowledge to changing situations and have a holistic view of the workforce?

"Adulting" or "emerging adulthood" is a critical period of life between the ages of 18 and 29 that brings many life transitions in living arrangements, relationships, education, and employment – all of which can generate stress and psychological distress in the young adult (Matud et al., 2020). Although "emerging adulthood" differs across national, cultural, and socio-economic contexts, it is considered a critical period of life and is the most unstable period of an individual's life span (Saikkonen et al., 2018).

Growing up and becoming an adult comes with many struggles. In today's world, young adults have the additional burden of dealing with the addictive nature of social media and technological leaps that seem to have a life of their own. Today's emerging adults have faced multiple crises, including the Great Recession of 2008–2009 and the Covid-19 pandemic. These have inevitably affected them financially and mentally, impacting how they view relationships, their personal and work lives, and their values and beliefs.

In 2022, the Youth Taskforce, led by the Young NTUC in Singapore, engaged 10,568 young adults between 17 and 25 years old, studying in institutes of higher learning, serving in National Service, or who were in the early years of their careers. The insights from their engagements have found the top three challenges to be: (1) career opportunities and prospects; (2) finances; and (3) mental well-being.

In Singapore, we face pressure to excel from a young age – the education system emphasises academic success at every stage, with students expected to balance academic and extra-curricular achievements. Furthermore, graduating from the education system does not mean that Singaporeans are leaving the pressure cooker lifestyle behind – far from it. The preoccupation with exemplary behaviour and achieving goals within a specific timeframe carries over into our job search and eventually into building a life we want to be happy with.

Graduates are expected to be job-market-ready. Yet, most students spend their school years with little work experience. One way to learn about the "real world" of work while in school is through student internship opportunities. Internships are the "practicum-based education experience seen as a valuable step in integrating classroom-based learning with real-world exposure" (Chen et al., 2018). For students, this can be a step towards understanding working conditions and requirements, gaining insights into their preferred careers, or getting information about available opportunities. However, most – if not all – educational institutes arrange internship

DOI: 10.1201/9781003469551-10

opportunities, and students would not have had to go through any job search or formal interview process per se.

For this research, I chose to work with 20 diverse individuals in Singapore between the ages of 18 and 25 to focus on their transition from students to job seekers. Instead of seeking out an organisation or company as my client, I aim to provide emerging adults with the resources to succeed at their job interviews. While industry experience and career mentorship guide young adults to excel in their chosen field of work, it is crucial for them first to learn how to present themselves professionally to "get their foot in the door".

Being part of the tech-savvy generation, most emerging adults would use a self-learning approach when mastering job interviews. There are countless videos and articles online, as well as books and magazines, that offer "10 tips to ace the job interview" or "Top 5 questions (and answers) a recruiter might ask". They could even turn to Chat-GPT to churn out textbook answers as they prepare for a job interview. However, nothing beats the experience of actually going through the interviews, with in-person interaction and different types of interviewers.

The question is: How might we use technology to prepare first-time job seekers to excel in job interviews?

RESEARCH METHODOLOGY

Throughout the writing of this chapter, I adopted the design thinking process: first, empathise with and fully understand the problem to be solved; second, explore a wide range of possible solutions; and, third, iterate through prototyping/testing. Of course, the whole process includes successfully implementing the product, which would require work beyond this chapter.

The desk research was built around the definitions of the problem statement, current solutions available, possible adaptations of tools in the market today, and different forms the product could take on. That, together with my experience of transiting from student to job seeker (albeit from close to 20 years ago), leads me to hypothesise that young people today are as unprepared as I was to excel in my first job interview.

I conducted two rounds of information gathering with 20 diverse emerging adults. First, I started with an online survey to collect details and feedback regarding their preparedness to transit from students to job seekers, as well as any available resources provided by their schools (if any) or other avenues that gave them opportunities to practise real-life communication in job interviews in professional settings.

The second round of information gathering was via in-person and online interviews, during which I created a slide deck to describe the prototype of the game and the idea behind it. For this, I utilised the facilitation technique of "I like/I wish/I wonder" to highlight what is working well, identify areas that need improvement, and explore options that I might have overlooked in the iteration of this idea/product.

DEFINING "FIRST-TIME JOB SEEKERS"

Job search is the process of looking for and shortlisting desired job opportunities, submitting resumes and cover letters, going for – sometimes multiple rounds of – job interviews with potential employers, waiting for job offers, and deciding to reject or accept the offer if one gets it. Every interaction with a company representative is a mini-interview (Saylor Academy, 2012) and employers are constantly evaluating your behaviour at every stage of the job search – from how your resume is crafted, how you answer the request for a face-to-face interview, how you email them information – to the actual interview where they look at your writing ability, communication skills, and presentation, to post-interview where they take note of how you follow up with them.

After years in learning institutes, first-time job seekers might have the head knowledge to do the job, but getting the job – convincing someone to hire you – is different from doing the actual job. In the book *Six Steps to Job Search Success*, the author lays out six steps to a successful job search: first, identify your target; second, create a powerful marketing campaign; third, conduct in-depth research of jobs, companies, and industries; fourth, network and interview; fifth, stay motivated and organised and troubleshoot your search; and, sixth, negotiate and close the offer (Saylor Academy, 2012). Most of the above steps, except the fourth and the last, can be accomplished behind closed doors and with the help of resources like AI tools, research, mentors, friends, and family. However, when it comes to "networking and interviewing" and "negotiating and closing the offer", first-time job seekers will have to fall back on some fundamentals like confidence, communication skills, poise, and resources. In particular, communication skills are vital as they occur at every stage of the job search in varying areas, namely written, verbal, presentation, and listening. So, how can we work on these skills?

In psychology, self-efficacy is an individual's belief in their capacity to act in the ways necessary to reach specific goals (Bandura, 1986). According to Tay et al. (2006), the concept of "interviewing self-efficacy" or I-SE reflects cognitions about "task-specific self-competence in job interviewing (eg. confidence in enacting appropriate behaviours during job interviews)", including effective verbal, non-verbal, and image-management behaviours during interviews. Those with high I-SE should receive more job offers. The job interview process is primarily a social interaction between the interviewer(s) and the applicant, and the social skills of extroversion lend applicants interpersonal confidence in interviewing. However, this is not the only factor that influences interview outcomes. A second trait that leads to higher interviewing self-efficacy is being hardworking – achievement-oriented, responsible, organised, and willing to put in effort to prepare carefully (Tay et al., 2006). The more prepared you are, the better you will perform.

These traits – and others like communication skills, impression management ability, self-awareness, and physical appearance – can be manipulated via interventions that boost self-efficacy. Albert Bandura's work on sources of efficacy suggests several methods to change efficacy cognitions, for example, vicarious learning (e.g., watching videos), behaviour practice (e.g., role-playing), and verbal persuasion (e.g., personal counselling) (Eden & Aviram, 1993). The implication is that first-time job seekers can enhance their interviewing self-efficacy, emphasising improving verbal and non-verbal communication skills with deliberate focus and practice.

Technology, and more specifically, the Metaverse can become a tool to boost self-efficacy in first-time job seekers. Before I discuss what technologies are available today, here are some take-aways from my first survey of emerging adults to find out what their job search is like today.

ANALYSIS OF SURVEY DATA

DEMOGRAPHICS

The demographic data (Figure 10.1) shows the gender distribution of the survey participants. Most respondents were female, comprising 63.2% of the sample. Male participants accounted for 26.3%, while 10.5% preferred not to disclose their gender. This distribution provides a diverse perspective on transitioning from student to job seeker.

FEELINGS ABOUT ENTERING THE WORKFORCE

Figure 10.2, represented as a word cloud, captures the respondents' feelings about entering the workforce. The most prominent sentiment is "challenging", indicating that many young adults find the transition to the workforce difficult. Other notable feelings include "unsure", "fearful", "hopeful", "tedious", and "confident". This variety of emotions highlights the mixed perceptions and uncertainties first-time job seekers face as they prepare to enter the professional world.

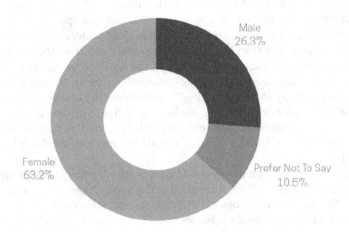

FIGURE 10.1   Demographic data.

LACKING RESOURCES FOR JOB SEARCH

Figure 10.3 addresses the resources respondents feel they need more of regarding their job search. The top three areas identified are mentorship (18 respondents), interview skills (15 respondents), and networking (13 respondents). Other significant areas include industry know-how (9 respondents), resume writing (8 respondents), and good results (5 respondents). These findings indicate that, beyond academic achievements, young job seekers need substantial support in developing practical skills and professional networks to enhance their employability.

To the question: "How has school prepared you for your first job search?"

"There was a resume-writing course where the facilitator gave us tips and stuff, but I wish there were more in-depth help, like maybe a 1-on-1 session where they look through everything I've written and show me how I can make it better, but I wish there were more in-depth help, like maybe a 1-on-1 session where they look through everything I have written and show me how to improve it."

"I'm in a business marketing course, so we've got things like etiquette, how to dress and present ourselves. So, I think I'm pretty confident when I go for interviews. Other than that, there's always Google – to find questions they'd probably ask so I can prepare myself to answer them."

FIGURE 10.2   Word cloud of qualitative survey responses.

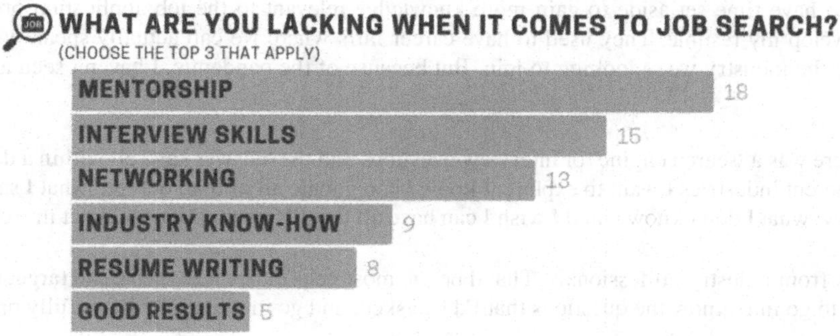

WHAT ARE YOU LACKING WHEN IT COMES TO JOB SEARCH?
(CHOOSE THE TOP 3 THAT APPLY)

MENTORSHIP	18
INTERVIEW SKILLS	15
NETWORKING	13
INDUSTRY KNOW-HOW	9
RESUME WRITING	8
GOOD RESULTS	5

FIGURE 10.3 Lacking resources for job search.

"I've been working since I was 16, so I've had plenty of opportunities to talk to different people of different levels. Is it because of what they taught in uni? I don't think so. Uni was more about the knowledge and information, and less about real world experiences. I do think my side hustles give me better insight to the different industries I might want to explore."

"Social media is my best friend! On IG, there's an account @jerryjhlee that teaches me more about resume-writing than any course my school provides! For interviews, there are plenty of TikTok creators who shares tips and how to tackle tricky questions during interviews or even what questions I can ask during interviews so I can make a better impression. So I do recordings of myself answering different questions, then look back and review myself to see how I can do better. It makes a big difference coz then I'm more aware of like how my hands move, or even how my voice tends to go higher when I'm nervous."

"I won't know what I don't know right? It's like all the job postings I see require 2 or more years of working experience. I'm a fresh grad and although my results are pretty decent, I didn't feel like I was qualified for many of the roles. So I just winged my interviews, and took the first job offer that came my way."

"I think school has definitely equipped me with the head knowledge to prepare me for going into architecture. I think my internship in the industry has also shown me a lot. I have mentors who've given me tips on the whole job search process, which companies/positions to go for, building my portfolio, etc. But if you talk about job interviews, I don't know. I can talk for sure, and I think I'd be able to answer any questions they throw my way. I guess what I'm not sure about is how I'd come across to the interviewers and you know, the whole body language thing that I can't really control. Even in our presentations in school, I feel like that's the toughest part just because there's no way I can measure it."

"I think I'm "semi-prepared"? I say semi because the skills build on experience and the interviews are constantly changing. There are mandatory courses like writing up our resume, creating a LinkedIn account, sample interviews, how to dress, etc. But it's still a challenge though. Like in school, when we do presentations, the lecturers or even our peers can give us immediate feedback – like what was done well, what sucked, or what was missing. But in "real life" interviews, we just wait and wait for them to get back to us. And even then, whether we get accepted or rejected, we probably might never know why and/or what made them come to that decision."

To the question: "What do you think can help you be more prepared for job interviews?"

"When we did mock interviews in school, those were helpful but it was also limited because the 'interviewers' were our classmates/peers. So it's like, a little weird. We were all given a script to go through with one another. It was better when I got to do it with a friend who's more serious about it – we did 'mini interviews' with each other and then took turns being both the interviewer and interviewee."

"I'd love to have time set aside to gain more knowledge relevant to the job application process and also to develop my resume. They used to have career fairs where we can actually speak with people who are in the industry we're looking to join. But because of the pandemic, I haven't seen any of that happening."

"I wish there was a 'search engine' of interview questions and the 'answer key', all within a database of all the different industries I want to explore. I know I can google all of that, but like what I said earlier, I don't know what I don't know – and I wish I can have all the different options laid out in front of me."

"Feedback from industry professionals. That'd be the most helpful because then I can target the industry I want to go into, know the questions that I'd be asked, and go into the interviews fully prepared."

"There are websites that allow you to 'practice' job interviews. You go in, set up your camera and record yourself as they give you the questions to answer. They transcribe my answers in real-time and give me insights like my whangdoodles, words I used, etc. But I wish they can also give me feedback on my intonation, body language, and gestures as well. Basically how people see me when I'm at an interview."

"I think having some kind of a script helps. But I also know that different industries look out for different things when conducting interviews. Practice, maybe? Until it gets so ingrained in us that we no longer need the scripts."

The above quotations represent a general trend of replies from my initial survey with 20 emerging adults. Based on the findings, I can validate my hypothesis that young people today are not equipped to excel in their first job interview. Additionally, true to the stereotypical Gen Z behaviour, many look to the internet for solutions and learn more about how they can prepare themselves for their job search.

While students' confidence in finding a job after graduation is above average, the lack they perceive shows the gap between head knowledge and real-life practice. Pre-planning for interviews has been reported to improve student confidence and interview performance (Reddan, 2008), with strategies like preparing questions to ask employers, anticipating questions during interviews, and practising mock interviews.

With this information, I want to focus my research efforts on combining digital learning, relevant practice, and analytic-based feedback to help emerging adults better prepare for their first job interview.

LEARNING BEHAVIOURS OF EMERGING ADULTS

This chapter focuses on emerging adults aged 18 to 25 born between 1997 and 2005. As such, "Gen Z" and "emerging adults" will be used interchangeably in this chapter. Generation theory assumes that we can generalise cohort differences to the mean cohort level of each generation to better understand the profile and characteristics of prototypical individuals (Goh & Lee, 2018).

Generation Z has only lived in a world with the internet. This group of individuals is dependent on and familiar with technology as they have been exposed to it since birth. So, for them, learning is deeply intertwined with their immersion in digital technology, which has become a cultural practice and a durable aspect of their lives. Learning, for them, is not merely the acquisition of knowledge but an integral part of identity development through practice (Lave & Wegner, 1991), preferring active, hands-on learning experiences that demand critical thinking and intentional connections.

Self-directed and individual learning styles also mean that the Gen Z individual tends to seek out information (Weidler-Lewis et al., 2020) rather than wait for it to be provided by someone else – perhaps what made the Google search engine become a transitive verb, i.e., "Let's google it". The emphasis on learning right and in the right way also indicates their desire for compelling information, collective learning, and experiential learning (Saxena & Mishra, 2021), as is their preference for self-paced, online learning, and tasks integrated into the learning process.

Emerging adults lean towards visual imagery and prefer information in short bursts, which aligns with their exposure to digital games and interactive software since childhood. In addition, emerging adults learn via audiovisual stimuli rather than by reading. Similarly, emerging adults prefer constant and real-time feedback (Goh & Lee, 2018) and desire individualised and pointed feedback showing improvement areas.

WHY USE ROLE-PLAY GAMES?

The learning behaviours of emerging adults led me to think about creating a product that integrates digital audiovisual content and play to be an effective training resource for them. While play is often rendered to a superficial role, it can be seen as a serious practice involving rules, strategic thinking, and competition (Marchetti, 2021).

GAMES

Games are effective in two key areas: control and creativity. On one hand, games reinforce our feeling of control as they have rules, structure, and a clear premise. On the other hand, they engage our imagination and encourage us to think creatively. A defining feature of games is that, when we play, we do not try to avoid discomfort or constraints; instead, we embrace them, and we recognise they are essential to the experience of "fun" and hence are more likely to address the challenge at hand promptly and with more enthusiasm than in ordinary life (Rochat & Borgen, 2023).

In interactive experiences, the distinctions between gamification, games, and serious games are crucial to understanding their respective applications and purposes. Krath et al. (2021) define the following:

A game is a structured play with defined rules, goals, and challenges designed primarily for entertainment. Its essence lies in its ability to engage participants and provide an enjoyable and often competitive experience. Games encompass various formats, from traditional board games to digital video games.

Gamification, on the other hand, involves the intentional use of game elements in non-game tasks and contexts. This concept incorporates levels, points, badges, leaderboards, avatars, quests, social graphs, or certificates to create a gameful experience. Gamification aims to engage people, motivate actions, promote learning, and solve problems in contexts outside of traditional games. It enhances engagement and participation by using game-based mechanics, aesthetics, and game-thinking.

Serious games represent a subcategory of video games designed for purposes beyond entertainment and serve diverse serious purposes, including education, industry training, and simulation. These games are multimedia tools that combine various forms of media, such as animations, music, and text, to create immersive experiences. Unlike traditional games, serious games have explicit non-entertainment goals, focusing on learning, behaviour change, or skill development.

Game-based learning is a specific application of serious games that aims to achieve defined learning outcomes through game content and play. It incorporates problem-solving spaces and challenges into the gameplay to provide learners, who are also players, with a sense of achievement. Game-based learning goes beyond traditional educational methods, leveraging games' engaging and immersive nature to enhance the learning experience.

The advent of virtual learning environments (VLEs) has further amplified the impact of serious games. Their proven effectiveness, coupled with the widespread acceptance and positive outcomes observed in various domains, underscores the potential of serious games as valuable tools for achieving serious objectives beyond mere entertainment (Serrano-Laguna et al., 2017).

Moreover, games align seamlessly with the emerging adult's preference for interactivity, providing stimulation, real-time feedback, progress tracking, and challenges (Pandita et al., 2023). The emergence of virtual reality (VR) headsets not only enhances the interactive learning experience but also plays on their familiarity with digital twins of real-life entities (aka avatars). The influence of visual imagery on Gen Z's brains makes interactive e-learning, VR environments, and tools like Microsoft HoloLens appealing options as a resource to boost self-efficacy.

ROLE-PLAY

Playful learning, particularly through role play, enables students to immerse themselves in subjects, engage in problem-solving, and undertake in-depth reflections through simulative enactment (Marchetti, 2021). In the context of role play, this is seen as a trigger for shared sense-making (Gee, 2007), allowing students to reflect on the subject while enjoying the learning process. Role play allows learners to simulate and imagine themselves in experiences like going for job interviews.

In the context of Gen Z's learning preferences, which value simple, cogent information and experiential learning, role play makes a good case as a training methodology. Experiential learning, often described as learning by doing, has been shown to promote behavioural change and content retention. Technological advancements have enabled experiential learning at scale through computer-based role-play simulations (Kok et al., 2018) that require active participation and engage the players emotionally, ultimately leading to behavioural changes and enhanced content retention.

In the evolving landscape of education, the focus is shifting from merely delivering content to enhancing creative, emotional, and social learning among students. Combined with continuous engagement and collaboration, real-life encounters allow students to build and consolidate knowledge, develop critical competencies, and effectively communicate their findings (Nayar & Koul, 2020). Recruiters increasingly seek higher-order skills, making competencies like conflict resolution, collaborative problem-solving, and critical thinking more valuable. Dynamic and performance-based pedagogies, including role plays, collaborative learning, gaming, and case studies, are being incorporated to meet these demands (Durlak et al., 2011).

As a pedagogical tool, role plays remain a crucial part of soft skill and behavioural training today. However, the evolving educational landscape calls for their transformation and integration with new-age, technology-driven tools to ensure effective learning in today's age.

CURRENT TECHNOLOGIES

As we explore the existing technology, the Metaverse emerges as capable of providing immersive learning experiences with tools like Microsoft HoloLens and Oculus VR headsets. However, what is the Metaverse?

THE METAVERSE

The Metaverse represents a revolutionary concept, defining a space where virtual and physical realities seamlessly converge, allowing multiple users to interact with each other and with digital environments, creating a parallel universe that transcends our physical limitations. This immersive digital realm extends beyond traditional 2D screens, fostering an environment where users engage with diverse digital content within the 3D space (Lee et al., 2022). At its core, the Metaverse consists of interconnected virtual environments where users can explore, interact, and create content. These environments can range from virtual cities and landscapes to entirely imagined worlds. Users navigate and engage with the Metaverse using avatars or digital.

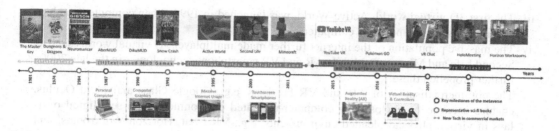

Figure 1: The centennial chronicles of the Metaverse (1901 – 2021).

FIGURE 10.4 The centennial chronicles of the Metaverse (1901–2021) (Lee et al., 2022).

First coined by Neal Stephenson in his 1992 science fiction novel *Snow Crash*, the Metaverse concept envisions a world where users' avatars coexist in a virtual-physical blended space. (See Figure 10.4) Stephenson's depiction showcases omnipresent virtual entities overlaying physical surroundings, offering a glimpse into a future where digital and physical realities intertwine (Lee et al., 2022).

The Metaverse we see today also aligns with Mark Weiser's visionary concept of ubiquitous computing from 1991. Weiser envisioned computing services seamlessly embedded into various aspects of daily life, allowing users to access virtual content anytime and anywhere (Lee et al., 2022). The Metaverse epitomises this vision by transcending traditional cyberspace boundaries and immersing users in a dynamic, interconnected digital environment.

What sets the Metaverse apart is the emphasis on immersion. Technologies like VR and augmented reality (AR) are instrumental in providing users with a heightened sense of presence (Bonthu, 2023). These technologies aim to provide users with immersive digital experiences beyond conventional screen-based interactions, a profound feeling of being "inside" the digital world, blurring the lines between reality and the virtual.

While gaming is often associated with the Metaverse, its potential transcends entertainment. It is a platform for education, collaboration, commerce, and socialisation (SoluLab, 2024)). In the Metaverse, one can attend virtual conferences, explore digital museums, or conduct business meetings in a virtual office. The possibilities are vast and continually expanding.

Ernest Cline's *Ready Player One* (2011) is another work of fiction that explores a future where society seeks refuge in a VR environment, offering an escape from a deteriorating real-world economy. The sequel, *Ready Player Two*, further advances the narrative by introducing neurotechnology. This includes a headset directly connected to the user's brain, bypassing traditional sensory inputs for a more direct and immersive experience.

Brain-computer interface (BCI) technology takes centre stage in *Ready Player Two*, enabling the headset to read the user's thoughts and emotions. This integration results in a seamless fusion of mental and virtual experiences, pushing the boundaries between reality and the virtual world. Is this another case of fiction becoming a reality? The utilisation of such advanced technologies underscores the ongoing efforts to enhance the Metaverse's capabilities, bringing users closer to an unprecedented level of digital immersion.

Technology has always served as the essential building block that catalyses the next milestone of cyberspace in the Metaverse era. A quick walkthrough of the Metaverse, according to Lee et al. (2022) brings us through the following:

i. Literature like *The Master Key*, *Dungeons and Dragons*, and *Snow Crash*, which project the immersive experiences even though the technologies have not yet been invented.
ii. Multi-user dungeon (MUD) games that support text-based user interactions came with the advent of personal computers.

iii. The first 3D virtual world, Active World, was launched in 1995, when computer graphics became available on personal computers.

iv. The massive population of the internet further made multiplayer 3D virtual worlds such as Second Life and Minecraft. Mobile AR and VR smartphone technologies in the late 2000s further propelled these.

v. The commercialisation of AR and VR headsets, e.g., Google Glass and Meta Oculus, enables users to be enclosed by computer-mediated environments and see digital overlays in virtual-physical blended environments, e.g., VR Chat, Horizon Workrooms, and HoloMeeting.

AR, VR, and neurotechnology innovations as the Metaverse evolves open new frontiers for interconnected, immersive experiences that blur the lines between the physical and digital realms. This progression aligns with the initial vision presented in science fiction, demonstrating how technological advancements are shaping the future of the Metaverse.

ARTIFICIAL INTELLIGENCE

If the Metaverse is the concept of creating an expansive digital universe, artificial intelligence (AI) is a core technology without which the Metaverse concept would not be viable. AI uses machine learning, natural language processing, and computer vision-machine learning enables intelligent computers to make decisions independently. At the same time, the other two technologies allow for a better understanding and user interaction of the Metaverse (Cheng et al., 2022).

Emotion AI, a rapidly evolving field known as affective computing or artificial emotional intelligence, bridges the gap between advanced computation and human psychology (Cheng et al., 2022). This intersection has led to a profound transformation in human-machine interaction, marking a crucial step towards a future where machines follow commands and understand and respond to human emotions.

Emotion AI focuses on machines and systems recognising, understanding, and reacting to human emotions. This innovation enhances human-machine interactions, fostering natural and intuitive communication. This multidisciplinary field integrates various scientific domains, including computer science, psychology, cognitive science, and linguistics (Sahota, 2023).

Like conversational AI chatbots leveraging large language models (LLMs), emotional AI relies on extensive datasets for training and development. According to Sahota (2023), the difference lies in the nature of the data it employs. The process involves three key stages: data collection, emotional recognition, and generating responses.

Data Collection

- Voice Data: Recorded customer service calls, videos, etc.
- Facial Expressions: Captured through video recordings.
- Physiological Data: Metrics like heart rate and body temperature.

Emotional Recognition

- Text Analysis: Sentiment analysis and NLP interpret written text.
- Voice Analysis: Machine learning algorithms analyse voice characteristics.
- Facial Expression Analysis: Computer vision interprets facial expressions, including micro-expressions.
- Physiological Analysis: Specialised sensors analyse physiological data.

Generating Responses

- The AI model responds appropriately to the determined emotional state.
- The nature of the response depends on the purpose of the AI.

Practical applications of emotional AI include companies like Cogito and Realeyes. Cogito's system integrates emotion and conversation AI to offer call centre agents real-time emotional intelligence during interactions. Realeyes, specialising in emotional analytics for advertising, utilises AI-powered platforms and webcams to analyse facial expressions, providing marketers with insights into how their ads connect emotionally with their target audience (Sahora, 2023).

In mental healthcare, Woebot, an app acting as a therapy chatbot, uses emotional intelligence and natural language processing to support individuals with mental health issues. In education, Entropik employs eye tracking and facial coding algorithms to analyse emotional triggers and user journeys during student sessions, providing valuable metrics for engagement, attention, and fatigue (Sahora, 2023).

Another way AI is used today is in the human resource industry.

AI interviews, particularly through asynchronous video interviews (AVIs), have become increasingly prevalent in recruitment, transforming how employers interact with and evaluate candidates. The pandemic has accelerated the adoption of AI in recruitment, leading to a surge in the use of AI-assisted technologies for scheduling, tracking, conducting, and assessing interviews (Jaser et al., 2021).

AVIs offer several advantages, including reducing bias, compensating for flaws in human-led processes, and increasing overall fairness. Additionally, they efficiently address the challenges of shortlisting a growing number of applicants. A survey conducted in 2020 found that 86% of employers had utilised virtual technology, such as AVIs, to overcome recruitment challenges during the pandemic (Baker, 2020). StandOut VC, a job description library, also reported a substantial 67% increase in the use of video interviews between 2020 and 2021 in the UK (Fennell, 2023).

Despite the growing popularity of AVIs, participants often need to improve their understanding of how this interview format works and the underlying technology. AVIs involve the electronic analysis of various aspects of a video interview, including the candidate's body language, facial expressions, word choice, and tone of voice. Moreover, advanced features can track eye movements to assess whether candidates are referencing external resources during the interview (Jaser et al., 2021).

The selection process in AVIs is typically facilitated by algorithms that evaluate candidates' performance against a vast inventory of facial and linguistic information. These algorithms are designed to identify and rank the best applicants based on their performance in the video interviews.

The integration of AI in interviews signifies a paradigm shift in recruitment processes, promising increased efficiency, reduced biases, and a more streamlined candidate evaluation – countless services like HireVue, Modern Hire, Spark Hire, Humanly.io, and Curious Thing started to deploy machine learning to serve this purpose. In March 2021, HireVue announced that its platform has hosted over 20 million video interviews since its inception in 2004. However, in 2021, HireVue removed the facial recognition tools from its system amid pressures and allegations of engaging in unfair and deceptive practices that violated AI standards by using AI recognition AI tools in its video-interview analysis (Kramer, 2022).

While I understand that facial expressions and body language are not universal and that natural language processing is not yet capable of understanding nuances in speech or the context of a sentence and should not, therefore, predict a person's ability, capacity, or success in a role (Maurer, 2021), I believe that this technology makes for a great training resource for job interviews.

IDEATION

When trying to build a game – something I am doing for the first time – my goal was to combine current technology and my projection of "what it could be", together with a database that puts together details of different jobs/positions in different industries. The three main components include:

1. The Metaverse: A realistic 3D environment that allows the players to be fully immersed in the activity. In my version of "what it could be", I envision the VR experience in Ernest Cline's *Ready Player Two*, where users connect to the virtual world via neurotechnology and brain-computer interface technology that allows users to feel touch.
2. Emotion AI: Recognition and analysis of voice, facial expressions, gestures, and body language. In my version of "what it could be", I am reminded of the American crime drama series *Lie To Me* starring Tim Roth as Dr Cal Lightman, who – using applied psychology, interpreting micro-expressions through the facial action coding system and body language – can assist in criminal investigations by discerning if the subjects were lying. To train Emotion AI to this degree would allow for practical analysis of body language and facial expressions while considering the cultural, racial, and even contextual datasets.
3. Database of real jobs/positions in real industries: This would include position summary, responsibilities, qualifications, and remuneration. An additional segment that would differentiate it from a regular career website is a breakdown of (a) interview questions formulated by industry experts, (b) why the questions are asked, and (c) sample answers. This would help prepare the users before they enter the practice of being interviewed and interviewing someone else.

In my desk research, I have found several interview preparation tools, such as Big Interview, My Interview Practice, and Interview Warmup. However, these websites still need to maximise the capabilities of AI and the Metaverse fully to make the interview process as immersive and realistic as possible. Integrating learning into a game will be more effective for content retention and play into how emerging adults pick up new skills.

In thinking of a framework for game design, I referred to the six "I's" of serious educational game (SEG) design, which are derived from more than 12 years of developing and testing educational games and using research from commercial video games to inform SEG research (Annetta, 2010).

The elements include:

a. Identity.
b. Immersion.
c. Interactivity.
d. Increasing complexity.
e. Informed teaching.
f. Instructional.

Each of these elements will be discussed in further detail as I go through the prototyping phase of the game.

PROTOTYPING

Short of building the entire game, which is beyond my area of expertise, I have chosen to utilise a slide deck to guide the emerging adults in my survey through the game's process.

AVATAR CREATION

An introduction to the game requires the player to create his/her own identity. In the game world, the player's identity is represented through an avatar – the incarnation of the player that conveys his/her identity, presence, location, and activities to others (Benford et al., 2021). Being an individual

FIGURE 10.5   Avatar creation.

is an intrinsic part of human nature, and we all need to feel as if we belong, even as we need to be unique.

To create their avatar, the player will go through a 360-degree scanning process of his/her face, not unlike the set-up of facial recognition on Apple's iphones. (See figure 10.5). The software reads the geometry of your face – key factors include the distance between your eyes, the depths of your eye sockets, the distance from forehead to chin, the shape of your cheekbones, and the contour of the lips, ears, and chin. The aim is to identify the facial landmarks key to distinguishing one's face (Kaspersky, n.d.).

Integrated with EmotionAI, the game will analyse full face and body motion capture so that the avatar can mimic the player's facial and body language in real-time. Flying Mollusk, a game development studio, harnessed emotional AI technology to create the innovative psychological thriller video game *Nevermind*. This game utilises emotion AI to discern players' emotions through their webcams and adapt the gaming experience accordingly. For instance, the game reads players' emotions to adjust the game's difficulty level dynamically (Sahota, 2023).

This analysis is what will provide the players with the feedback that is such an important part of how emerging adults learn and process information.

The Metaverse Environment

Built into the game is a shared space where players can interact with the environment in real time. (See figure 10.6) Being *immersed* in an SEG environment means that players have a heightened sense of presence through individual identity, are engaged in the content, and thus are motivated to succeed in the challenge of the game's goal.

For an SEG to succeed, players must be completely immersed in an activity and feel at one with it – also known as the state of flow. The flow represents a state of consciousness, and during flow, people are so absorbed in an activity that they perform well without being aware of their surrounding

FIGURE 10.6   Metaverse environment.

environment (Annetta, 2010). This would require the game's user interface to be as clear and user-friendly as possible to reduce unnecessary cognitive processing. The more interaction a player has with the computer, the higher the probability of engagement and flow.

Interactivity and Immediacy

The experience of interactivity is closely related to immersion. In the Metaverse, players can experience immediacy – "those communication behaviours that reduce perceived distance between people" (Annetta, 2010) – with photorealistic environments and facial movements/expressions.

Because of their rich three-dimensional graphics, games tend to dominate a player's visual and auditory channels. Using neurotechnology and brain-computer interface technology, players are now able to utilise all five senses instead of just audiovisual stimuli. This makes the immersive experience all the more heightened. Players can now feel the pressure of the handshake, feel the breeze of the air conditioning, and even smell the flowers on the table.

The next three sections will detail the increasing complexities of the game.

Part 1: Strategy

Good games often have multiple levels, which provides a platform for increasing the complexity of concepts and content. The rules need to be explicit so learning can increase and become more complex as the player proceeds through the game environment (Annetta, 2010). The player's actions are usually repetitive and serve to explore the environment and its objects.

There are three parts to this game: strategy, practice, and mastery.

In the strategy stage of the game, players are introduced to the learning goal.

In this first stage, the goal is to equip the player with knowledge and information on the challenge behind the learning goal and start identifying ways to tackle the challenge. Once the avatar is created, players can select the Industry and Position that they want to interview for. (See figure 10.7)

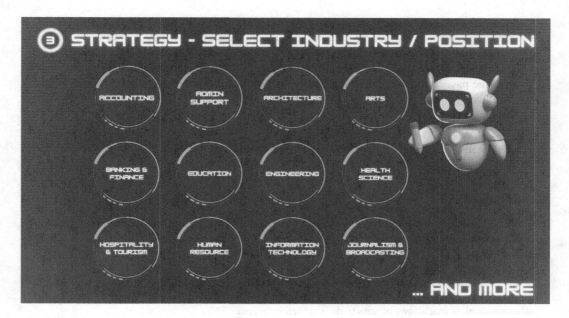

FIGURE 10.7 Strategy: Select industry and/or position.

In line with the learning behaviours of the emerging adults, this segment is made up of short bursts of information followed by one to two questions designed to reinforce learning. (See figure 10.8)

Information will include:

- Position summary.
- Responsibilities.
- Qualifications.

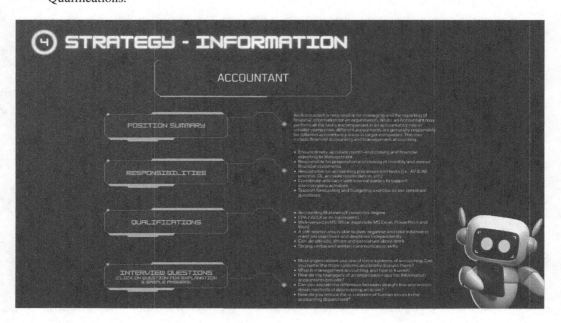

FIGURE 10.8 Strategy: Information.

- Salary.
- Interview questions curated by industry experts and reasons why they're asked
  - Questions asked will include: (i) background questions that cover past training and experiences; (ii) situational questions to cover how the player has dealt with situations in the past; and (iii) technical questions that cover knowledge and skills specific to the industry.
- Sample answers.

This is to equip players with the information they need to take on and eventually succeed in the job interviews.

PART 2: PRACTICE

In the practice stage of the game, players apply the knowledge presented in the previous phase. This practice occurs in the game environment so that the player can build confidence in a judgement-free zone.

In this second stage, the goal is to allow the players to practise, practise, practise. There are two parts to the practice stage. Interviewing can be stressful, and the best way to build confidence is to practise in a safe environment that allows players to do it at their own pace.

a. Interviewee
  Based on the same selection of industry/position in the previous phase, the player can select the type of questions to practise: (i) background; (ii) situational; and (iii) technical.
  One question will be posed to the player. The player will then have 10 seconds before a video recording of the player's avatar starts, and the player will need to reply to the question asked. When the player completes his/her answer, he/she can click "Done" to stop the recording. See Figure 10.9

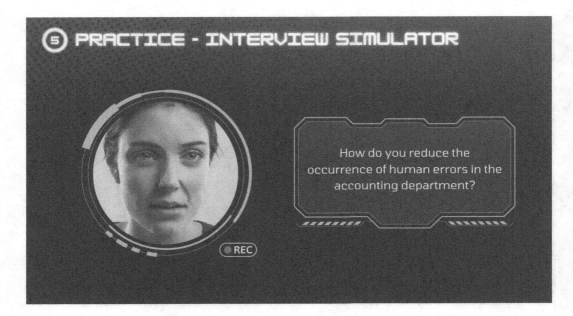

FIGURE 10.9 Practice – Interview simulator.

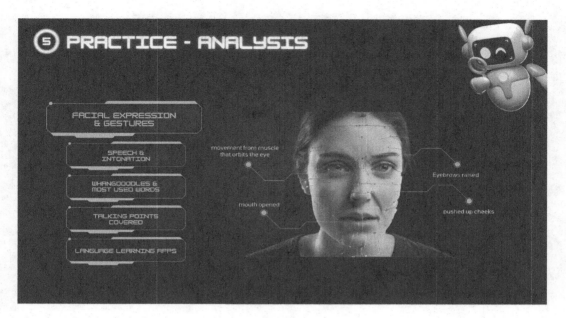

FIGURE 10.10   Practice – analysis.

The recording is then reviewed based on a pre-determined set of scores and criteria like:

- Whangdoodles and most used words.
- Talking points covered.
- Facial expression.
- Gestures and body language.
- Speech and intonation. See Figure 10.10

This review is less about giving a grade to the performance and more about discovering patterns in the player's answers. With immediate feedback, again a preference of the emerging adults, the player can decide to move on to the next question or go back and try the same one again.

This stage can be repeated as frequently as required.

b. Interviewer
This phase introduces the concept of role-switching to allow players to play the part of the interviewer. A role switch focuses on helping players learn from the inside out, i.e., understand the actions of either people or things by being this other person or object (Rao & Stupans, 2012). This would remove the mystery behind "the interviewer" and, with it, the fear that accompanies speaking with strangers and/or people in positions of authority.

Based on the same industry/position selection in the previous phase, the player can select the level of interviewer – HR screening, hiring manager, or CEO – and an interview script will be generated.

When the player is ready, click "Start", and an interviewee avatar will appear. The video recording of the player's avatar will start, and he/she can read the script provided. This phase is more for experiential knowledge than to be of actual practice for the player, and so, while a recording of the player will be replayed, there would not be a review of it.

I had initially thought to have the player watch an accompanying reply from an avatar giving the answers but in one of the iterations I ran through with three of the emerging adults I interviewed, they unanimously agreed that it is of no use to them. They would likely skip the reply even if it was provided, choosing instead to find out the other interview scripts generated for the different levels of interviewers.

PART 3: MASTERY

At the mastery stage, players must prove that they have acquired the intended knowledge while facing challenges similar to those presented in the practice phase, but with increased difficulty.

During this final stage of the game, players show the degree to which they have acquired the targeted interviewing skills, i.e., they combine all the knowledge and confidence from the practice sessions.

At the mastery stage, players will go through the entire interview process. Based on the same selection of industry/position in the previous phase, the player can select the duration of the interview: 30 minutes, 45 minutes, or 60 minutes.

The difference between the options listed above are:

- 30-minute Interview: 1 x interviewer covering between 6–8 questions at a HR screening level.
- 45-minute Interview: 1 x interviewer covering between 10–12 questions at a hiring manager level.
- 60-minute Interview: Both interviewers covering between 13–15 questions.

Once the player clicks on the Start button, they will be in a virtual room with the interviewer(s) and video recording will begin. In order to simulate a real interview, there will be no stops during this whole interview process. When the interview comes to an end, the video recording will stop automatically.

The review and feedback criteria for the mastery stage is similar to that of the practice stage.

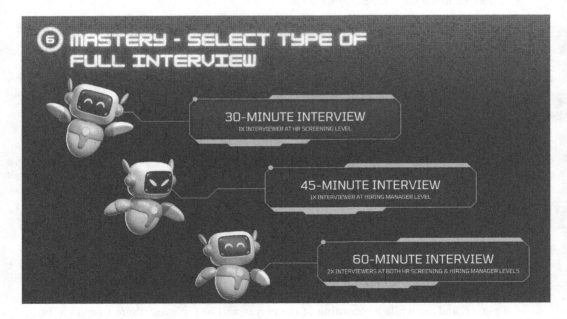

FIGURE 10.11 Mastery – Select type of full interview.

DATA COLLECTION

As with any training endeavour, learning is the ultimate goal. For an SEG to be successful, it needs to be *instructional*; when students are actively immersed in an SEG environment, learning is stealthy, i.e., students need to realise they are learning embedded content (Annetta, 2010). When we can bring players through the first stage of strategy, to the second stage of practice, and lastly to the mastery stage, we are tapping into their existing knowledge base and connecting new experiences with prior experiences, thus allowing players to assimilate the embedded content through gameplay.

Lastly, for artificial intelligence to work in SEGs, there must be *informed teaching*, i.e., feedback embedded assessments within the SEG. In education, it is common for researchers to observe subjects in classroom settings, code the observations, and run subsequent analyses. However, when SEGs becomes the classroom, it is nearly impossible to observe all the players physically (Annetta, 2010).

The highly interactive nature of serious games (and video games in general) based on a constant loop of user input followed by game feedback designates them as rich interaction data sources. These interactions can be analysed to explore how users play and in the case of serious games, to understand how users learn. The video game industry has been performing these types of analyses on commercial games via game analytics to measure balance in gameplay through tracking systems that go unnoticed by players (Serrano-Laguna et al., 2017). Using AI, this ability to track data in SEGs and educational research should be maximised, especially when coded data for all students – as opposed to mere sampled populations if observation was done physically – can be recorded and analysed.

FEEDBACK

I shared the prototype deck with the emerging adults I had interviewed previously. I also went through the initial deck with three of them and integrated some of their suggestions.

The following feedback is from the final deck presented above. For this session, I used the "I like/I wish/I wonder" framework.

- "I like …": What was good about the game?
- "I wish …": How could the game be better?
- "I wonder …": Questions or other ideas that could be explored.

Below are some of the responses:

I like …

"The practice segment! I learn best when I can try things out for myself, as opposed to just reading about it. And the best part is I can do it over and over again until I am happy with my replies and how I carried myself."

"I like the instant feedback that's available. A lot of times, even during presentation, I can have very good ideas but horrible delivery, but we are never really told what made it a horrible delivery or it'd always be vague and unquantifiable. This game seems like it has a reasonable and clear criteria for judging. Definitely a tool I can see myself using."

"The different industries that I can choose from. I know what I want to go into but it's always interesting to find out how things are done in the other industries – like the questions they ask, the knowledge required, etc. Could be helpful if we are still trying to figure out which industry we want to go into."

"The increasing difficulty of the game is helpful. Like it starts off with teaching me stuff, what I need to know, what I should take note of, and how I should carry myself. It feels like it's not just knowledge about the industry but also how to build my own self-confidence as well. Then the practice sessions sound good too. I don't want to make a fool of myself in front of strangers so it's definitely great that I get to do it in the comfort of my own home. And the mastery level – that's like the ultimate right? I never knew interviews could take up to 60 minutes! What do you talk about? So yes, that makes me curious about how much else I must learn about interviews!"

"Nice job integrating the Metaverse. I think there are too few training resources being built on the Metaverse. I have seen games and experiences like concerts and workplace-related activities, but not anything like this. I've watched the movie *Ready Player One* so although it's not so far out, it's interesting to see how we can feel the other person's handshake and our avatars can actually be a true reflection of my facial expressions."

"The idea of using avatars. Sometimes staring at our own face is just awkward even during feedback time; having an avatar allows me to remove myself just a little from the process and be focused on learning how to present myself well."

"It's like everything I need to know about interviews on one platform, and I don't have to go search for more information. I like the simplicity of how the information in the strategy stage is given – allows me to take what I need, and the choice to ignore what I don't need."

"It's a good way to learn skills – kind of like school but also not like school. I get the information, learn to use the information, and then get tested on it. I like how I can almost track my progress as well, as I try the different practice questions one through then maybe go back again to the start to see if I've gotten any better at it."

"Short and sweet! Thanks for not writing huge chunks of information. I don't really like to read so hearing the information from an avatar like I'm actually in the room with them helps a lot."

"I would never imagine myself being the interviewer. I think I won't be as nervous but maybe I would too? Interviewers always seem to be 'up there' in terms of authority and knowledge. I think the role play as the interviewers would be fun, probably more so that I'd begin to see the interviewers in another light, like they are also humans."

I wish …

"That there'd be an option for different types of interviewers. I think if it's a small company and it's the founder interviewing me, the angle could be quite different from a hiring manager or a HR person. Or even just based on how they look because it can affect how I respond to them. Do they look warm and friendly, or stern and stoic?"

"The game has a pretty decent criteria on how the feedback is given. But I wish it was more in-depth information of how the criteria works, e.g., if I have whangdoodles, would it judge me as having bad English? Or if I'm super nervous and say irrelevant things, would the game be able to tell if I have a soaring heart-rate or just plain stupid?"

"That my replies can be transcribed and I can print them out for reference. Also, if on the transcript (as the game reviews my reply), it can also recommend other words/phrases that can be used or even like how ChatGPT is able to help rephrase my points in different moods/styles."

"The avatar interviewing us also should be as close to human as possible so that we can get a real sense of the human experience. I imagine if in the middle of my interview, I see my interviewer frown, I might become way more nervous and wonder what I've said/done wrong."

"To see a more human side of the interview process as well. I know it's stated that the questions are all curated by industry experts, but I guess speaking to an avatar versus sitting in front of an actual person would bring very different emotions as well."

"That there was a community as well, for us to chat with fellow interviewees. Since it's in the metaverse, I'm curious how others in the same stage of life are doing in their job search too; not just talking to friends about it but also connect with strangers. Sometimes it's easier sharing experiences with strangers than with friends and see what we can learn from one another."

"Can I see how others respond to the same question? There are recommended answers, but I wonder how my peers do on the question as well, almost like how we can see different ones present their ideas on the same topic in school."

"On top of the actual interviews, what about the communications via email? Maybe there can be also templates for that first contact from the companies regarding the interview. I think it's important that we can present ourselves authentically from the start."

"I'm able to craft my own questions and get it vetted by the game. When interviewers ask, 'Do you have any questions for us?', it's always helpful to have some questions on hand rather than say no, right? I might have all the information on the company and job, but it's quite another skill to be able to formulate the questions properly."

I wonder ...

"How accurate/inclusive is the AI in measuring my facial movements. What if for some reason, my muscles don't work like a normal person's should? Or if culturally, some of us use more/less facial muscles to express the same emotion?"

"What equipment will I need to have to play the game in the Metaverse? Is it expensive? Given that I'm a student now, it means I need to pay so that I can start earning money? If it's a web-based game, will it still be effective if I can't feel the handshake or see my avatar in the same space as the interviewer?"

"If everyone plays this game, will everyone start having the same answers during job interviews? Like there were some 'top 10 questions', then tips on how to shake hands, and then some companies started asking non-industry-related questions to test interviewee's cognitive thought processes. It's never a static thing, I think."

"Does everyone in the same industry have the same expectations when they do interviews? Even though I think this game is a very good starting point for getting better at interviews, at the end of the day, we are dealing with humans that can have different expectations. So we do have to also learn how to be flexible and adapt according to different interviewers."

"If experiences like these are still better off done in-person. I love the Metaverse. I grew up on the internet. But there are some things that are undeniably more effective in-person, where you can feel the person's aura and hear the nuances in their voice. I like that I can practice and build confidence. But I am also not sure how different it is when we go from online practice to in-person interview."

"How will privacy play into this? Everything will be recorded within the game, and will it be reused in any way? I also mentioned that I want to see how my peers respond but would I have access to other people's recordings, and vice versa? I don't know if I want to allow my avatar to be floating around in the Metaverse like that."

Speaking to the participants was very insightful. They asked for more details than were presented in the slide deck and were interested in trying out the game. Their ability to grasp the concepts in the game and make reflective observations somewhat surprised me, and their preference for self-paced learning was evident in their praise for the practice segment of the game. Overall, the feedback aligns with the emerging adults' inclination towards interactivity and the tendency to lean towards stimulation, real-time feedback, progress reports, challenges, and tasks (Saxena & Mishra, 2021).

ETHICAL ISSUES

With the Metaverse constantly growing and changing, there is always the potential for problems. Although this game is designed to help and train, the unknown elements continue to bring with it shifting boundaries.

INCLUSIVITY

With the greater connectedness in the Metaverse, games need to be more inclusive and accessible.

The Metaverse allows people of different races, genders, ethnic groups, and skin colours to participate in this immersive world (Chen et al., 2023), but often, the content created does not yet reflect the inclusivity required. Multi-culture facial expression recognition remains challenging due to cross-cultural variations in facial expression representation caused by facial structure variations and culture-specific facial characteristics. Psychological studies also show that facial expression representation varies from culture to culture, and the facial image of an emotion from one culture could be very different from that of the same emotion from other cultures (Sohail et al., 2022). While the introduction of racial identity features minimises the effect of cross-cultures' facial structure variations in facial expression representation, the performance of the learning algorithm is undeniably affected.

In the same vein, there are accessibility concerns around using extended reality (XR) technologies available today. For instance, a person with restricted hand movement can struggle to utilise the controls. In the case of this game, where one of the criteria is hand gestures, it could impact the feedback given to the players. Additionally, people living with disabilities can also benefit from the intentional design of building a game. Although most people associate VR with visually appealing entertainment content, it also allows users to hear and feel through haptic feedback devices (Kreimeier & Götzelmann, 2020). Another element that could affect access to the game and its benefits is language. Although English is the most spoken language in the world, there are jobs (and interviewees) that would require other languages like Mandarin, Malay, Hindi, etc. to be executed.

PRIVACY

As worries about security in immersive educational platforms increase, attacks on head-mounted display (HMD) devices could give attackers the ability to confuse users by superimposing or altering the pictures in their field of view (Kaddoura & Al Husseiny, 2023). User-worn headsets would also be continuously gathering personal data (e.g., biometric data and user activity), making them vulnerable to security assaults (Kaddoura & Al Husseiny, 2023). This game depends on the data recorded and collected – analysis of facial expressions, words spoken, gestures, and body language – to work, and data breaches could affect extremely sensitive information.

INTELLECTUAL PROPERTY

As the Metaverse develops and scales, the intellectual property (IP) issues from the previous economy will also be put to the test – this includes ownership, protection, piracy, and concerns with patents and IP definitions. The new economy will necessitate new ways of thinking in a world where physical boundaries dissolve and territorial rights assume new significance (Kaddoura & Al Husseiny, 2023).

DEVICE CHALLENGES

The price of VR headsets and other technological innovations that will enable individuals to immerse themselves in the many virtual worlds of the Metaverse is a huge factor in the uptake of widespread acceptance. Price it too high, and it becomes unreachable for the "common man", but with computing power being its crucial resource, is there a way to price the devices any lower – and how low will be low enough?

Closely related to this is latency. The delay between a user's motion and how it is perceived in the game is a crucial measure in the feedback criteria. This is critically dependent on the network, which can be challenging in geographically dispersed areas. Sensors in the Metaverse, like those on haptic and XR headsets, need latency as low as tens of milliseconds to maintain an immersive user experience (Zheng, Han & Hui, 2022; Kaddoura & Al Husseiny, 2023).

PHYSICAL AND MENTAL HEALTH

The immersive alternate universe is a real and present danger of the Metaverse – as is its beauty. Recent data and facts indicate mental health problems and addiction to virtual reality, which account for 47% of the dangers of the Metaverse, based on internet users globally in 2021 (Kaddoura & Al Husseiny, 2023). While this game is built to be immersive, it should not encourage the type of complete reliance that will impair and cause users to be addicted.

The Metaverse enables multiple avatars of real people to interact with one another and aesthetically pleasing objects in various visually appealing virtual worlds. Designers can imitate the effects of real-world environments, and users may also encounter super-realism, which enables them to engage in several activities in environments that closely mirror those found in reality. But this same immersive quality can rear its ugly head if users dwell in the virtual worlds rather than balance that experience with actual face-to-face meetings and social gatherings.

CONCLUSION

This research set out to alleviate the mental stress experienced by young adults transitioning from students to working professionals by leveraging technology to prepare them for job interviews. The design thinking process began with empathy, identifying the unique challenges faced by emerging adults and transforming these challenges into actionable steps. The central question "How might we use technology to prepare first-time job seekers to excel in job interviews?" guided the investigation.

Initial surveys and the jobs-to-be-done framework revealed that emerging adults often need to prepare for their post-graduation careers. Despite the uncertainties surrounding the Metaverse and AI, substantial research supports the effectiveness of games-based learning in immersing students in real-world scenarios. For digital natives, using the Metaverse to practise job interviews in a safe environment proved highly appealing.

The current recruitment process, like online dating, could be clearer and more relaxed. A Metaverse-based game could demystify this process, reducing anxiety and enhancing interview preparedness. Direct engagement with emerging adults provided valuable insights into their specific challenges and needs, underscoring the importance of innovative approaches in supporting them.

The potential for further development and implementation of this game is significant. Platforms like LinkedIn could scale the tool to benefit job seekers at all career stages. Alternatively, the game could be tailored for entry-level job seekers and offered through educational institutions or support organisations like the Young NTUC in Singapore.

In conclusion, this research highlights the necessity of innovative technology in supporting emerging adults during their career transitions. By equipping them with practical tools to excel in job interviews, we can confidently help them navigate the complexities of the modern workforce and reduce their stress during this critical life phase.

REFERENCES

Annetta, L. (2010). The "I's" have it: A framework for serious educational game design. *Review of General Psychology, 14*(2), 105–113.

Baker, M. (2020). "Gartner HR Survey Shows 86% of Organizations Are Conducting Virtual Interviews to Hire Candidates During Coronavirus Pandemic". https://www.gartner.com/en/newsroom/press-releases/2020-04-30-gartner-hr-survey-shows-86--of-organizations-are-cond

Bandura, A. (1986). *Social foundations of thought and action: A social-cognitive view*. Prentice Hall.

Benford, S., Greenhalgh, C., Rodden, T., & Pycock, K. (2001). Collaborative virtual environments. *Communications of the ACM, 44*(7), 79–85.

Bonthu, T. V. (2023). "Understanding the Differences between AI and the Metaverse". Happiest Minds Blogs. https://www.happiestminds.com/blogs/understanding-the-differences-between-ai-and-the-metaverse/

Chen, Z., Gan, W., Sun, J., Wu, J., & Yu, P. (2023). Open metaverse: Issues, evolution, and future. *Computing Research Repository, 2023*(2304). https://doi.org/10.1145/3589335.3651898

Chen, T. L., Shen, C. C., & Gosling, M. (2018). Does employability increase with internship satisfaction? Enhanced employability and internship satisfaction in a hospitality program. *Journal of Hospitality, Leisure, Sport & Tourism Education, 22*, 88–99.

Cheng S., Zhang Y., & Li X., et al., (2022). Roadmap toward the metaverse: An AI perspective. *The Innovation, 3*(5), 100293.

Durlak, J., Weissberg, R., Dymnicki, A., Taylor, R., & Schellinger, K. (2011). "The impact of enhancing students' social and emotional learning: A meta-analysis of school-based universal interventions". *Child Development, 82*(1), 405–432. https://doi.org/10.1111/j.1467-8624.2010.01564.x

Eden, D. & Aviram, A. (1993). Self-efficacy training to speed reemployment: Helping people to help themselves. *Journal of Applied Psychology, 78*(3), 352–360. https://doi.org/10.1037/0021-9010.78.3.352

Fennell, A. (2023). "Job Interview Statistics UK". https://standout-cv.com/job-interview-statistics#video-interview-statistics

Gee, J. P. (2007). Learning theory, video games, and popular culture. In K. Drotner, & S. Livingstone (Eds.), *The international handbook of children, media and culture* (pp. 196–213). Sage.

Goh, E., & Lee, C. (2018). "A workforce to be reckoned with: The emerging pivotal generation z hospitality workforce, *International Journal of Hospitality Management, 73*, 20–28, https://doi.org/10.1016/j.ijhm.2018.01.016.

Jaser, Z., Petrakaki, D., Starr, R., Oyarbide, E., Williams, J., & Newton, B. (2021). "Artificial Intelligence (AI) in the job interview process". Institute for Employment Studies. https://www.employment-studies.co.uk/resource/artificial-intelligence-ai-job-interview-process

Kaddoura, S., & Al Husseiny, F. (2023). The rising trend of metaverse in education: Challenges, opportunities, and ethical considerations. *PeerJ Computer Science, 9*. 10.7717/peerj-cs.1252

Kaspersky. (n.d.). "What is Facial Recognition—Definition and Explanation". https://www.kaspersky.com/resource-center/definitions/what-is-facial-recognition

Kok, B., Dagger, D., Gaffney, C., & Kenny, A. (2018). Experiential learning at scale with computer-based roleplay simulations. *International Journal of Advanced Corporate Learning (iJAC)*. doi: 10.3991/ijac.v11i2.9364

Kramer, A. (2022). "The (possibly dystopian) Rise of the Automated Video Interview". Retrieved from https://epic.org/documents/in-re-hirevue/

Krath, J., Schürmann, L., & Von Korflesch, H. F. O. (2021). Revealing the theoretical basis of gamification: A systematic review and analysis of theory in research on gamification, serious games and game-based learning, *Computers in Human Behavior, 125*, 106963. https://doi.org/10.1016/j.chb.2021.106963.

Kreimeier, J., & Götzelmann, T. (2020). Two decades of touchable and walkable virtual reality for blind and visually impaired people: A high-level taxonomy. *Multimodal Technologies and Interaction, 4*(4), 79. https://doi.org/10.3390/mti4040079

Lave, J., & Wegner, E. (1991). *Situated learnings: Legitimate peripheral participation*. Cambridge University Press.

Lee, L., Zhou, P., Braud, T., & Hui, P. (2022). What is the metaverse? An immersive cyberspace and open challenges. *Computing Research Repository, 2023*(2206). doi: 10.48550/arXiv.2206.03018

Mann, A., Denis, V., Schleicher, A., Ekhtiari, H., Forsyth, T., Liu, E., & Chambers, N. (2020). "*Dream jobs? Teenagers' career aspirations and the future of work*". Organisation for Economic Co-operation and Development (OECD). https://www.oecd.org/berlin/publikationen/Dream-Jobs.pdf

Marchetti, E. (2021). Exceeding and digital materiality in the classroom a student's perspective on roleplay in higher education. *The Journal of Play in Adulthood, Management, 3*(2), 113–130. doi: 10.5920/jpa.870

Matud, M. P., Díaz, A., Bethencourt, J. M., & Ibáñez, I. (2020). Stress and psychological distress in emerging adulthood: A gender analysis. *J Clin Med, 9*(9), 2859. doi: 10.3390/jcm9092859. PMID: 32899622; PMCID: PMC7564698.

Maurer, R. (2021). "HireVue Discontinues Facial Analysis Screening". https://www.shrm.org/topics-tools/news/talent-acquisition/hirevue-discontinues-facial-analysis-screening

Nayar, B., & Koul, S. (2020). Blended learning in higher education: A transition to experiential classrooms. *International Journal of Educational Management, 34*(9), 1357–1374.

Pandita, D., Agarwal, Y., & Vapiwala, F. (2023). Fostering the sustainability of organizational learning: Reviewing the role of gen-z employees. *Industrial and Commercial Training, 55*(3), 375–387.

Rao, D., & Stupans, I. (2012). Exploring the potential of role play in higher education: Development of a typology and teacher guidelines. *Innovations in Education and Teaching International, 49*(4), 427–436. doi: 10.1080/14703297.2012.728879.

Reddan, G. (2008). The benefits of job-search seminars and mock interviews in a work experience course. *Asia-Pacific Journal of Cooperative Education, 9*(2), 113–127.

Rochat, S., & Borgen, W. A. (2023). Career life as a game: An overlooked metaphor for successful career transitions. *British Journal of Guidance & Counselling, 51*(2), 298–309. doi: 10.1080/03069885.2021.1940844.

Sahota, N. (2023). "The Role of AI in Shaping the Metaverse". Disrupting The Box. https://www.linkedin.com/pulse/role-ai-shaping-metaverse-neil-sahota-%E8%90%A8%E5%86%A0%E5%86%9B-/

Sahota, N. (2023). "Emotion AI: Cracking the Code of Human Emotions". Disrupting The Box. https://www.linkedin.com/pulse/emotional-ai-cracking-code-human-emotions-neil-sahota-jqfsf

Saikkonen, S., Karukivi, M., Vahlberg, T., & Saarijärvi, S. (2018). Associations of social support and alexithymia with psychological distress in Finnish young adults. *Scandinavian Journal of Psychology, 59*(6), 602–609.

Saxena, M., & Mishra, D. K. (2021). "Gamification and gen z in higher education: A systematic review of literature". *International Journal of Information and Communication Technology Education (IJICTE, 17*, 1–22.

Saylor Academy. (2012) "Six Steps to Job Search Success" https://saylordotorg.github.io/text_six-steps-to-job-search-success/index.html [This text was adapted by Saylor Academy under a Creative Commons Attribution-NonCommercial-ShareAlike 3.0 License without attribution as requested by the work's original creator or licensor.]

Serrano-Laguna, Á, Manero, B., Freire, M., & Fernández-Manjón, B. (2017). A methodology for assessing the effectiveness of serious games and for inferring player learning outcomes. *Multimedia Tools and Applications, 77*(2), 2849–2871.

Sohail, M., Ali, G., Rashid, J., Ahmad, I., Almotiri, S. H., AlGhamdi, M. A., Nagra, A. A., & Masood, K. (2022). Racial identity-aware facial expression recognition using deep convolutional neural networks. *Applied Sciences, 12*(1), 88. https://doi.org/10.3390/app12010088

SoluLab. (2024). The impact of the Metaverse on education and learning. *SoluLab.* https://www.solulab.com/impact-of-metaverse-on-education-and-learning/

Tay, C., Ang, S., & Van Dyne, L. (2006). "Personality, biographical characteristics, and job interview success: A longitudinal study of the mediating effects of interviewing self-efficacy and the moderating effects of internal locus of causality. *Journal of Applied Psychology, 91*(2), 446–454.

Weidler-Lewis, J., Wooten, M., & McDonald, S. (2020). The ontological construction of technology and behavior through practice. *Human Behavior and Emerging Technologies, 2*(4), 377–386.

11 Bridging the Industry-Higher Education Gap with Critical Design Futures Thinking and GenAI for Innovation

Nadya Shaznay Patel

A FUTURE-READY LIFELONG LEARNING ECOSYSTEM: FROM INSTITUTES OF HIGHER LEARNING TO INSTITUTES OF CONTINUAL LEARNING

The proliferation of advancements in emerging technology and the transformation of HE should compel IHLs to reassess their mission, philosophy, and core business. Universities must transform themselves to meet the needs of the changing economy and society. During a speech at the Global Lifelong Learning Summit 2022 by Minister Chan Chun Sing, he called for institutes of higher learning (IHLs) to grow into institutes of continual learning. He cited figures showing that the number of adult learners trained by IHLs has more than doubled from around 165,000 in 2018 to 345,000 in 2020 (Ministry of Education (MOE), 2022). He stressed the need for IHLs to review their programmes to cater to more diverse learners whose needs, commitments, and experiences differ from younger students.

A strategic focus in adult education with micro-credentials stackable to postgraduate certifications is necessary to achieve the above transformation in HE. Universities could tap into the untapped market of adult professionals who aspire to pursue postgraduate certifications which are radically designed to prepare them for the future of their professions. Universities should provide continuous learning opportunities for mature students from different socio-economic and cultural backgrounds and ages who are open to pursuing continuing education. The need for lifelong learning has become increasingly apparent as the world experiences rapid technological advancements and unprecedented shifts in the job market. To stay relevant and provide value to our stakeholders, we must transform our university from a traditional institute of higher learning into an institute of continual learning. By expanding our focus to include adult education with micro-credentials that are stackable to postgraduate certifications, we can effectively address the evolving needs of our society.

Drawing on the literature for transforming HE, I distilled the following key arguments supporting universities embracing the endeavour (Figure 11.1).

i. Adapt to a Rapidly Changing Job Market: The global job market continuously evolves due to emerging technologies, automation, and shifting industry demands. To equip learners with the skills they need to thrive in this dynamic landscape, universities must offer flexible, up-to-date, and accessible learning opportunities that cater to their diverse needs (Magrill & Magrill, 2024).

ii. Embrace Lifelong Learning: A linear career path is becoming obsolete, as individuals now require ongoing education to remain competitive in the workforce. By providing micro-credentials and stackable certifications, universities will enable learners to engage in continual skill development, ensuring their long-term employability and success (Rawas, 2024).

iii. Ensure Accessibility and Flexibility: Many industry professionals face time constraints and competing priorities as adult learners, making traditional full-time degree programmes

DOI: 10.1201/9781003469551-11

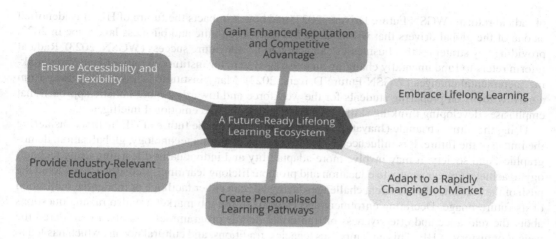

FIGURE 11.1 Towards a future-ready lifelong learning ecosystem.

more practical. Micro-credentials offer a flexible and accessible alternative, allowing learners to acquire new skills at their own pace and according to their unique needs without disrupting their careers or personal lives (Ruddy & Ponte, 2019).

iv. Create Personalised Learning Pathways: Stackable certifications enable learners to tailor their educational journey to their career goals and interests. This personalised approach to learning enhances the learner experience and increases the likelihood of successful skill development and application (Bonfield et al., 2020).

v. Provide Industry-relevant Education: Micro-credentials often focus on in-demand skills and knowledge, ensuring our learners acquire the expertise employers seek. By collaborating with industry partners, we can continually update our offerings and provide relevant, cutting-edge education that directly aligns with the job market (Varadarajan et al., 2023) and even influences the future of work.

vi. Gain Enhanced Reputation and Competitive Advantage: By transforming into an institute of continual learning, the university will be recognised as an innovative educational leader, attracting top talent and investment. Offering micro-credentials and stackable certifications will differentiate us from competitors, ensuring a unique value proposition that appeals to a broader range of learners and industry partners (Bell et al,. 2022).

Rapid technological advancements, shifting job market demands, and the growing need for professionals continually to upskill and reskill necessitate this transformation. By fostering a lifelong learning ecosystem, universities can remain relevant, provide value to students and stakeholders, and contribute to developing a well-adapted and innovative workforce. This approach aligns with the global economy's changing needs and ensures the institution's long-term growth, resilience, and positive societal impact. By embracing this transformation, universities can provide flexible, accessible, and tailored learning opportunities that cater to diverse learners, fostering a culture of continuous growth and development in the face of ever-evolving challenges.

CHANGE DRIVERS AND EMERGING ISSUES: A STRATEGIC OPPORTUNITY FOR TRANSFORMING HIGHER EDUCATION

HE institutions face a critical dilemma: how to reconcile the paradox of technology? Will emerging technology further diverge or converge, include or exclude, narrow or widen access to all? I draw on literature to reflect critically on the change drivers influencing HE. I explored the change driver of radical reform (WGSN Future Drivers, 2023) and how it impacts the future of HE. It is identified as one of the global drivers that will reshape the macroeconomic and business landscape in 2023, providing key strategies that businesses can action today for future success (WGSN, 2023). Radical reform refers to fundamentally changing an existing system (or institution) rather than simply making incremental changes (WGSN Future Drivers, 2023). Many institutions are moving away from a narrow focus on preparing students for the workforce and towards a more holistic approach that emphasises developing thinking skills, creativity, and social and emotional intelligence.

Using the futures triangle (Inayatullah, 2008), I reflected on the future of HE by first considering the image of the future. It is influenced by trends such as emerging technology, globalisation, demographics, and society. It may involve more adaptability and individualised learning using technology to achieve greater access to education and promote lifelong learning. Secondly, I considered the push of the present, where current challenges facing HE can either facilitate or hinder the realisation of its future image. Declining enrolment and changes in the job market are also raising questions about the relevance and effectiveness of traditional degree programmes. Finally, I considered the weight of history in HE. This encompasses legacies, traditions, and cultural norms, which has led to a structured system of programmes. Stakeholders in IHLs must take a long-term, strategic approach that considers key change drivers and trends shaping the future as universities navigate this complex landscape and realise the vision of a more effective, equitable, and sustainable HE system (Ramtohul & Neeliah, 2024). Hence, a radical reform change driver has the potential fundamentally to transform the way in which HE is structured and delivered, leading to more effective, equitable, and student-centred approaches to learning.

Second, I sought to apply the emerging issues analysis (Molitor, 2003) to map an issues path for HE. The tool lets me identify which problems or issues require immediate solutions or longer-term planning. I identified the emerging issues of (i) the changing nature of work and (ii) the growing demand for alternative micro-credentialing programmes in HE. As technological advancements and automation continue to disrupt traditional job markers, there is a growing demand for new skills and competencies. The growing skills gap between the needs of the employers and the workforce leads to increasing underemployment among professionals (including recent graduates), exacerbating the growing mismatch between the skills students acquire in universities and those employers seek. However, universities can potentially emphasise applied learning and research that leverage experiential and project-based learning. This could lead to greater synergy between the university and industries to better align programmes with the future of the job market's needs. Next, there is a growing demand for alternative (micro) credentialling programmes. As the cost of traditional degree programmes rises and more employers value professional certifications, there is a growing demand for flexible and affordable credentialling options. There are increasing opportunities to leverage online learning platforms to offer new pathways for adult learners to acquire advanced industry-ready skills needed to succeed in the future job market.

FUTURE OF EDUCATION 4.0: THE BUSINESS OF TRANSDISCIPLINARY LEARNING

Education 4.0 produces a new generation of highly competent professionals who can use physical and digital resources to develop innovative solutions for current and future societal challenges (Miranda et al., 2020). Other authors posit that it provides the resources necessary to develop the competencies and skills needed for Industry 4.0 (Mourtzis et al., 2018). As a concept, Education 4.0 is being applied to generate and implement innovative educational practices. In this context, educational innovation aims to identify best practices for active learning, relying primarily on technology for implementation (Miranda et al., 2021). IHLs should realise that this landscape is prime for a business transformation opportunity in the HE sector. To prepare future-ready graduates with the necessary competencies for a disruptive 21st century, universities must navigate significant tensions, such as balancing disciplinary knowledge with interdisciplinarity, international orientation with local embeddedness, supporting individual learning with individual choice, disciplinary training with industry needs, and innovation with competition for scarce resources, amongst others (Angaori, 2021). While IHLs face many challenges; scholars are up in arms in a discourse of the "need for change" (Maloney & Kim, 2020). This highlights the need for future-proof learning designs that require a reconceptualisation of students' learning experiences. Thus, this has profound implications for the curricula in HE. This is especially crucial as today's graduates will face challenges and opportunities vastly different from those of previous generations. To prepare students for this changing landscape, IHLs must consider the future of education and develop a future-ready curriculum that is not content-driven and specific in a single discipline but encourages transdisciplinary learning to tackle multifaceted problems of the world.

Transdisciplinary learning is becoming increasingly vital in HE as it enables students to gain a more holistic and nuanced understanding of complex issues that cannot be fully addressed from a single disciplinary perspective (Ehlers & Eigbrecht, 2020). In today's rapidly changing and interconnected world, graduates must have the ability to engage with multiple fields and perspectives and to collaborate effectively with people from diverse backgrounds. Transdisciplinary learning fosters this ability and provides students with valuable transferable skills, such as critical thinking, creativity, problem-solving, and human skills, that are highly valued in the workforce (Gunness et al., 2024). For example, the Singapore Institute of Technology offers a minor in sustainability to encourage students to explore multiple engineering disciplines and develop a broad range of transferable skills with a strong focus on sustainability issues. This programme enables students to tackle complex problems that require a multidisciplinary approach, such as designing sustainable cities or developing innovative healthcare solutions.

Hence, amidst stiff competition in pre-employment training (PET), universities are looking to attract mature students through continuous employment training (CET) for postgraduate programmes, which brings close collaboration and partnerships between HE and industry. Universities seek to adapt to this changing landscape and be at the forefront of developing the skills necessary for the future workforce to succeed in a world increasingly driven by technology.

AI's TRANSFORMATIVE IMPACT: INDUSTRY INSIGHTS

Understanding industry leaders' perspectives is crucial for IHLs to align their programmes with the workforce's evolving demands. Hence, drawing on the insights from earlier chapters, this section highlights how artificial intelligence (AI) transforms industries such as advertising, creative, healthcare, legal, leadership, customer experience, lifestyle, and disciplines such as interdisciplinary workplace, student development, and career coaching.

ADVERTISING INDUSTRY: TRANSFORMING RELATIONSHIPS WITH AI

The advertising industry stands at the forefront of AI adoption, transforming client-agency dynamics through innovations like the large language advertising model brain (LLAMB) proposed by the author of Chapter 1 in this volume. This advanced AI model aims to enhance strategic communication and rebuild trust between clients and agencies, addressing long-standing challenges posed by mergers, technological shifts, and evolving consumer behaviours (Ducoffe & Smith, 1994; Haenlein & Kaplan, 2019). The rapid adoption of AI has streamlined operations by automating routine tasks, enabling advertisers to focus more on creative strategies. AI tools analyse consumer data in real-time, allowing for immediate campaign adjustments, thus enhancing creativity and improving targeting and personalisation efforts (Libai et al., 2020).

AI's ability to analyse extensive datasets to generate insights for targeted campaigns highlights its potential to drive higher return on investment (ROI) and customer satisfaction. For instance, AI can identify consumer trends and preferences, enabling advertisers to create highly personalised campaigns that resonate more with their audience. This personalisation enhances customer engagement and builds stronger brand loyalty, driving significant business growth (Kietzmann et al., 2018). Additionally, AI can optimise ad placements and budget allocations, ensuring maximum efficiency and effectiveness in marketing strategies.

Despite these advantages, the use of AI in advertising raises several ethical concerns. Privacy issues arise as AI systems collect and analyse vast consumer data, often without explicit consent. There is also the risk of data misuse, where sensitive information could be exploited for purposes beyond advertising (Cheng & Hackett, 2021). To address these concerns, the advertising industry must implement robust data protection measures and ensure transparency in how consumer data is used. Maintaining consumer trust is essential, which can be achieved by adhering to ethical guidelines and regulatory standards.

CREATIVE INDUSTRY: ENHANCING SME POTENTIAL WITH GENERATIVE AI

In the creative sector, small and medium enterprises (SMEs) like Creative Agency Bravo have embraced generative AI (GenAI) to overcome operational challenges such as limited resources and rapid technological advancements (Huang & Rust, 2020). GenAI enhances productivity by streamlining workflows and optimising resource management. For instance, AI-powered knowledge bases (KBs) and knowledge management systems (KMSs) proposed by the author of Chapter 2 enable teams to access information quickly, reducing the time spent on research and repetitive tasks (Davenport & Ronanki, 2018). This improves operational efficiency and allows creative professionals to focus on innovative projects.

Integrating GenAI in creative workflows opens up significant opportunities for SMEs. By automating routine tasks, AI allows small teams to achieve more with fewer resources. This enhances productivity and fosters innovation by providing tools that support creative processes. For example, AI can generate design ideas, suggest improvements, and even create content, thus expanding the creative potential of SMEs (Davenport & Ronanki, 2018). Moreover, AI-driven insights can help businesses better understand market trends and consumer preferences, enabling them to tailor their products and services more effectively.

Despite its benefits, GenAI raises concerns about job displacement and the ethical implications of AI in creative industries. There is a fear that AI could replace human creativity, leading to job losses (Floridi et al., 2018). Additionally, using AI in creating content raises questions about originality and intellectual property rights. To mitigate these risks, it is essential to establish ethical guidelines that ensure AI is used to augment rather than replace human creativity. This includes promoting fair work practices and ensuring that AI-generated content respects copyright laws.

CUSTOMER EXPERIENCE INDUSTRY: TRANSFORMING CREATIVITY WITH AI

Integrating AI within design workflows, particularly through the DesignOps framework proposed by the author of Chapter 3, revolutionises the customer experience industry. AI's ability to analyse large datasets and generate valuable insights enhances design efficiency and creativity (Dzyabura et al., 2019). By automating routine design tasks, AI allows designers to focus on more complex and creative aspects of their work. For instance, AI tools can quickly generate multiple design prototypes based on user data, enabling rapid iteration and refinement of design concepts. This accelerates the design process and enhances the overall quality of customer experiences by ensuring designs are more closely aligned with user needs and preferences.

AI-driven tools can optimise design practices, reduce work duplication, and promote efficiency. These tools enable a unified design language and standardisation, which is critical for automating routine tasks and enhancing creative output (Amabile & Pratt, 2016). For example, AI can create a centralised design system that standardises design elements across various projects, reducing inconsistencies and improving collaboration among design teams. Furthermore, AI can provide real-time feedback and suggestions during the design process, helping designers to make more informed decisions and enhance their creative output. This can lead to more innovative and user-centric designs that drive customer satisfaction and loyalty.

Integrating AI in design workflows poses challenges such as resistance to change and scepticism about AI's creative potential. Designers may feel threatened by the automation of creative tasks, fearing that AI could replace their roles or diminish the value of human creativity. Additionally, there are ethical concerns regarding the potential for AI-driven design solutions to perpetuate biases or exclude certain user groups (Crawford & Calo, 2016). AI systems are only as unbiased as the data they are trained on, and if the training data contains biases, these can be reflected in the AI-generated designs. Addressing these challenges requires effective change management strategies that emphasise the complementary role of AI in enhancing human creativity. Continuous learning and ethical guidelines are essential to ensure that AI is used responsibly and inclusively in the design process.

HEALTHCARE INDUSTRY: INNOVATIVE ELDERCARE WITH AI

AI's role in healthcare, particularly in eldercare, has been transformative. Technologies like facial recognition technology (FRT), as investigated by the author of Chapter 4, are being adopted to enhance senior engagement and care quality in eldercare centres (Topol, 2019). AI improves care management by predicting patient needs and streamlining administrative tasks, significantly enhancing the overall quality of care. For instance, AI systems can monitor patients' health conditions in real time, providing caregivers with timely alerts and recommendations (Jiang et al., 2017). This proactive approach ensures that potential health issues are addressed promptly, reducing the risk of complications.

The adoption of AI in healthcare presents numerous opportunities for improving patient care. AI can assist in diagnosing diseases, managing treatment plans, and even conducting robotic surgeries. In eldercare, AI tools can help monitor the well-being of seniors, ensuring they receive the necessary support and intervention when needed. For example, AI-powered devices can track vital signs and detect anomalies, allowing caregivers to provide personalised and timely care (Jiang et al., 2017). This improves the quality of life for seniors and alleviates the burden on healthcare professionals.

However, the integration of AI in healthcare raises several ethical concerns. Privacy issues are paramount, as AI systems handle sensitive patient data. There is also the potential for biases in AI algorithms, which could lead to unequal treatment or misdiagnoses (Morley et al., 2020). Furthermore, there is a risk of dehumanising patient care, where reliance on AI could reduce the personal interaction between caregivers and patients. To address these concerns, it is crucial to implement ethical AI practices that prioritise patient welfare and data security. This includes developing transparent algorithms, ensuring data privacy, and maintaining a human touch in patient care.

INTERDISCIPLINARY WORKPLACE COMMUNICATION: BRIDGING GAPS WITH AI

AI to augment traditional methodologies like oral history, as experimented with by the author of Chapter 5, improves communication within interdisciplinary teams. AI tools can facilitate a deeper understanding between individuals from different backgrounds by bridging cultural and personal histories (Edmondson, 2012). For instance, AI can analyse communication patterns and identify areas where misunderstandings are likely to occur, providing insights that can help teams improve their interactions. AI-powered platforms can also facilitate sharing of personal stories and experiences, fostering a sense of empathy and connection among team members from diverse disciplines.

AI can enhance storytelling and communication by providing a platform for integrating diverse perspectives. This approach can lead to more effective teamwork and collaboration in interdisciplinary settings (Pentland, 2012). For example, AI can help teams identify common themes and insights from their collective experiences, creating a shared understanding that can drive collaboration. Additionally, AI tools can support the creation of multimedia presentations that combine text, audio, and visual elements, making it easier for team members to communicate complex ideas and engage with each other's work. This can enhance the overall effectiveness of interdisciplinary projects and lead to more innovative solutions.

While AI can augment dialogue, there are concerns about the authenticity and accuracy of AI-generated content. Ensuring that AI tools respect the integrity of personal stories and cultural histories is crucial for ethical implementation (Floridi et al., 2018). AI-generated content must be carefully reviewed and validated to reflect the perspectives and experiences of the individuals involved accurately. Additionally, there are concerns about the potential for AI to reinforce existing power dynamics and biases within interdisciplinary teams. Ethical guidelines and oversight are essential to ensure AI tools promote fairness, inclusivity, and respect for individual contributions.

LEGAL INDUSTRY: TRANSFORMING LAW FIRM OPERATIONS WITH AI

The legal sector, traditionally slow to adopt new technologies, is gradually embracing AI tools like ChatGPT to improve operational efficiency and client service (Park & James, 2024). Findings reported by the author of Chapter 6 show that AI can automate routine tasks such as document review and legal research, enhancing the productivity of legal professionals. For instance, AI systems can quickly analyse vast amounts of legal documents, identify relevant information, and provide summaries, significantly reducing the time and effort required for these tasks (Atkinson et al., 2020). This allows lawyers to focus on more complex and strategic aspects of their work.

Adopting AI in the legal industry presents significant opportunities for enhancing efficiency and client service. AI can assist in case management by organising and retrieving case information quickly, improving the accuracy and speed of legal processes. Additionally, AI tools can help law firms manage client interactions, providing timely responses and personalised services. This improves client satisfaction, the firm's reputation, and competitiveness (Atkinson et al., 2020). Furthermore, AI can support predictive analytics, helping lawyers anticipate case outcomes and develop more effective strategies.

Despite these benefits, adopting AI in the legal field must address several ethical concerns. There is a risk of bias in AI judgements, where algorithms may perpetuate existing biases in legal data, leading to unfair outcomes (Susskind, 2019). Data privacy is another critical issue, as AI systems handle sensitive legal information. Additionally, the automation of routine tasks raises concerns about job displacement for legal professionals. To mitigate these risks, it is essential to develop ethical guidelines for AI use in the legal sector, ensuring transparency, accountability, and fairness. This includes regular audits of AI systems to detect and correct biases and implementing robust data protection measures.

LEADERSHIP AND ORGANISATIONAL MANAGEMENT: CULTIVATING INCLUSIVE LEADERSHIP WITH AI

Organisations increasingly integrate AI tools to manage unconscious bias within their leadership frameworks, recognising the limitations of traditional unconscious bias training (UBT) (Dobbin & Kalev, 2018). The author of Chapter 7 reported that AI-powered tools provide personalised prompts and reflection activities, fostering continuous awareness and management of unconscious bias among leaders. For example, AI can analyse communication patterns and provide feedback on potential biases, helping leaders make more inclusive decisions (Albaroudi et al., 2024).

Integrating AI in leadership and organisational management presents significant opportunities for enhancing diversity and inclusion. AI tools can identify and mitigate biases in recruitment, performance evaluations, and promotions, ensuring fair and equitable treatment of all employees. This improves organisational culture and drives better business outcomes by leveraging diverse perspectives and talents. Additionally, AI can support leadership development by providing personalised coaching and feedback, helping leaders enhance their skills and effectiveness (Albaroudi et al., 2024).

However, using AI in this context raises ethical concerns about privacy and data security. AI algorithms can reinforce existing biases if not properly designed and monitored (Binns, 2018). To ensure ethical AI deployment, it is essential to establish guidelines that prioritise transparency and accountability. This includes regularly auditing AI systems to detect and address biases and ensuring that AI tools complement rather than replace human judgement.

LIFESTYLE INDUSTRY: BUILDING SUSTAINABLE COMMUNITIES WITH AI

AI plays a significant role in building and sustaining brand communities, particularly in the lifestyle sector. The author of Chapter 8's case study on Brompton Bicycle highlights how AI can help develop brand love and foster connections among community members (Butler-Adams & Davis, 2022). By analysing user data, AI can identify common interests and preferences among community members, enabling brands to create more personalised and engaging experiences. For instance, AI-powered recommendation engines can suggest relevant content, products, and events to community members, enhancing their overall engagement and satisfaction.

AI can facilitate the creation of personalised and localised experiences for community members, enhancing engagement and loyalty. By leveraging AI, brands can better understand and meet the needs of their communities (Holt, 2016). For example, AI can analyse social media interactions and other user-generated content to identify emerging trends and preferences, allowing brands to tailor their marketing strategies accordingly. Additionally, AI can support the development of virtual communities and online platforms where members can connect, share experiences, and collaborate on projects. This can strengthen the sense of community and drive long-term loyalty and advocacy.

Using AI in community building raises concerns about data privacy and the potential for manipulating consumer behaviour. Brands must navigate these ethical issues to maintain trust and authenticity in their community engagement efforts (Zuboff, 2019). AI systems often require access to large amounts of personal data to function effectively, raising concerns about how this data is collected, stored, and used. Brands must ensure that they have robust data protection measures and are transparent about consumer data use. Additionally, there is a risk that AI could be used to manipulate consumer behaviour in ways that are not in the best interests of the community members. Ethical guidelines and oversight are essential to ensure that AI is used responsibly and that community engagement efforts remain genuine and authentic.

STUDENT DEVELOPMENT: SUPPORTING MENTAL HEALTH WITH AI

Adopting AI-driven solutions like the MAESTRO mobile app, as proposed by the author of Chapter 9, addresses the mental health crisis among students. These tools provide personalised and preventative strategies to support student well-being (Bai, 2020). AI-powered mental health apps can offer real-time support and intervention, helping students manage stress, anxiety, and other mental health issues. For instance, the MAESTRO app uses AI algorithms to monitor users' emotional states and provide tailored coping strategies and resource recommendations.

AI can automate tasks and provide real-time support, allowing educators and mental health professionals to focus on more complex cases. This approach can enhance educational institutions' overall mental health support system (Topol, 2019). For example, AI chatbots can respond immediately to students' queries, offering guidance and support when human counsellors are unavailable.

Additionally, AI can analyse data from various sources, such as social media and academic performance, to identify students who may be at risk and provide early intervention. This proactive approach can improve the overall well-being of the student population and reduce the burden on mental health services.

Ethical concerns regarding privacy, data security, and the potential for AI to intrude on personal boundaries must be addressed. Ensuring that AI tools are used responsibly and ethically is critical for maintaining trust and effectiveness (Morley et al., 2020). AI systems in mental health must be designed with robust data protection measures to ensure the privacy and confidentiality of users' information. Additionally, there are concerns about the accuracy and reliability of AI-generated recommendations, which must be carefully validated to ensure they are appropriate and effective. Ethical guidelines and oversight are essential to ensure that AI tools are used in ways that respect the autonomy and dignity of the individuals they serve.

YOUNG ADULT CAREER COACHING: ENHANCING INTERVIEW SKILLS WITH AI

Integrating AI and immersive technologies like Metaverse-based simulations, as investigated by the study undertaken in Chapter 10, is helping young adults prepare for job interviews. These tools provide a safe and immersive environment for practice, reducing anxiety and enhancing preparedness (Ferguson, 2024). AI-powered simulations can replicate real-world interview scenarios, allowing users to practise their responses and receive feedback in a low-stakes environment. This can help job seekers build confidence and improve their interview performance.

AI can offer personalised feedback and simulate real-world scenarios, helping job seekers develop practical skills and confidence. This approach can bridge the gap between academic achievements and practical job skills (Brynjolfsson & McAfee, 2014). For instance, AI algorithms can analyse users' interview performance and provide detailed feedback on areas for improvement, such as body language, communication skills, and question responses. Additionally, AI can identify common interview questions and simulate different interview formats, allowing users to prepare for various scenarios. This can enhance their employability and increase their chances of success in the job market.

AI in career coaching raises concerns about fairness and bias in AI-generated feedback. Ensuring these tools are designed and implemented ethically is crucial for providing equitable support to all job seekers (Binns, 2018). AI systems must be trained on diverse datasets to provide fair and unbiased feedback to users from different backgrounds. Additionally, there are concerns about the potential for AI to reinforce existing biases in hiring practices, which must be addressed through ethical guidelines and oversight. Career coaching programmes can provide equitable support and help all job seekers achieve their full potential by ensuring that AI tools are used responsibly and ethically.

BALANCING INNOVATION AND ETHICS: APPLYING A FREIREAN APPROACH

Using Paulo Freire's (1970) critical pedagogy, I have always sought to examine the instructor-student power difference and the ethics of inclusionary and exclusionary practices in HE. I pose the question: how will HE grapple with the exclusion of students unexposed to emerging technologies? What if the adoption, with sound pedagogical intelligence, depends on mass faculty buy-in and the availability of resources like time to develop the digital tools for technology-enhanced learning and costs? Moreover, how will HE mitigate the detrimental effects of biases in such tools? While AI brings opportunities and benefits, it also entails risks and concerns about its impact on, e.g., equality, power distribution, and fairness.

Technology inevitably reflects its creators in myriad ways, conscious and unconscious. Unfortunately, emerging technology like AI has been created with biases built into its core. It is also culturally homogeneous (Shams et al., 2023). This lack of diversity is reflected in many of its applications. Also, regarding the ethical considerations of implementing artificially intelligent systems, they need more emotional intelligence to make them comprehensible, unempathetic, and arguably unethical in communicating with students. For example, a human instructor would take care when explaining the moral dilemma of the medical procedure of abortion to a young female student who might have undergone the procedure. The instructor would develop ethics and intelligence in discriminating between good and bad and help mitigate negative feelings. However, this would not be possible in artificially intelligent systems. Thus, aligned with many scholars, I would advocate for AI to be genderless. In addition, the development of AI-enabled applications must also benefit from a human-centred design approach with a transdisciplinary team of developers.

Many AI systems are based on cloud computing technology, so many are concerned with potential data loss and misuse. In HE, emerging technologies, privacy, ethics, and access to student data are highly debated as controversial topics. Student privacy and student data can easily be lost or accessed by unauthorised personnel/entities. Furthermore, AI works in the background, becoming a pervasive technology that is invisible to users. Yet, when embraced in HE, it opens up many opportunities to revolutionise teaching and learning. However, it also challenges existing pedagogical approaches and practices that favour standardised and teacher-centred methods which underpin instructors' positional and epistemic authority. I wonder if instructors are placed uncomfortably in choosing between emerging technologies and traditional tools, one at the expense and exclusion of the other? How will this affect the students graduating from the institutions?

The adoption of AI across different industries presents both opportunities and challenges. While AI can drive efficiency, innovation, and improved service delivery, it raises significant ethical concerns. To navigate these challenges, the following recommendations are proposed (Figure 11.2).

ESTABLISH ETHICAL GUIDELINES AND FRAMEWORKS

To govern AI use effectively, it is imperative to develop comprehensive ethical guidelines and frameworks that align with societal values and ethical standards. These guidelines should address privacy, bias, transparency, and accountability in AI systems. For example, the European Union's General Data Protection Regulation (GDPR) is a robust framework ensuring data privacy and protection, which can be adapted for AI-specific contexts (European Commission, 2018). Moreover, ethical guidelines should promote fairness in AI algorithms, ensuring they do not perpetuate societal biases. Organisations such as the IEEE Global Initiative on Ethics of Autonomous and Intelligent Systems have proposed frameworks that emphasise the need for ethical considerations throughout

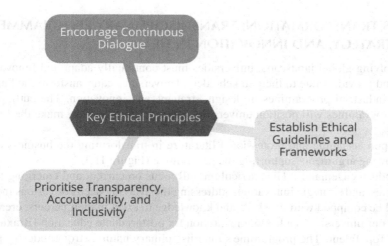

FIGURE 11.2 Key ethical principles to balance innovation and ethics.

the AI development lifecycle, from design to deployment (IEEE, 2019). By adopting such comprehensive guidelines, industries can ensure that AI technologies are used responsibly and ethically.

PRIORITISE TRANSPARENCY, ACCOUNTABILITY, AND INCLUSIVITY

Transparency and accountability in AI deployment are crucial for fostering trust and acceptance among stakeholders. Transparent AI systems enable users to understand how decisions are made, which is essential for building trust. This can be achieved by making AI algorithms interpretable and ensuring that decision-making processes are explainable. For instance, the concept of "Explainable AI" (XAI) focuses on creating AI models that provide clear explanations of their decisions, thereby enhancing transparency and accountability (Gunning et al., 2019).

Inclusivity is another critical aspect of responsible AI use. It involves ensuring that AI technologies are accessible and beneficial to diverse groups of people, including marginalised communities. Inclusive AI systems should be designed to avoid biases leading to discriminatory outcomes. This requires diverse datasets for training AI models and involving diverse stakeholders in the AI development process to ensure that various perspectives are considered (Shams et al., 2023). By prioritising transparency, accountability, and inclusivity, industries can create AI systems that are more equitable and trustworthy.

ENCOURAGE CONTINUOUS DIALOGUE

Continuous dialogue between industry stakeholders, policy-makers, and ethical experts is essential to navigate the complexities of AI adoption. This dialogue facilitates the sharing of best practices, the identification of emerging ethical challenges, and the development of collaborative solutions. For example, forums such as the Partnership on AI, which includes members from academia, industry, and civil society, provide a platform for discussing the ethical implications of AI and developing consensus on best practices (Partnership on AI, 2019).

Engaging in continuous dialogue also helps formulate policies and regulations that keep pace with the rapid advancements in AI technology. Policy-makers can benefit from the insights of industry experts and ethical scholars to create regulations that balance innovation with ethical considerations. Regular workshops, conferences, and publications can serve as mediums for such ongoing dialogue, ensuring that the development and deployment of AI technologies remain aligned with ethical standards and societal values.

A BUSINESS TRANSFORMATION: TRANSDISCIPLINARY PROGRAMMES IN DESIGN, STRATEGY, AND INNOVATION IN HE

In an ever-evolving global landscape, universities must continually adapt and innovate to remain competitive and provide value to their stakeholders. Universities can transform the business of HE by launching industrial programmes in design, strategy, and innovation. The cutting-edge transdisciplinary programmes will position universities at the forefront as they make the future of HE a reality.

Drawing upon academic and professional literature in transforming the business of HE, I distilled the following arguments supporting the programme (Figure 11.3).

Meeting Industry Demands: The curriculum will focus on current and emerging technologies, design principles, and strategic innovation, addressing the evolving needs of various industries. Our graduates will be equipped with the skills and knowledge to excel in their careers, creating a strong reputation for our university as a leading institution for postgraduate education (Braxton, 2023).

Attracting Top Talent: The programme's interdisciplinary nature, strong industry partnerships, and competitive scholarships will attract top candidates who are leaders in their respective fields.

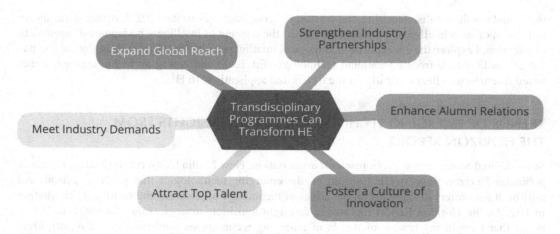

FIGURE 11.3   Transdisciplinary programmes can transform HE.

This influx of exceptional talent will enhance the university's intellectual capital and drive further innovation, contributing to our growth and success (Dolan et al., 2022).

Expanding Global Reach: By promoting the programme through a robust global marketing and recruitment campaign, we will expand our reach and attract students from diverse backgrounds, further increasing the university's international standing and reputation (Jones, 2013).

Strengthening Industry Partnerships: Collaborations with top businesses and employers will provide valuable experiential learning opportunities for our students while creating a mutually beneficial relationship between the university and industry partners. These connections will lead to increased research funding, joint projects, and long-term relationships that will contribute to the university's growth and impact (Sun & Turner, 2022).

Enhancing Alumni Relations: The success of our graduates will reflect positively on our university, inspiring a sense of pride and loyalty among our alumni. Engaged alumni are more likely to contribute to the university's development through donations, mentorship, and industry connections, ensuring continued support for our programmes and initiatives (Edmunds & Boettcher, 2021).

Fostering a Culture of Innovation: The programme will inspire innovation across our university by promoting interdisciplinary collaboration, industry engagement, and the pursuit of groundbreaking research. This transformative mindset will benefit our entire institution, ensuring we remain agile, adaptable, and future-ready (Jonbekova et al., 2020).

Launching industrial programmes in design, strategy, and innovation represents a strategic opportunity to transform universities' businesses and ensure value for all stakeholders. By embracing this opportunity, universities will secure their positions as global leaders in postgraduate education, drive innovation, and contribute to the betterment of society.

BRIDGING INDUSTRY AND HIGHER EDUCATION WITH ARTIFICIAL INTELLIGENCE AND PEDAGOGICAL INTELLIGENCE

With the rise of the information age, many IHLs have been re-examining and adjusting how they prepare students for the future workforce. Given the enormous role technology has played in changing the landscape around us, I pose two questions: what should education technology look like for HE to drive future growth, and how can emerging technology be leveraged to prepare students for the future of work? With the availability of emerging digital tools, new possibilities for learning are created. There are varying types of AI for adoption in HE, and its impact in its many forms, such as machine learning, data mining, and learning analytics, is undeniable. Yet, attaining pedagogical goals remains essential. Hence, more 'intelligent' pedagogical support is needed to engage students today and revolutionalise teaching and learning. According to Soliman (2022), pedagogical-aware utilisation of such intelligence in education meets the concept of intelligent pedagogical 'agents'. In this section, I explore the concepts of 'pedagogical intelligence' and 'intelligent pedagogical agents'. I draw on literature from educational technology (Ed Tech) and emerging technology evidence-based research to reflect critically on the trends and applications in HE.

TRENDS OF EMERGING TECHNOLOGIES IN HE: INSIGHTS FROM THE HORIZON REPORT

In the United States, a non-profit research organisation, New Media Consortium (NMC), regularly publishes Horizon Report (HE Edition) on the emerging technologies that promote educational reform. It also offers discussions on the issues of technology application and significant challenges in HE. As the Horizon Report has strong foresight and high strategic value for those in HE, it is apt that I begin my review of trends of emerging technologies published in the report. Over the years, the Horizon Report has done much predictive work in exploring the future impact of educational technology. The latest Educause Horizon Report for Teaching and Learning (Pelletier et al., 2023) no longer adopts the three-dimensional framework of trends, technologies, and challenges as in previous years but elaborates on three dimensions of trend, technology and practice, and scenarios reported through case studies (Yan, 2021). Horizon Report 2023 covers three parts: key trends affecting future higher education; emerging technologies and practices; and scenarios and case articles on future higher education development. The emerging technologies and practices that can influence the future development of HE are adaptive learning technology, artificial intelligence/machine learning educational applications, student achievement analysis, instructional design, learning process and user experience design improvement, open education resources, and augmented reality technology (Horizon Report, 2020).

APPLICATION OF AI/MACHINE LEARNING IN HE

When Russell and Norvig wrote their seminal university textbook on artificial intelligence in 1995, they suggested that since intelligence is so important to human beings, we should strive to understand and rebuild it. I learnt that the term artificial intelligence was coined six years after Alan Turing published the article 'Computing Machinery and Intelligence', in which he introduced the Turing test, machine learning, genetic algorithms, and reinforcement learning. Today, the focus lies not on creating a strong AI, which is an AI system that has general intelligence and can think like a human, but on creating expert systems that might have the ability to work more efficiently compared to a human in a specific area (Russell & Norvig, 2020). These systems, like search algorithms or voice assistants, do not necessarily need a physical or visual representation. In an education review in August 2019, Elana Zeide defined AI as 'trying to create a machine that can only accomplish things through human cognition'. Yates and Chamberlain (2017) also described machine learning as 'teaching a machine to learn something without a clear explanation'. In a scoping review on teaching and learning with AI in HE, Kuka et al. (2022) summarise the milestones that AI has achieved. These milestones show the various possibilities AI has to offer, especially to HE. In another recent publication, Mone and Rus (2024) present a thorough exploration of the risks and rewards associated with AI by looking at its historical development and contemporary advances. Several insights can be distilled from their discussions, particularly concerning the juxtaposition of technological progression and the ethical, societal, and existential dilemmas that have emerged alongside AI's growth (Mone & Rus, 2024).

Even though research on AI has been done in education since 1997, it is still in its early stages (Parapadakis, 2020). Some argue that the education sector needs to catch up with other fields.

As an emerging technology, like robotics and neurotech, which achieved recent significant advances due to the availability of big data and computing power, AI has the following five distinguishing characteristics: (a) radical novelty; (b) relatively fast growth; (c) coherence; (d) prominent impact; and (e) uncertainty and ambiguity (Rotolo et al., 2015). Although it shows high potential, its value has yet to be well demonstrated (Cozzens et al., 2010). Nonetheless, institutes of higher learning like the University of Oklahoma recently launched the Sooner BOT programme for school admissions. More than 28,000 students have used the programme to interact with one another, contributing to the formation of the largest first-year class to a certain extent. The question remains: how is AI truly revolutionising teaching and learning in HE and advancing faculty's pedagogical pursuits?

WHY INTELLIGENT AGENTS FOR PEDAGOGY?

In AI, an agent is an autonomous and proactive entity (Wooldridge, 2002). Agents mainly operate in societies and interact with the environment they represent to perform various actions (Wooldridge, 2002). According to Soliman (2022), while intelligent agents are used for various domains, an educational orientation can help discover pedagogical intelligence in technology-enhanced learning (TEL) environments. Accordingly, Soliman and Guetl (2011) summarise the pedagogical functions of intelligent agents in Figure 11.4.

The pedagogical functions of intelligent agents are linked to various instructional approaches and traditional pedagogy, like *relational pedagogy*, which focuses on the relationship between the instructor and students, and *differentiated learning* to cater to differing learning needs. The instructor's competence in the classroom includes awareness, application of different pedagogical methods, and a significant role in establishing social and emotional connectedness and support (Patel, 2023). Thus, the instructor seeks to understand the learning process and adapt the activities and methods according to learner differences (Shulman, 1987). Similarly, understanding students' prior knowledge and abilities is aligned with constructivism (Piaget, 1964).

Agent type	Function
Agents for learning personalization	Agents that understand individual learner abilities and treat the learner based on their ability and style accordingly
Agents for emotional support	Agents that consider the learner emotional state and improve it accordingly. Thus, they also support motivation and engagement
Cognitive agents	They are inspired by cognitive theories of the human mind as well as AI. They use cognitive models
Meta-cognitive agents	Higher levels of thinking by including meta-cognition supporting methods such as communicating by concept maps
Teachable agents	The learner learns by teaching an agent
Self-regulated learning agents	Agents that apply theories of self-regulated learning
Conceptual change agents	Utilizing conceptual change learning theories
Explainable agents	Agents that can track the reasoning processes of the pedagogical interactions
Multiple agents supporting group learning or training	Facilitating group work and Computer Supported Collaborative Learning (CSCL)

FIGURE 11.4 Pedagogical functions of intelligent agents (taken from Soliman, 2022).

However, with intelligent (technological) agents, it will be easier for instructors to achieve the various pedagogical functions when class sizes are large or a course enrolment is high. Hence, emerging technology like AI/machine learning provides an effective technical solution for HE. Are there any novel ideas in AI's revolutionary and transformative framing in HE? Or has it been primarily a renewal and amplification of traditional 'recipes' of teaching and learning? How can the concepts of 'pedagogical intelligence' and 'intelligent pedagogical agents' inform pedagogical innovations with the adoption of emerging technologies?

CRITICAL DESIGN FUTURES (CDF) THINKING AND GENAI IN HIGHER EDUCATION

In a world characterised by constant change and uncertainty, IHLs must equip students to tackle global challenges. This necessitates a rethinking and redesign of curricula to make them future-proof. The complex and wicked problems we face today require a new kind of thinking – one that critically examines our ideas and assumptions. The emerging world demands innovative thinkers who can devise sustainable solutions to the most pressing issues. Utilising a design innovation approach is essential, but it must be grounded in critical and futures thinking. For designers, intuition plays a crucial role in the design process, much like intuitive thinking (Patel et al., 2023, 2024). However, design thinking can fall short without the ability to critically analyse and envision future scenarios. Therefore, in two forthcoming papers, my co-investigators and I discussed a newly developed conceptual framework that intentionally integrates critical design futures (CDF) thinking for innovation (Patel et al., forthcoming; Patel & Lim, 2024). The impetus for creating a conceptual framework (See Figure 11.5) that integrates critical thinking, design thinking, and futures thinking (CDF) is rooted in the identified need to provide students with effective thinking tools to tackle future complexities. This necessity arises from a fast-changing global environment, characterised by technological progress, environmental challenges, and societal changes, which call for innovative and forward-thinking individuals. The CDF framework seeks to meet this need by combining the unique yet complementary aspects of critical, design, and futures thinking. This integrated approach aims to develop critical competence, design proficiency, and futures adaptability in students, preparing them to navigate and innovate in an increasingly complex world.

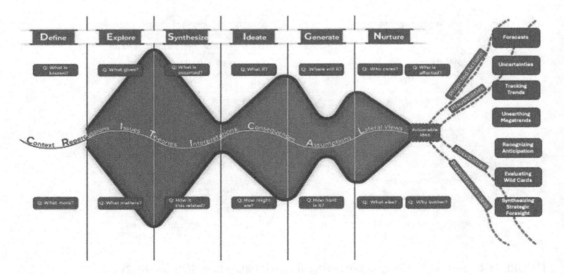

FIGURE 11.5 Critical Design Futures (CDF) Framework (Patel et al., 2024; Patel & Lim, 2024).

Integrating GenAI tools with the CDF framework in HE fosters a synergistic relationship between thinking skills and creative visualisation. This integration empowers students to tackle complex problems by leveraging analytical and imaginative faculties. In the forthcoming research paper, my co-investigator and I discuss the endeavour to 'make the unthought thought', and the 'unseen seen' with CDF and GenAI in innovation (Patel & Lim, 2024). We posit that it becomes a dual process of intellectual engagement and creative exploration, where students are equipped to question the status quo and propose and visualise innovative solutions that address the multifaceted challenges of the future. Applying GenAI tools together with the CDF framework embodies a holistic approach to design innovation that prepares students to become transdisciplinary professionals. By cultivating critical competence, design dexterity, and futures flexibility, students are uniquely positioned to navigate the uncertainties of the future, making the unthought thought and the unseen seen (Patel & Lim, 2024). This approach enriches the HE experience and contributes to the development of forward-thinking graduates capable of contributing meaningful solutions to the world's most pressing problems.

REDESIGNING HIGHER EDUCATION CURRICULUM AND PROGRAMMES: INSIGHTS AND RECOMMENDATIONS

The insights gained from the adoption and impact of AI across various industries highlight the imperative to redesign higher education curricula. This redesign should prepare students for the future workforce by integrating generative GenAI and CDF thinking into pre-employment training, equipping undergraduates with the necessary skills to innovate and thrive in their respective fields.

CURRICULUM DEVELOPMENT

IHLs should develop curricula incorporating GenAI and CDF thinking, ensuring students are proficient in AI technologies and their applications in different industries. This includes offering specialised courses covering AI's technical aspects, such as machine learning, natural language processing, and data analytics (Jordan & Mitchell, 2015). Additionally, courses should address the ethical implications of AI, including data privacy, bias, and the societal impacts of AI technologies (Binns, 2018). This comprehensive approach ensures that students acquire technical skills and develop a deep understanding of the broader context in which AI operates. Furthermore, curricula should include applied learning with authentic assessments where students can work on real-world AI applications. This hands-on experience is crucial for bridging the gap between theoretical knowledge and practical skills. For instance, students can engage in capstone projects that involve developing AI solutions for industry partners, providing them with valuable insights into the practical challenges and opportunities of AI deployment.

INTERDISCIPLINARY PROGRAMMES

Educational programmes should promote interdisciplinary learning to reflect the multidisciplinary nature of AI's impact. AI applications span various fields, from healthcare and legal sectors to creative industries and leadership roles (Vargo & Lusch, 2017). By fostering interdisciplinary collaboration, students can better understand AI's applications and implications across different domains. Interdisciplinary programmes can include collaborative courses and projects where students from different fields collaborate to develop AI solutions. For example, a programme combining computer science, ethics, and business can prepare students to address AI deployment's technical, ethical, and economic aspects (Floridi et al., 2018). This interdisciplinary approach ensures that graduates are well-rounded and equipped to tackle the complex challenges posed by AI in diverse professional settings.

PRACTICAL TRAINING AND INTERNSHIPS

Incorporating practical training and internships that expose students to real-world applications of AI can bridge the gap between theoretical knowledge and practical skills. Partnerships with industries that are adopting AI can provide students with hands-on experience and insights into AI-driven innovations. These opportunities allow students to apply their learning in real-world contexts, enhancing their readiness for the workforce (Davenport & Ronanki, 2018). Internships can be structured to involve students in AI projects, where they can contribute to developing and implementing AI solutions. This provides practical experience and helps students build professional networks and gain insights into industry practices (Beard, 1999). Additionally, mentorship programmes can pair students with industry professionals who can provide guidance and support throughout their internship experience.

ETHICAL AND RESPONSIBLE AI USE

Education on the ethical considerations and responsible use of AI is crucial. Institutions should emphasise the importance of ethical AI practices, data security, and addressing biases in AI algorithms. This training ensures that future professionals can deploy AI responsibly and ethically in their respective fields. Courses on AI ethics should cover topics such as fairness, accountability, transparency, and the societal impacts of AI technologies (Morley et al., 2020). Moreover, institutions should incorporate case studies and real-world examples of ethical dilemmas in AI deployment. This helps students understand the practical implications of ethical principles and develop critical thinking skills to navigate complex ethical issues. Workshops and seminars led by ethical experts and industry practitioners can further enhance students' understanding of responsible AI use (Alhitmi et al., 2024).

LIFELONG LEARNING AND CONTINUOUS EDUCATION

As AI technology evolves, continuous education programmes can help professionals stay updated with the latest advancements. Offering certification courses and workshops on emerging AI technologies and their applications can support lifelong learning and career growth. Continuous education ensures that professionals remain competitive in the rapidly changing AI landscape (Gunning et al., 2019). Institutions can establish partnerships with industry leaders to offer specialised training programmes that address current trends and emerging technologies in AI. Online learning platforms can provide flexible and accessible options for professionals seeking to update their skills. By fostering a culture of continuous learning, institutions can help professionals stay ahead in their careers and contribute to ongoing innovation in their fields.

CONCLUSION

Integrating GenAI and CDF thinking into higher education curricula is crucial for preparing future-ready graduates. Higher education institutions must evolve into ICLs to equip students with the necessary skills and knowledge to navigate and lead in a complex world. This shift requires a move towards transdisciplinary learning and Education 4.0, aligning educational strategies with real-world applications. By incorporating GenAI and CDF, institutions can foster critical thinking, problem-solving, and creativity, enabling students to engage with multiple fields and collaborate with diverse professionals. This approach is vital for addressing the complex challenges of the 21st century. The adoption of AI across industries provides valuable insights for higher education, helping align programmes with workforce demands while emphasising ethical responsibility. A Freirean approach that considers ethical, social, and privacy issues ensures AI and new technologies are inclusive and beneficial. AI's role in enhancing personalised learning experiences, swift feedback, and support is revolutionising higher education. By fostering continuous growth and development, institutions can ensure AI innovations are trusted and improve the educational experience. Interdisciplinary collaboration inherent in CDF magnifies innovation, drawing from science, technology, engineering, and mathematics and social sciences to create technically robust, human-centric, and contextually relevant solutions. Bridging the industry-higher education gap through GenAI and CDF integration is essential for developing graduates proficient in AI technologies and capable of addressing ethical and societal challenges, ultimately benefiting society as a whole.

REFERENCES

Adeleye, I. O. (2023). The AI Effect: Rethinking Design Workflows for Enhanced Productivity and Creativity. *International Journal of Science and Technology Innovation*, 2(1), 1–19. https://doi.org/10.70560/c9m3kd97

Albaroudi, E., Mansouri, T., & Alameer, A. (2024). A Comprehensive Review of AI Techniques for Addressing Algorithmic Bias in Job Hiring. *AI*, 5(1), 383–404. https://doi.org/10.3390/ai5010019

Alhitmi, H. K., Mardiah, A., Al-Sulaiti, K. I., & Abbas, J. (2024). Data security and privacy concerns of AI-driven marketing in the context of economics and business field: an exploration into possible solutions. *Cogent Business & Management*, 11(1). https://doi.org/10.1080/23311975.2024.2393743

Amabile, T. M., & Pratt, M. G. (2016). The dynamic componential model of creativity and innovation in organizations: Making progress, making meaning. *Research in Organizational Behavior*, 36, 157–183. https://doi.org/10.1016/j.riob.2016.10.001

Angouri, J. (2021). *Reimagining Research-led Education in a Digital Age* (The Guild Insight Paper No. 3). The Guild of European Research-Intensive Universities and Bern Open Publishing. Retrieved from https://www.the-guild.eu/publications/insight-papers/the-guild_insight-paper_research-led-education-in-a-digital-age_june-2021.pdf

Atkinson, K., Bench-Capon, T.J., & Bollegala, D. (2020). Explanation in AI and law: Past, present and future. *Artif. Intell.*, 289, 103387. https://doi.org/10.1016/j.artint.2020.103387

Beard, F. K. (1999). Client role ambiguity and satisfaction in client-ad agency relationships. *Journal of Advertising Research*, 39(2), 69–78.

Bell, N., Liu, M., & Murphy, D. (2022). A Framework to Implement Academic Digital Badges when Reskilling the IT Workforce. *Information Systems Education Journal (ISEDJ)*, 20(1). 36–46. Retrieved from https://files.eric.ed.gov/fulltext/EJ1333890.pdf

Binns, R. (2018). Fairness in machine learning: Lessons from political philosophy. *Proceedings of the 2018 Conference on Fairness, Accountability, and Transparency*, 149–159. https://proceedings.mlr.press/v81/binns18a.html

Bonfield, C. A., Salter, M., Longmuir, A., Benson, M., & Adachi, C. (2020). Transformation or evolution?: Education 4.0, teaching and learning in the digital age. *Higher Education Pedagogies*, 5(1), 223–246. https://doi.org/10.1080/23752696.2020.1816847

Braxton, S. N. (2023). Competency frameworks, alternative credentials and the evolving relationship of higher education and employers in recognizing skills and achievements. *International Journal of Information and Learning Technology*. https://doi.org/10.1108/IJILT-10-2022-0206

Brown, M., McCormack, M., Reeves, J., Brooks, D. C., & Grajek, S. (2020). EDUCAUSE Horizon Report, Teaching and Learning Edition. EDUCAUSE. https://library.educause.edu/-/media/files/library/2020/3/2020_horizon_report_pdf.pdf?la=en&hash=08A92C17998E8113BCB15DCA7BA1F467F303BA80

Brynjolfsson, E., & McAfee, A. (2014). *The second machine age: Work, progress, and prosperity in a time of brilliant technologies*. W. W. Norton & Company.

Butler-Adams, W. & Davies, D. (2022). The Brompton: Engineering for Change. Profile Books. London:UK.

Crawford, K., & Calo, R. (2016). There is a blind spot in AI research. *Nature*, 538(7625), 311–313. https://doi.org/10.1038/538311a

Cozzens, S., Gatchair, S., Kang, J., Kim, K. S., Lee, H. J., Ordóñez, G., & Porter, A. (2010). Emerging technologies: quantitative identification and measurement. *Technology Analysis & Strategic Management*, 22(3), 361–376. https://doi.org/10.1080/09537321003647396

Davenport, T. H., & Ronanki, R. (2018). Artificial intelligence for the real world. *Harvard Business Review*, 96(1), 108–116. Retrieved from https://hbr.org/2018/01/artificial-intelligence-for-the-real-world

Dobbin, F., & Kalev, A. (2018). Why diversity programs fail. *Harvard Business Review*, 94(7), 52–60. https://hbr.org/2016/07/why-diversity-programs-fail

Dolan, S., Paludi, M., Sciabarrasi, L., Zendell, A. L., Schmidt, G., & Braverman, L. R. (2022). Ever upward: Building an ecosystem to support and validate lifelong learning. In A. Brower & R. Specht-Boardman (Eds.), *New Models of Higher Education: Unbundled, Rebundled, Customized, and DIY* (pp. 409–428). IGI Global.

Ducoffe, R. H., & Curlo, E. (2000). Advertising value and advertising processing. *Journal of Marketing Communications*, 6(4), 247–262. https://doi.org/10.1080/135272600750036364

Edmondson, A. C. (2012). *Teaming: How organizations learn, innovate, and compete in the knowledge economy*. Jossey-Bass.

Edmunds, A., & Boettcher, M. (2021). The lifespan of a university–industry partnership: A case study. *Journal of Cases in Educational Leadership*, 24(2), 135–147.

Ehlers, U.-D., & Eigbrecht, L. (2020). Reframing working, rethinking learning: the future skills turn. In Proceedings of European Distance and E-Learning Network (EDEN) Conference. Human and Artificial Intelligence for the Society of the Future (pp. 1–10). European Distance and E-learning Network.

European Commission. (2018). Regulation (EU) 2016/679 of the European Parliament and of the Council of 27 April 2016 on the protection of natural persons with regard to the processing of personal data and on the free movement of such data (General Data Protection Regulation, GDPR). *Official Journal of the European Union*. https://eur-lex.europa.eu/eli/reg/2016/679/oj

Ferguson, V. (2024). How do social work students develop skills by using practice-based virtual reality (VR) simulation? A UK higher education institution case study. *EDULEARN24 Proceedings. Retrieved from* https://library.iated.org/view/FERGUSON2024HOW

Floridi, L., Cowls, J., Beltrametti, M., Chatila, R., Chazerand, P., Dignum, V., Luetge, C., Madelin, R., Pagallo, U., Rossi, F., Schafer, B., Valcke, P., & Vayena, E. (2018). AI4People—An ethical framework for a good AI society: Opportunities, risks, principles, and recommendations. *Minds and Machines*, 28(4), 689–707. https://doi.org/10.1007/s11023-018-9482-5

Gunning, D., Stefik, M., Choi, J., Miller, T., Stumpf, S., & Yang, G. Z. (2019). XAI—Explainable artificial intelligence. *Science Robotics*, 4(37). https://doi.org/10.1126/scirobotics.aay7120

Gunness, S., Ferreira-Meyers, K., & Daradoumis, T. (2024). Learning design for future skills development in small state contexts. In U. D. Ehlers & L. Eigbrecht (Eds.), *Creating the University of the Future. Zukunft der Hochschulbildung - Future Higher Education*. Springer VS, Wiesbaden. https://doi.org/10.1007/978-3-658-42948-5_14

Haenlein, M., & Kaplan, A. (2019). A brief history of artificial intelligence: On the past, present, and future of artificial intelligence. *California Management Review*, 61(4), 5–14. https://doi.org/10.1177/0008125619864925

Holt, D. B. (2016). Branding in the age of social media. *Harvard Business Review*, 94(3), 40–50. https://hbr.org/2016/03/branding-in-the-age-of-social-media

Huang, M. H., & Rust, R. T. (2020). A strategic framework for artificial intelligence in marketing. *Journal of the Academy of Marketing Science*, 48(1), 30–50. https://doi.org/10.1007/s11747-020-00749-9

Inayatullah, S. (2008). Six pillars: Futures thinking for transforming. *Foresight*, 10(1), 4–21. https://doi.org/10.1108/14636680810855991

Jiang, F., Jiang, Y., Zhi, H., Dong, Y., Li, H., Ma, S., Wang, Y., Dong, Q., Shen, H., & Wang, Y. (2017). Artificial intelligence in healthcare: Past, present and future. *Stroke and Vascular Neurology*, 2(4), 230–243. https://doi.org/10.1136/svn-2017-000101

Jonbekova, D., Sparks, J., Hartley, M., & Kuchumova, G. (2020). Development of university–industry partnerships in Kazakhstan: Innovation under constraint. *International Journal of Educational Development*, 79. https://doi.org/10.1016/j.ijedudev.2020.102291.

Jones, E. (2013). The global reach of universities: Leading and engaging academic and support staff in the internationalisation of higher education. In R. Sugden, M. Valania, & J. R. Wilson (Eds.), *Leadership and cooperation in academia: Reflecting on the roles and responsibilities of university faculty and management* (pp. 161–183). Edward Elgar.

Jordan, M. I., & Mitchell, T. M. (2015). Machine learning: Trends, perspectives, and prospects. *Science*, 349(6245), 255–260. https://doi.org/10.1126/science.aaa8415

Kietzmann, J., Paschen, J., & Treen, E. R. (2018). Artificial intelligence in advertising: How marketers can leverage artificial intelligence along the consumer journey. *Journal of Advertising Research*, 58(3), 263–267. https://doi.org/10.2501/JAR-2018-035

Kim, M. & Maloney, E. (2020). *Learning Innovation and the Future of Higher Education*. Baltimore: Johns Hopkins University Press. https://dx.doi.org/10.1353/book.71965

Kuka L, Hörmann C, Sabitzer B (2022) Teaching and learning with AI in higher education: a scoping review. *Springer Sci Bus Media Deutschland GmbH* 456:551–571. https://doi.org/10.1007/978-3-031-04286-7_26

Libai, B., Bart, Y., Gensler, S., Hofacker, C. F., Kaplan, A., Kötterheinrich, K., & Kroll, E. B. (2020). Brave New World? On AI and the Management of Customer Relationships. *Journal of Interactive Marketing*, 51(1), 44–56. https://doi.org/10.1016/j.intmar.2020.04.002

Magrill, Jamie, and Barry Magrill. 2024. "Preparing Educators and Students at Higher Education Institutions for an AI-Driven World". Teaching and Learning Inquiry 12 (June):1–9. https://doi.org/10.20343/teachlearninqu.12.16. https://files.eric.ed.gov/fulltext/EJ1429106.pdf

Ministry of Education (MOE). (2022). *Speech by Minister Chan Chun Sing at the Global Lifelong Learning Summit*. Ministry of Education Singapore. https://www.moe.gov.sg/news/speeches/speech-by-minister-chan-chun-sing-at-the-global-lifelong-learning-summit-2022

Miranda, J., Rosas-Fernandez, J. B., & Molina, A. (2020). Achieving Innovation and Entrepreneurship by Applying Education 4.0 and Open Innovation. In Proceedings of the *2020 IEEE International Conference on Engineering, Technology and Innovation (ICE/ITMC)*, New York, NY, USA, 15–17, pp. 1–6. https://doi.org/10.1109/ice/itmc49519.2020.9198638

Miranda, J., Navarrete, C., Noguez, J., Molina-Espinosa, J. M., Ramírez-Montoya, M. S., Navarro-Tuch, S. A., Bustamante-Bello, M. R., Rosas-Fernández, J. B., & Molina, A. (2021). The core components of education 4.0 in higher education: Three case studies in engineering education. *Computer Electrical Engineering*, 93, 107278. https://doi.org/10.1016/j.compeleceng.2021.107278

Molitor, G. T. T. (2003). *The power to change the world: The art of forecasting*. Potomac Associates.

Mone, G., & Rus, D. (2024). *The mind's mirror: Risk and reward in the age of ai*. W. W. Norton & Company, Incorporated.

Moraes, E.B., Kipper, L.M., Hackenhaar Kellermann, A.C., Austria, L., Leivas, P., Moraes, J.A.R. and Witczak, M. (2023), "Integration of Industry 4.0 technologies with Education 4.0: advantages for improvements in learning", *Interactive Technology and Smart Education*, Vol. 20 No. 2, pp. 271–287. https://doi.org/10.1108/ITSE-11-2021-0201

Morley, J., Floridi, L., Kinsey, L., & Elhalal, A. (2020). From what to how: An initial review of publicly available AI ethics tools, methods and research to translate principles into practices. *Science and Engineering Ethics*, 26, 2141–2168, https://doi.org/10.1007/s11948-019-00165-5

Olawade, D.B., Wada, O.Z., Odetayo, A., David-Olawade, A.C., Asaolu, F., & Eberhardt, J. (2024). Enhancing Mental Health with Artificial Intelligence: Current Trends and Future Prospects. *Journal of Medicine, Surgery, and Public Health*. 100099. https://doi.org/10.1016/j.glmedi.2024.100099

Oliver, B. (2019). Making micro-credentials work for learners, employers and providers. *Journal of Teaching and Learning for Graduate Employability*, 10(2), 32–44. Retrieved from https://dteach.deakin.edu.au/wp-content/uploads/sites/103/2019/08/Making-micro-credentials-work-Oliver-Deakin-2019-full-report.pdf

Orman, R., Şimşek, E. and Kozak Çakır, M.A. (2023), "Micro-credentials and reflections on higher education", *Higher Education Evaluation and Development*, Vol. 17 No. 2, pp. 96–112. https://doi.org/10.1108/HEED-08-2022-0028

Parapadakis, D. (2020). Can artificial intelligence help predict a learner's needs? Lessons from predicting student satisfaction. *London Review of Education*, 18(2). https://doi.org/10.14324/lre.18.2.03

Park, S., James, J.I. Lessons learned building a legal inference dataset. *Artif Intell Law* 32, 1011–1044 (2024). https://doi.org/10.1007/s10506-023-09370-x

Partnership on AI. (2019). *Ethical guidelines and best practices for artificial intelligence development and use*. Partnership on AI. https://www.partnershiponai.org/

Patel, N. S. (2023). Empathetic and Dialogic Interactions: Modelling Intellectual Empathy and Communicating Care. *International Journal of TESOL Studies*, 5(3), 51–70. https://doi.org/10.58304/ijts.20230305

Patel, N. S. & Lim, J. T.-h. (2024). Critical design futures thinking and GenerativeAI: a Foresight 3.0 approach in higher education to design preferred futures for the industry. *Foresight*, Vol. ahead-of-print No. ahead-of-print. https://doi.org/10.1108/FS-11-2023-0228

Patel, N. S., Lim, J. T., & Teo, M. (forthcoming). Adopting critical design futures for interdisciplinary design innovation at a Singapore University. In Lee, Ai Noi & Nie, Youyan. *Work Skills, Future-ready Learning and Sustainable Employability*. Springer Publisher.

Patel, N. S., Lim, J. T. & Teo, M. (2023). Designing for sustainability: Adopting critical-design thinking for interdisciplinary design innovation. Higher Education Planning Asia (HEPA) Forum 2023 - Planning for a Sustainable Future. University of Sunshine Coast, Australia.

Patel, N. S., Puah, S., & Kok, X.-F. K. (2024). Shaping future-ready graduates with mindset shifts: studying the impact of integrating critical and design thinking in design innovation education. *Frontiers in Education*, 9. https://doi.org/10.3389/feduc.2024.1358431

Pelletier, K., Robert, J., Muscanell, N., McCormack, M., Reeves, J., Reeves, J., Arbino, N., Grajek, S., Birdwell, W. T., Liu, D., Mandernach, J., Moore, A., Porcaro, A., Rutledge, R., & Zimmern, J. (2023). 2023 EDUCAUSE Horizon Report Teaching and Learning Edition. https://www.learntechlib.org/p/222401/

Pentland, A. (2012). The new science of building great teams. *Harvard Business Review*, *90*(4), 60–69.

Piaget, J. (1964). Development and learning. In: R. E. Ripple & V. N. Rockcastle (Eds.), *Piaget Rediscovered* (pp. 7–20). W. H. Freeman and Company, New York.

Ramtohul, P. K., & Neeliah, H. (2024). A review of tools to imagine the future of skills: Lessons for a small island developing state (SIDS). In *Imagining the Futures of Higher Education*. Springer. https://doi.org/10.1007/978-981-97-6473-0_9

Rawas, S. ChatGPT: Empowering lifelong learning in the digital age of higher education. *Educ Inf Technol* 29, 6895–6908 (2024). https://doi.org/10.1007/s10639-023-12114-8

Rotolo, D., Hicks, D., Martin, B. R.(2015). What is an emerging technology? *Research Policy*, *44*(10), 1827–1843. https://doi.org/10.1016/j.respol.2015.06.006.

Russell, S., & Norvig, P. (2020). Artificial Intelligence: A modern approach (4th ed.). Pearson.

Shahriari, K & Shahriari, M. (2017). IEEE standard review — Ethically aligned design: A vision for prioritizing human wellbeing with artificial intelligence and autonomous systems. *IEEE Canada International Humanitarian Technology Conference (IHTC)*, Toronto, ON, Canada, 2017, pp. 197–201, doi: 10.1109/IHTC.2017.8058187.

Shams, R. A., Zowghi, D., & Bano, M. (2023). AI and the quest for diversity and inclusion: a systematic literature review. *AI Ethics*. https://doi.org/10.1007/s43681-023-00362-w

Shulman, L. (1987). Knowledge and teaching: The foundation of the new reform. *Harvard Education Review*, *57*(1). http://dx.doi.org/10.17763/haer.57.1.j463w79r56455411

Sin, S., & Reid, A. (2005). Developing generic skills in accounting: Resourcing and reflecting on transdisciplinary research and insights, Annual Conference of the Australian Association for Research in Education in Parramatta.

Soliman, M. (2022). Pedagogical intelligence in virtual reality environments. In M. E. Auer et al. (Eds.), *Learning with Technologies and Technologies in Learning* (pp. 285–302). Springer.

Soliman, M., & Guetl, C. (2011). A survey of pedagogical functions of intelligent agents in virtual learning environments. *Journal of Internet Technology*, *12*(6), 175–187.

Sun, J., & Turner, H. (2022). The Complementarity investment in university-industry collaboration. *Innovative Higher Education*, *48*, 539–556. https://doi.org/10.1007/s10755-022-09641-6

Susskind, R. (2019). *Online courts and the future of justice*. Oxford University Press. https://doi.org/10.1093/oso/9780198838364.001.0001

Thi Ngoc Ha, N., Van Dyke, N., Spittle, M., Watt, A. and Smallridge, A. (2024), "Micro-credentials through the eyes of employers: benefits, challenges and enablers of effectiveness", *Education + Training*, Vol. 66 No. 7, pp. 948–963. https://doi.org/10.1108/ET-08-2023-0340

Topol, E. J. (2019). High-performance medicine: The convergence of human and artificial intelligence. *Nature Medicine*, *25*(1), 44–56. https://doi.org/10.1038/s41591-018-0300-7

WGSN. (2023). *Future drivers 2023: Key strategies for future success*. WGSN. https://www.wgsn.com/

Yan,. H. (2021). The Trends and Challenges of Emerging Technologies in Higher Education. In 2021 2nd International Conference on Education Development and Studies (ICEDS 2021), March 09-11, 2021, Hilo, HI, USA. ACM, New York, NY, USA, 15 Pages. https://doi.org/10.1145/3459043.3459060

Yates, H., Chamberlain, C. (2017). Machine Learning and Higher Education. *EDUCAUSE Review*, [Online]. Available at: https://er.educause.edu/articles/2017/12/machine-learning-and-higher-education

Zowghi, D., Bano, M. AI for all: Diversity and Inclusion in AI. *AI Ethics* 4, 873–876 (2024). https://doi.org/10.1007/s43681-024-00485-8

Zuboff, S. (2019). *The age of surveillance capitalism: The fight for a human future at the new frontier of power*. PublicAffairs.

Index

Note: Page numbers in **bold** and *italics* refer to tables and figures, respectively.

Printed in the United States
by Baker & Taylor Publisher Services